Cell Physiology
of Blood

Society of General Physiologists Series • **Volume 43**

Cell Physiology
of Blood

Society of General Physiologists • 41st Annual Symposium

Edited by
Robert B. Gunn
Emory University
and
John C. Parker
University of North Carolina
at Chapel Hill

Marine Biological Laboratory
Woods Hole, Massachusetts

9–12 September, 1987

©**The Rockefeller University Press**
New York

Contents

Preface   *ix*

Keynote Speech

Chapter 1   **How Does the Sodium Pump Pump?**
             I. M. Glynn   **1**

**Hematopoiesis: The Ontogeny and Regulation of Red Cells, White Cells,
and Platelets**

Chapter 2   **Introduction to the Regulation of Hematopoiesis**
             David G. Nathan   **19**

Chapter 3   **Growth and Differentiation of Hematopoietic Stem Cells**
             *T. M. Dexter, I. L. O. Ponting, R. A. Roberts, E. Spooncer,
             C. Heyworth, and T. Gallagher*   **25**

Chapter 4   **Stem Cell Functions Assessed in Clonal Culture**
             Makio Ogawa   **39**

Chapter 5   **Human Colony-stimulating Factors and Stromal Cell Function**
             Colin A. Sieff, Charlotte M. Niemeyer, and Douglas V. Faller   **47**

Chapter 6   **Erythropoietin: In Vitro and In Vivo Studies of the Regulation of
             Erythropoiesis**
             John W. Adamson   **57**

Chapter 7   **The Role of Purified and Recombinant Hematopoietic Growth
             Factors in the Regulation of Various Stages of Human Mega-
             karyocytopoiesis**
             *Ronald Hoffman, John Straneva, Li Lu, Bruce Roth, Edward
             Bruno, and Robert Briddell*   **67**

Chapter 8   **Transfer of Genes in Hematopoietic Cells with Retroviral Vec-
             tors**
             *Arthur Nienhuis, David Bodine, Timothy Browder, Siu-Wah
             Chung, Cynthia Dunbar, Stefan Karlsson, and Peter Wong*   **79**

Organization and Function of Membrane Skeletal Proteins

Chapter 9   **The Red Cell Membrane Skeleton: A Model with General Biologi-
             cal Relevance but Pathological Significance for Blood**
             *Peter Agre, Barbara L. Smith, Ali M. Saboori, and Andrew
             Asimos*   **91**

Chapter 10   **The Spectrin-based Membrane Skeleton: Extensions of the Current Paradigm**
                Vann Bennett, Jonathan Davis, Kevin Gardner, and Joseph P. Steiner   101

Chapter 11   **Structure and Function of the Platelet Membrane Skeleton**
                Joan E. B. Fox and Janet K. Boyles   111

Chapter 12   **The Cortical Actin Gel of Macrophages**
                John H. Hartwig, Ken S. Zaner, and Paul A. Janmey   125

Anion Transport

Chapter 13   **The Red Cell Anion–Transport System: Kinetics and Physiological Implications**
                Jesper Brahm   141

Chapter 14   **Structure and Tissue-specific Expression of the Mouse Anion-Exchanger Gene in Erythroid and Renal Cells**
                Ron R. Kopito, Mona M. Andersson, Doris A. Herzlinger, Qais Al-Awqati, and Harvey F. Lodish   151

Chapter 15   **Functional Roles of Carboxyl Groups in Human Red Blood Cell Anion Exchange**
                Michael L. Jennings, Matthew P. Anderson, and Sally J. McCormick   163

Chapter 16   **Mechanisms of Anion Net Transport in the Human Erythrocyte**
                O. Fröhlich and P. A. King   181

Chapter 17   **Properties of the Principal Anion-Exchange Mechanism in Human Neutrophils**
                Louis Simchowitz   193

Chapter 18   **Mechanisms of Membrane Potential Probes**
                Alan S. Waggoner   209

Dye Indicators of Membrane Potentials in Blood Cells

Chapter 19   **Membrane Potential and the Cytotoxic Ca Cascade of Human Red Blood Cells**
                Jeffrey C. Freedman, Ellen M. Bifano, Lynn M. Crespo, Promod R. Pratap, Ronald Walenga, R. Eugene Bailey, Susan Zuk, and Terri S. Novak   217

Chapter 20   **Measurement of Membrane Potential Responses Elicited from Blood Cells: Effect of the Dye/Cell Ratio and the Presence of an Intracellular Calcium Probe**
                Bruce Seligmann, William O. Haston, James S. Wasvary, and John R. Rediske   233

Chapter 21 The Dynamics and Relationship of K^+ Efflux and Ca^{++} Influx in B Lymphocytes after Antigen-Receptor Cross-Linking
John T. Ransom and John C. Cambier 241

Chapter 22 Optical Measurement in Single Cells of Membrane Potential Changes Linked to T and B Lymphocyte Antigen Receptors
H. Alexander Wilson, David Greenblatt, Fred D. Finkelman, and Thomas M. Chused 253

Chapter 23 Platelet Membrane Potentials and Their Significance in Monitoring Stimulus Response Coupling
Elizabeth R. Simons, Theresa A. Davies, Sheryl M. Greenberg, Judith M. Dunn, and William C. Horne 265

Ion Channels and Activation of Blood Cells

Chapter 24 Role of Potassium and Chloride Channels in Volume Regulation by T Lymphocytes
Michael D. Cahalan and Richard S. Lewis 281

Chapter 25 Intracellular pH Regulation in Neutrophils: Properties and Distribution of the Na^+/H^+ Antiporter
Sergio Grinstein, Wendy Furuya, and Ori D. Rotstein 303

Chapter 26 Potassium Conductances in Macrophages
Elaine K. Gallin and Leslie C. McKinney 315

Chapter 27 The Events Leading to Secretory Granule Fusion
Guillermo Alvarez de Toledo and Julio M. Fernandez 333

Phosphoinositide Cycle and Protein Kinase C

Chapter 28 Inositol Lipids and Phosphates in Erythrocytes and HL60 Cells
R. H. Michell, C. E. King, C. J. Piper, L. R. Stephens, C. M. Bunce, G. R. Guy, and G. Brown 345

Chapter 29 Protein Kinase C and Its Associated Substrates in the Human Erythrocyte
H. Clive Palfrey and Ahmad Waseem 357

Chapter 30 Biochemical and Molecular Events Controlled by Lymphokine Growth Factors
William L. Farrar, Annick Harel-Bellan, and Douglas K. Ferris 371

Chapter 31 Initiation and Termination of Ca^{2+} Signals: Studies in Human Platelets
Timothy J. Rink and Stewart O. Sage 381

List of Contributors 395

Subject Index 401

Preface

This book contains the written contributions of speakers invited to the 41st Annual Meeting of the Society of General Physiologists, held in Woods Hole, Massachusetts, during September 9–12, 1987. Abstracts for the 95 contributed poster presentations at the meeting are published in the December 1987 issue of *The Journal of General Physiology*.

As organizers of this symposium on Cell Physiology of Blood we had to make a difficult judgement: would it be better to have the presentations cover a subject in depth, or should we convene scientists with a wide range of interests who happen to work on blood cells? It would not have been difficult, for example, to fill the entire program with talks exclusively about membrane transport, or cytoskeletal proteins, or intracellular signaling, or growth and differentiation. Each of these fields is moving rapidly; each would have commanded a devoted audience. We opted for diversity. Blood cells are the subject of such a variety of research that we thought it would be constructive to bring together the leading investigators in a number of areas, gambling that they would find subjects of mutual interest to discuss. In addition, we both felt that as members of the transport workers' union we needed a broadening experience. We were not disappointed.

Many people contributed to the success of the meeting. We would like especially to acknowledge the session chairmen, to whom we delegated the invitation of speakers: David Nathan and Arthur Nienhuis for the session on hematopoiesis, Peter Agre for the membrane skeletal program, Jeffrey Freedman for the workshop on dyes that measure membrane potential, Michael Cahalan for the section on ion channels, and Clive Palfrey for the presentations on intracellular signaling. Joseph Hoffman in his introduction of the keynote speaker gave a historical account of blood cell research at the Marine Biological Laboratories, mentioning M. H. Jacobs, A. K. Parpart, and F. R. Hunter. Daniel Tosteson wound up the proceedings with a summary and reflections. To these individuals, to the invited speakers and the contributors of abstracts, to Maureen Bodo of The Rockefeller University Press, Jane Leighton of the Society of General Physiologists, and to the officers of the society we extend our thanks.

Robert B. Gunn
John C. Parker

Keynote Speech

Chapter 1

How Does the Sodium Pump Pump?

Physiological Laboratory, University of Cambridge, Cambridge, England

Cell Physiology of Blood © 1988 by The Rockefeller University Press

Introduction

During the last decade, there have been very substantial advances in our understanding of both the structure and the mechanism of the sodium pump, though we are still a long way from fully understanding either. Curiously, despite the advances on both fronts, there is as yet little interaction between them. Those who study structure and those who study function are aware of each other's work and happy that it should be going on, but they have not yet received much insight from each other's experiments. The situation is one which, I believe, the child psychologists call "parallel play."

The Structure of the Na,K-ATPase

Let me say something about structure first. It is not an area in which I have been directly involved, so I can be presumed to have fewer prejudices.

It is now clear that the Na,K-ATPase molecule contains two kinds of subunit, an α chain of molecular weight ~100,000, and a β chain, which is a glycopolypeptide whose protein portion has a molecular weight of ~40,000. The two kinds of chain exist in equal numbers, but there has been a good deal of controversy about whether the Na,K-ATPase molecule in situ consists of one α and one β chain, which it probably does, or two α-β protomers. Two years ago, two groups of workers, Shull et al. (1985) in Cincinnati and Kawakami et al. (1985) in Kyoto, succeeded in cloning and sequencing DNA complementary to the messenger RNA coding the α subunit. This was the α subunit from sheep kidney in the American work, and from *Torpedo* electric organ in the Japanese work. The deduced amino acid sequences were in very good agreement with each other and with what was already known from direct analysis of proteolytic digestion products, and they showed strong homologies with the sequences found in the calcium-transporting ATPase of sarcoplasmic reticulum and in the H,K-ATPase responsible for acid secretion in the gastric mucosa. There are also striking homologies with the H-transporting ATPase of yeasts and the K-transporting ATPase of *Escherichia coli*. More recently, the sequence of the β subunit of the Na,K-ATPase has also been determined (Shull et al., 1986).

By looking at the relative hydrophobicity or hydrophilicity of the amino acid side chains as one moves along the sequence, it is possible to derive so-called hydropathy plots (Kyte and Doolittle, 1982). (Hydropathy is derived from Greek roots meaning "feeling for water.") For the α chain, the plots show six to eight regions made up mainly of amino acids with hydrophobic side chains, and these regions are separated by more hydrophilic regions. This arrangement suggests that the α chain crosses the lipid membrane at least six and probably eight times. In contrast, hydropathy plots for the β unit suggest only a single crossing. The sugars are attached to the portion of the β chain projecting into the extracellular medium.

Interpretation of hydropathy plots is not entirely foolproof, since hydrophobic regions are not necessarily confined to the parts of the polypeptide crossing the lipid bilayer. However, there is independent evidence that the α chain crosses the lipid bilayer at least six times. Under defined conditions, trypsin and chymotrypsin cut the α chain at three particular sites, all of which are accessible only from the intracellular surface. It is therefore possible to divide the chain into four regions and to know that the junctions between these regions are exposed at the intracellular surface. If the intact chain in situ is labeled with reagents that act solely from the extracellular surface or solely from the intracellular surface, then the position of the labels will identify the regions of the chain that project outwards or inwards. Lipid-soluble labels can also be used to identify the transmembrane portions. By building up the jigsaw puzzle in this way, it is possible to derive a rough picture of the polypeptide chain as it winds backwards and forwards through the lipid bilayer (for references, see Glynn, 1985). This picture fits moderately well with the picture derived from hydropathy plots (though there are some contradictions), and it also shows the approximate positions of the ATP-binding and phosphorylation sites at the intracellular surface, and of the region or regions adjacent to the ouabain-binding site at the extracellular surface.

A more direct approach to structure is, of course, to use the electron microscope. The most successful procedure has been to study Na,K-ATPase–rich membranes that have been exposed to vanadate or to orthophosphate for several weeks. The enzyme has

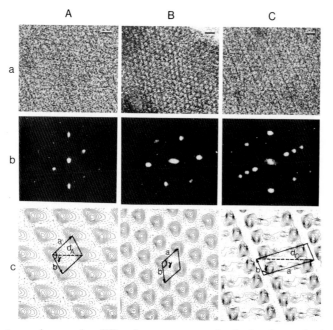

Figure 1. Electron micrographs, diffraction patterns, and calculated protein-density contour maps for three preparations of two-dimensional Na,K-ATPase crystals. Preparations *A* and *B* had been exposed to vanadate, preparation *C* to phosphate. (Reproduced with permission from Hebert et al., 1982.)

some lateral mobility in the plane of the membrane, and under these conditions the individual enzyme molecules come together, first as linear arrays and eventually as confluent areas of two-dimensional crystal. These crystalline areas can then be investigated using the standard techniques of negative staining, thin-sectioning, and freeze-fracture, and the repetitive images can be studied using sophisticated techniques of image analysis.

Fig. 1, from Hebert et al. (1982), shows electron micrographs, diffraction patterns, and calculated protein density contour maps for three preparations. The first two had been prepared with vanadate, and there is one α-β protomer per unit cell. The right-hand preparation had been exposed to phosphate, and the unit cell corresponds to the $\alpha_2\beta_2$ dimer. By studying negatively stained dimeric two-dimensional crystals at a whole series of tilt angles, and then subjecting each projection to Fourier analysis, Hebert and his colleagues were able to produce a three-dimensional model of Na,K-ATPase in dimeric form. Each unit cell contained two rod-like protein regions corresponding to α-β protomers. The units protruded ~40 Å on the cytoplasmic side of the bilayer and ~20 Å on the extracellular side. Unfortunately, the resolution in the plane of the membrane was only ~20–25 Å, so detailed structural analysis was not possible.

The Nature of the Pump Cycle

In the normal working of the pump, three sodium ions are transported outwards and two potassium ions are transported inwards for each molecule of ATP hydrolyzed. We are therefore dealing with an enzyme that has four substrates (ATP, water, intracellular sodium, and extracellular potassium) and four products (ADP, inorganic phosphate, extracellular sodium, and intracellular potassium). In considering an enzymic reaction involving this number of reactants and products, one has to ask the question: do all the reactants combine with the enzyme before any of the products are released, or do some combine only after some of the products have been released? If the latter is true, the enzyme is known as a "ping-pong" enzyme (Cleland, 1963) because it must

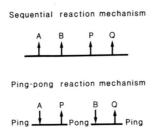

Figure 2. Multisubstrate enzymic reaction mechanisms classified according to Cleland (1963). *A* and *B* are substrates; *P* and *Q* are products.

exist in at least two stable states: a "ping" state that combines with the first set of reactants and releases the first set of products, and a "pong" state that combines with the second set of reactants and releases the second set of products (see Fig. 2).

It turns out that the sodium pump is a ping-pong enzyme (Fig. 3). Combination with ATP and intracellular sodium is followed by phosphorylation, release of ADP to the cell interior, and, after a conformational change, release of the sodium at the outside surface of the cell. The enzyme is now in the pong state. Combination with extracellular potassium is followed by hydrolysis of the phosphoenzyme, release of the inorganic phosphate to the cell interior, and, after a change in conformation of the dephosphoenzyme, release of the potassium at the intracellular surface of the membrane. The enzyme is now back again in the ping state.

What is the evidence for this story? Instead of giving a strictly historical account, I propose to take a walk, or rather a series of walks, around the enzyme cycle, attempting to justify its various features on the way. And though I shall discuss in some detail the kinds of evidence that support the scheme, I shall not show any experimental results. This is partly because this address was originally given after dinner at the end of a busy day, and partly because I wanted my listeners then—and I want my readers now—to appreciate the woods and not be overconcerned with the trees.

First, let us look at the right-hand side of the cycle, the part involving sodium ions.

In 1965, Post et al. showed that Na,K-ATPase from guinea pig kidney was phosphorylated by ATP in the presence of magnesium and sodium ions, and that the phosphoenzyme was hydrolyzed in the presence of potassium ions. The phosphorylated group was later shown to be a β-aspartyl carboxylic group (Post and Kume, 1973), which we now know to be residue 376 in the α chain. Although these experiments told one nothing about the movements of the catalyzing ions, it was economical to suppose that the phosphorylation was associated with an outward movement of sodium and the hydrolysis was associated with an inward movement of potassium. In other words, it looked as though the Na,K-ATPase might be a ping-pong enzyme, in which the ping state was unphosphorylated and reacted with ATP and intracellular sodium and the pong state was phosphorylated and reacted with water and extracellular potassium.

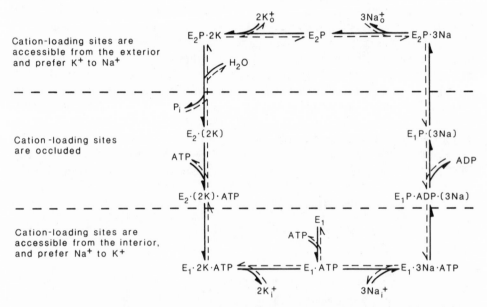

Figure 3. The normal Na,K-ATPase cycle. The brackets in $E_1P(3Na)$ and $E_2P(2K)$ denote that the ions are occluded. For simplicity, (*a*) magnesium ions have been ignored, (*b*) the slow conversion of $E_2(2K)$ to $E_1 \cdot 2K$ that occurs when ATP is not bound to the enzyme has been omitted, (*c*) the binding by E_1 of sodium ions and ATP in that order has been omitted. (Modified from Karlish et al., 1978.)

The need to postulate the existence of two different forms of the phosphoenzyme came about in this way. It was shown early on that the incorporated phospho group was an acyl phosphate. An acyl phosphate would be expected to react with ADP to form ATP, but Albers and his colleagues (Fahn et al., 1966*a*, *b*) found that the phosphoenzyme would do this only if the enzyme had been pretreated in certain ways: with *N*-ethylmaleimide, for example, or, as it later turned out, with oligomycin (Siegel and Albers, 1967), or with α-chymotrypsin (Jørgensen et al., 1982). They therefore suggested that there were two forms of phosphoenzyme: E_1P, which was formed first, and which could react with ADP; and E_2P, which was formed spontaneously from E_1P and which did not react with ADP but was rapidly hydrolyzed in the presence of potassium ions. If the enzyme had not been treated with *N*-ethylmaleimide (or oligomycin or α-chymotrypsin), the phosphoenzyme was mostly E_2P; if it had, the spontaneous conversion of E_1P to E_2P was blocked and the enzyme was mostly E_1P. The difference between E_1P and E_2P is thought to be simply a difference in conformation, since Post et al. (1969) could detect no difference between the electrophoretic behavior of the proteolytic digestion products of the two forms of phosphoenzyme. Very recently, it has been shown that there is probably also an intermediate form, $E*P$ (for references and discussion, see Glynn, 1988), but that is a complication we shall not go into here.

Now let us look at the left-hand side of the cycle, the part that is involved with potassium. The notion that there were two different dephospho forms came about in this way. Although the sodium pump is rather specific for sodium (only hydrogen ions and lithium ions can substitute for sodium ions, and they do so poorly), it is much less specific for potassium. Lithium, rubidium, cesium, ammonium, and thallous ions can all substitute for potassium ions. In 1972, Post et al. reported experiments in which they measured the rate of rephosphorylation by ATP of enzyme that had just been dephosphorylated. They found that the rate of rephosphorylation differed depending on whether lithium or rubidium ions had been present during the hydrolysis, and that this was true even when the experiments were done in such a way that the conditions during the rephosphorylation were identical. In other words, the enzyme appeared to remember which ions had catalyzed the hydrolysis. Post et al. suggested that this was because the catalyzing ions became occluded within the enzyme at the moment of hydrolysis and were released only later after a slow conformational change. Furthermore, since the enzyme became available for rephosphorylation sooner if higher ATP concentrations were used, they suggested that ATP accelerates the conformational change that precedes the release of the occluded ions.

Flux Studies

We have now looked at both halves of the cycle, but so far I have not discussed any evidence that bears on the sidedness of the events we are considering. The phosphorylation studies I have been talking about gave no information on this point and, clearly, another kind of approach was needed.

While the phosphorylation studies just discussed were going on, other investigators were looking at the fluxes catalyzed by the Na,K-ATPase in a variety of different conditions. These studies made use of nerves, muscles, and, above all, red cells, for what was important was the integrity of the membrane containing the pumps, not the concentration of pumps in the membrane.

Consider again the scheme in Fig. 3. Under physiological conditions, the cycle runs counterclockwise, and energy from the hydrolysis of ATP is used to move three sodium ions outwards and two potassium ions inwards. If we arrange it so that the concentration gradients of sodium and potassium are so steep that the free energy available from the hydrolysis of one molecule of ATP is insufficient to move the ions, the whole system runs backwards, synthesizing ATP at the expense of energy from the downhill movements of the ions (Garrahan and Glynn, 1967e; Glynn and Lew, 1970). This was, of course, an exciting result to get. It is as if you took a motorcar with an empty petrol tank, let it run downhill, and found petrol spurting from the filler cap. It will not work with a motorcar because the combustion of petrol in air is too far from thermodynamic equilibrium, but it did work with the sodium pump. All the experiment reveals in relation to mechanism, though, is that all of the steps in the cycle must be reversible and that in physiological conditions none can be very far from equilibrium.

Some of the other abnormal flux modes were more informative. When red cells, or resealed ghosts, were incubated in media containing no potassium, it was possible to detect a one-for-one exchange of internal and external sodium ions, as if the right-hand part of the cycle were shuttling backwards and forwards (Garrahan and Glynn, 1967a). If that suggestion is correct, then: (a) the exchange should require the presence of both ADP and ATP; we found that it did; (b) the exchange should not be accompanied by the hydrolysis of ATP; no hydrolysis was observed; (c) the exchange should have a high affinity for sodium ions inside and a low affinity outside; it had; and (d) the exchange should be blocked by oligomycin; it was. (For references, see Garrahan and Glynn, 1967a–d; Sachs, 1970; Glynn and Hoffman, 1971; Garay and Garrahan, 1973; Cavieres and Glynn, 1979. See also Glynn, 1985, for references to experiments in cells other than red cells.)

Under other conditions, it was possible to demonstrate a one-for-one exchange of internal and external potassium ions, as if the left-hand part of the scheme were shuttling backwards and forwards (Glynn et al., 1970). If the scheme is right, one would predict that the shuttling would require both inorganic phosphate and a nucleotide—it did—that it should not be accompanied by hydrolysis of ATP—it was not—and that it should have a high affinity for potassium outside and a low affinity inside—it had (Glynn et al., 1970; Glynn and Hoffman, 1971; Simons, 1974; Sachs, 1981). It will be noticed that the scheme gives two separate roles to ATP: phosphorylating the E_1 form of the dephosphoenzyme, and accelerating the conformational change of the E_2 form of the dephosphoenzyme. Since the accelerating effect does not depend on phosphorylation, non-phosphorylating ATP analogues—i.e., ATP molecules in which the terminal phosphate is attached by a methylene or imido group—might be expected to act just as well. Simons (1975) showed that compounds of this kind do in fact support potassium-potassium exchange.

Now we come to a deficiency in the scheme. If red cells are incubated in media lacking both sodium and potassium, according to the scheme the cycle should stop at the E_2P stage and there should be neither fluxes nor ATP hydrolysis. In fact, what is observed is a slow outward movement of sodium accompanied by a slow hydrolysis of ATP, the stoichiometry being the usual 3Na:ATP (Garrahan and Glynn, 1967a, b; Beaugé and Ortiz, 1973; Lew et al., 1973; Karlish and Glynn, 1974; Glynn and Karlish, 1976). To accommodate this finding, we had to add a slow return pathway to the scheme (Fig. 4). We called the outward movement of sodium ions "uncoupled sodium efflux," though at the time of its discovery there was no way of knowing that the outward movement of sodium was not accompanied by an inward movement of hydrogen ions. Very recently, Marin and Hoffman (1988) have shown that it is accompanied by an outward movement of anions.

Now, examination of the scheme with this central return pathway shows that, just as it is possible to go outwards by the right-hand pathway and back by the central pathway, so it is possible to go outwards by the left-hand pathway and back by the central pathway. Arguing in this way, Sachs (1986) predicted that, under appropriate conditions, an uncoupled efflux of potassium ions should occur, and he was able to demonstrate that it did.

So far, then, we have seen that the scheme can account for six flux modes: the normal forward mode, a reverse mode, sodium-sodium exchange, potassium-potassium exchange, uncoupled sodium efflux, and uncoupled potassium efflux. For good measure, we can add two more. First, because hydrogen ions can replace sodium ions weakly, the pump can catalyze a slow hydrogen-potassium exchange when sodium is absent at the intracellular face of the pump (Hara and Nakao, 1986). Second, because sodium ions have an extremely weak potassium-like effect, it is possible under some conditions to detect a slow 3:2 sodium-sodium exchange accompanied by the hydrolysis of ATP (Lee and Blostein, 1980). Unlike the classic sodium-sodium exchange, this does not require ADP because the inward pathway for sodium is the left-hand side of the scheme not the right-hand side.

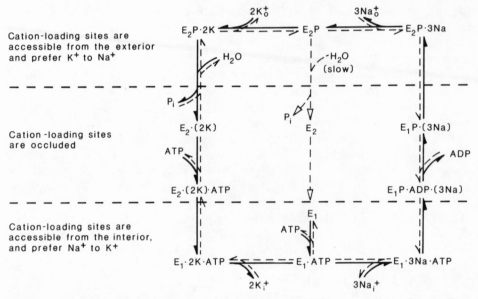

Figure 4. The normal Na,K-ATPase cycle with the extra central pathway necessary to account for uncoupled Na efflux and uncoupled K efflux; see text. (Modified from Karlish et al., 1978.)

Occluded-Ion Forms

It is obviously very satisfactory that the scheme can account for the existence of all these flux modes, and that the flux studies complement the phosphorylation studies, but can we get more direct evidence?

Examination of the scheme shows that the processes involved in the outward movement of sodium and the processes involved in the inward movement of potassium share certain common features. The first common feature is that, in each case, there is a step in which the transfer of a phospho group is accompanied by the trapping or occlusion of the ions catalyzing the transfer. In the sodium limb of the cycle, the transfer of the phospho group is from ATP to the enzyme, the catalyzing ions and the ions that are occluded are sodium ions, and they are occluded in E_1P. In the potassium limb, the transfer of the phospho group is from the phosphoenzyme to water, the catalyzing ions and the ions that are occluded are potassium ions, and they are occluded in E_2. The second common feature is that, in each case, the trapping of the ions is followed by a conformational change that allows the trapped ions to be released at the opposite surface of the membrane. In the sodium limb, the conformational change is the change from E_1P to E_2P, and it can be blocked by pretreating the enzyme with N-ethylmaleimide or with oligomycin or with α-chymotrypsin. In the potassium limb, the conformational change is from E_2 to E_1. There is no suitable blocking agent, but the change is very slow anyway, unless ATP is bound to a low-affinity site on E_2.

An obvious and crucial test of the scheme is to see whether it is possible to get independent evidence of the existence of these occluded-ion forms, and, if it is, whether they have the properties expected of them. It turns out that it is possible and that they do have the expected features, though they also have some features that are unexpected.

Let us take the occluded-sodium form first. In principle, the test is very straightforward. We must make E_1P in the presence of radioactive sodium, stabilize it somehow, force it through a cation-exchange column, and see whether the enzyme carries radioactivity into the effluent. The only step that looks difficult is the stabilization, but we already know how to do that. If we pretreat the enzyme with N-ethylmaleimide or with α-chymotrypsin, the conversion of E_1P to E_2P is blocked. If we also make sure that ADP is absent, then the E_1P has nothing to transfer its phospho group to, so the sodium ions cannot escape by a reversal of the reaction that led to their trapping. We did the experiment and found that E_1P contained three occluded sodium ions (Glynn et al., 1984). If we added ADP, the occluded ions escaped. If we failed to pretreat the enzyme with N-ethylmaleimide or α-chymotrypsin, the occlusion could not be detected. The occluded form was behaving just as it should. (A curious feature of this experiment is that it was not done much earlier. The notion that sodium ions were occluded in E_1P was first put forward 16 years ago [Glynn and Hoffman, 1971], and enzyme preparations active enough for the experiment to be practicable have been around since Jørgensen's paper of 1974.)

Let us now turn to the demonstration of occluded potassium ions in the E_2 form of the dephosphoenzyme. The story here is a little more complicated and involves a slight digression. We have already seen how the rephosphorylation experiments of Post et al. (1972) led them to suggest that potassium ions were occluded transiently when phosphoenzyme in the E_2P form was hydrolyzed in the presence of potassium ions. It has been known for a long time that the unphosphorylated enzyme changes its properties depending on whether it is suspended in sodium media or potassium media. In sodium media, it has a very high affinity for ATP and it is attacked one way by trypsin; in potassium media, it has a low affinity for ATP and it is attacked in a different way by trypsin (Hegyvary and Post, 1971; Nørby and Jensen, 1971; Jørgensen, 1975). It has been conventional to refer to the dephosphoenzyme in sodium media as being in the E_1 form and dephosphoenzyme in potassium media as being in the E_2 form, but there was initially no reason to identify E_2—the stable form of enzyme in potassium media—with the hypothetical transient form of dephosphoenzyme containing occluded potassium ions. In 1978, Karlish et al. described experiments in which stopped-flow fluorimetry was used to study the interconversion of E_1 and E_2 as the enzyme was switched between a sodium medium and a potassium medium. They made use of a fluorescent ATP analogue that had the useful property of increasing its fluorescence on binding to the enzyme. Because the enzyme has a high affinity for ATP in a sodium medium and a low affinity in a potassium medium, provided the concentration of analogue is low, there will be net binding and net release of analogue as the enzyme is switched between potassium and sodium media. And because the fluorescence of the analogue increases on binding, it is possible to follow the interconversion of E_1 and E_2 forms by observing the changes in fluorescence. Using this technique, Kárlish et al. (1978) showed that the conversion of E_1 to E_2 was rapid, but the conversion of E_2 to E_1 was astonishingly slow, with a time constant of between 5 and 10 s at room temperature. Furthermore, the rate of the slow conversion could be increased by increasing the concentration of the nucleotide analogue, which seemed to act with a rather low affinity. Now, the slow change in conformation and the acceleration by ATP acting at a low-affinity site are precisely the features that Post et al. (1972) had to postulate for their transient occluded-potassium form. So the question was: are the two forms identical? In other words, when one puts unphosphorylated enzyme into a potassium medium, does it spontaneously change into a form containing occluded potassium ions?

Luis Beaugé and I thought we would test this by seeing whether the enzyme would carry the occluded ions through a cation-exchange resin. A simple calculation showed that there was no hope of doing the experiment with ^{42}K, because the specific activity was much too low, but, as we have seen, the enzyme treats rubidium very much like potassium, and ^{86}Rb is available with very high specific activity. Since the time constant for the conversion of E_2 to E_1, in the absence of nucleotide, was of the order of 10 s, it was obviously necessary to force the suspension through the resin in a much shorter time, and at first we were doubtful whether, with such high flow rates, the resin could remove all of the free ion. It turned out, however, that the resin worked well even if the suspension was in contact with it for only ~ 1 s, and we were able to demonstrate that enzyme suspended in sodium-free rubidium media contained about two occluded rubidium ions per phosphorylation site (Beaugé and Glynn, 1979). Furthermore, by flowing the enzyme suspension through the column at different rates and comparing the amounts of occluded rubidium appearing in the effluent with the times spent on the column, it was possible to estimate the rate of release of occluded rubidium. It agreed reasonably well with the rate of the conformational change E_2 to E_1, and, like that rate, it was greatly increased by nucleotides acting with a low affinity and without phosphorylation (Glynn and Richards, 1982).

By this time we were convinced that we had succeeded in making the occluded-potassium form described by Post et al. (1972), but we certainly had not made it by the same route: the potassium-catalyzed hydrolysis of the E_2P form of phosphoenzyme. If we were right, we ought to be able to make an occluded-potassium form with identical properties by what we had by then begun to call the Postal route, but which (if these notions of the way the pump works are sound) is better referred to as the physiological route. We therefore suspended enzyme in a high-sodium medium containing a little rubidium chloride, and forced it first through a thin layer of Sephadex containing a low concentration of MgATP, and then through a cation-exchange resin. We expected that the ATP would phosphorylate the enzyme, the rubidium ions would catalyze the hydrolysis of the phosphoenzyme and become occluded, and the occluded rubidium would appear in the effluent. And this is what happened (Glynn and Richards, 1982). When the ATP concentration in the Sephadex was increased, the binding of ATP to low-affinity sites on $E_2(Rb)$ accelerated the conformational change to E_1 and less radioactivity appeared in the effluent.

Taken together, the experiments just described show that we can reach the occluded-potassium form by two routes, the physiological route and a direct route (see Fig. 5). There is, however, an important difference between the two routes in both the affinity and the sidedness of the sites that combine with the potassium ions. Using vesicles made from red cell membranes, Blostein and Chu (1977) showed that the potassium ions that hydrolyze E_2P act with a high affinity and come from the extracellular surface. The potassium ions that enter the $E_2(K)$ form by the physiological route must therefore come from the outer surface and they combine with high-affinity sites. The stopped-flow experiments of Karlish et al. (1978), using fluorescent ATP analogues, showed that the rate of conversion of E_1 to E_2 increased linearly with potassium concentration over quite a wide range, which suggests that in the direct route potassium ions combine at low-affinity sites on E_1. Karlish and Pick (1981) incorporated Na,K-ATPase into the membranes of artificial lipid vesicles and used the pattern of tryptic digestion to distinguish between the E_1 and E_2 forms. They found that only potassium ions at the originally intracellular surface were able to convert E_1 into E_2. It follows that in the direct route potassium ions combine at low-affinity sites at the intracellular surface of the membrane. Coupling the two routes back to back, as shown in Fig. 5, we therefore have a pathway for potassium ions that extends from high-affinity extracellular sites to low-affinity intracellular sites. That is, of course, just what is wanted to carry potassium ions into the cell.

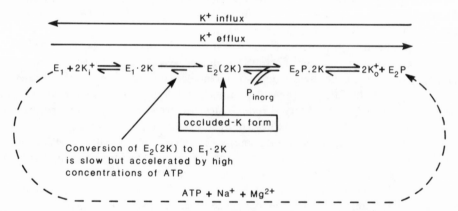

Figure 5. The two routes to the occluded-potassium form of the Na,K-ATPase: (*a*) the physiological route, i.e., hydrolysis of the E_2P form of the phosphoenzyme, catalyzed by potassium ions bound to high-affinity extracellular sites; (*b*) a change in conformation of the dephosphoenzyme following the binding of potassium ions to low-affinity intracellular sites. (Modified from Glynn et al., 1985*b*.)

At this stage, we did one experiment too many and discovered a complication. If the scheme in Fig. 5 is correct, we ought to be able to occlude rubidium ions by the direct route and release them quickly by adding magnesium and inorganic phosphate. When we did the experiment (Glynn et al., 1985*a*), we were surprised to find that the addition of magnesium and phosphate caused the rapid release of only 50% of the occluded rubidium. We thought that there must be something wrong with our mixing technique, and we were still more doubtful about the result when Joseph Hoffman, in the course of a visit, told us that Bliss Forbush, who used a rapid-filtration technique much more sophisticated than our ion-exchange columns, found that the addition of magnesium and inorganic phosphate released 100% of the occluded rubidium.

Further experiments, however, showed us that both sets of results were correct. Whether magnesium plus phosphate released 50 or 100% of the occluded rubidium depended on whether the suspending medium did or did not contain free rubidium ions. If free ions were present, only 50% of the occluded ions escaped quickly; if no free ions were present, all of the occluded rubidium escaped quickly. This not only meant that honor was satisfied all round, but it also suggested strongly that the release of the occluded ions was ordered. If the first ions to be released are replaced quickly (and repeatedly) by ions from the medium, then the remaining ions cannot escape. If this explanation is correct, the ions preventing the release of the second 50% of occluded ions should themselves become occluded. Forbush showed that they were, and we have confirmed that (Forbush, 1985; Glynn et al., 1985a).

This brings us to an interesting question. If the enzyme is an $\alpha_2\beta_2$ dimer, the ordered release of the rubidium ions could occur in two ways. The rapidly released ions might represent one of the two ions in each α-β protomer. Alternatively, the two halves of the dimeric enzyme might release their contents in an ordered sequence, so that the rapidly released ions represented all of the occluded ions in one of the α-β protomers of each dimeric enzyme molecule.

If we want to understand the mechanism of the pump, it is important to know which explanation is correct. How can we decide between them? If we could knock out one α-β protomer in each supposedly dimeric enzyme molecule, we would presumably prevent the 50% effect if that depended on interaction between the two halves of the dimer, but we should still expect to see it if it depended on the ordered release of the two ions in each α-β protomer. There is, of course, no way in which we can knock out one α-β protomer in each dimeric enzyme molecule. However, if we treat the enzyme with some agent that attacks the α-β protomers randomly, and we stop the reaction when the Na,K-ATPase activity is reduced to half, we should have a preparation in which a quarter of the enzyme will have been completely knocked out, a quarter will be normal, and half will have one functioning α-β protomer. If we allow this enzyme to occlude rubidium ions through the route that involves phosphorylation and dephosphorylation, then only the functional protomers will contain occluded rubidium. Now consider what happens when we add magnesium and phosphate in conditions in which we expect a 50% release. If the independence hypothesis is correct, each functioning α-β protomer will quickly release one of its two rubidium ions, so we should still find that half of the rubidium is released rapidly. On the alternative, "half-of-the-dimer" hypothesis, however, it is easy to show that we should expect either 25 or 75% to be released rapidly; the uncertainty comes about because we cannot predict whether the enzyme with only one unit working will release its rubidium or not.

We have done the experiment in two ways, using either α-chymotrypsin or fluorescein isothiocyanate to knock out half of the Na,K-ATPase activity. Whichever technique we used, we found that the fraction of occluded rubidium lost quickly from enzyme whose Na,K-ATPase activity had been inhibited to 50% or more was still close to 50%, and very significantly different from the fractional losses expected on the "half-of-the-dimer" hypothesis (unpublished experiments by D. E. Richards and I. M. Glynn). In other words, it is the release of the two rubidium ions in each α-β protomer that is ordered, not the emptying of the two halves of the dimer.

Conclusion

I should have liked to finish this discussion by coming back to our picture of the enzyme structure and pointing out the sites at which sodium ions and potassium ions are occluded. That is not possible. A conspicuous failure of the structural studies, so far, is that we have no idea of the identity or position of the sites that bind the ions at the membrane surface or take part in the occlusion. Kawakami et al. (1985) have pointed out that at least one, and possibly three, of the eight largely hydrophobic transmembrane segments of the α chain contain one or more acidic residues, and they suggest that these might be involved in the occlusion of ions within the enzyme molecule. Since sodium ions are occluded in a phosphorylated form of the enzyme, one might guess that the negative charges on the phosphate group were involved. But one would probably guess wrong, since Esmann and Skou (1985) have shown that, even without phosphorylation, oligomycin-treated enzyme can occlude sodium ions.

If we do not know the nature of the groups surrounding the occluded ions and we do not know the nature of the groups binding the ions at the membrane surfaces, we are not in a strong position to say how the ions get from the surface to the interior and on to the opposite surface. Analogy with better-understood enzymes would suggest that the gating, as it is sometimes called, is caused by small angular movements of rigid domains about a hinge, rather than by large movements of small and very mobile portions of the protein, but that is a vague statement even if it is correct.

There is, however, one feature of the movement of the ions through the membrane that we do know about and which I ought to mention. It has been known for a long time that, as might be expected from the stoichiometry, each cycle of the pump is accompanied by the outward movement of one positive charge. It follows that at least one step in the cycle must be sensitive to the membrane potential. Recently, Rephaeli et al. (1986) looked at the effects of potential on the major conformational changes. What they did was label Na,K-ATPase with a suitable fluorescent probe, incorporate it into the membranes of artificial lipid vesicles, generate potentials across the vesicle membrane with a suitable combination of ionophores and ion gradients, and observe the effects of these potentials on the conformational changes, which they followed by stopped-flow fluorimetry. It turned out that the conversion of $E_2(K)$ to $E_1 \cdot K$ was insensitive to potential, but the conversion of $E_1P(3Na)$ to $E_2P \cdot 3Na$ was accelerated fourfold by making the cytoplasmic face of the membrane positive to the extracellular face by ~ 170 mV.

The sensitivity to membrane potential of the steps in the sodium limb of the cycle has recently been confirmed by a novel technique used by Fendler et al. (1988) and by Borlinghaus et al. (1987). They apply fragments of Na,K-ATPase–rich membranes to one side of a black lipid membrane and then follow the potential changes generated across the composite membrane when ATP is suddenly released from caged ATP. It turns out that phosphorylation is accompanied by a sudden outward movement of positive charge, but this is not seen if the enzyme has been pretreated with α-chymotrypsin (Borlinghaus et al., 1987). It follows that the phosphorylation step is electrically silent, but the subsequent conformational change involves a net movement of positive charge in an outward direction.

That is as far as the story goes. Obviously, there is still an enormous gap between our knowledge of structure and our knowledge of mechanism. Just how that gap is to be filled it would be rash to predict. Presumably, as our knowledge of structure becomes more detailed, and Na,K-ATPase preparations with site-directed mutations become available, the parallel play of the child psychologists will be replaced by a more interesting and more creative game. But to know what that game is like we shall have to wait and see.

References

Beaugé, L. A., and I. M. Glynn. 1979a. Occlusion of K ions in the unphosphorylated sodium pump. *Nature.* 280:510–512.

Beaugé, L. A., and O. Ortiz. 1973. Na fluxes in rat red blood cells in K-free solutions. *Journal of Membrane Biology.* 13:165–184.

Blostein, R., and L. Chu. 1977. Sidedness of (sodium, potassium)-adenosine triphosphatase of inside-out red cell membrane vesicles. Interactions with potassium. *Journal of Biological Chemistry.* 252:3035–3043.

Borlinghaus, R. T., H.-J. Apell, and P. Läuger. 1987. Fast charge-translocations associated with partial reactions of the Na,K-pump. I. Current and voltage transients after photochemical release of ATP. *Journal of Membrane Biology.* 97:161–178.

Cavieres, J. D., and I. M. Glynn. 1979. Sodium-sodium exchange through the sodium pump: the roles of ATP and ADP. *Journal of Physiology.* 297:637–645.

Cleland, W. W. 1963. The kinetics of enzyme-catalyzed reactions with two or more substrates or products. I. Nomenclature and rate equations. *Biochimica et Biophysica Acta.* 67:104–137.

Esmann, M., and J. C. Skou. 1985. Occlusion of Na^+ by the Na,K-ATPase in the presence of oligomycin. *Biochemical and Biophysical Research Communications.* 127:857–863.

Fahn, S., M. R. Hurley, G. J. Koval, and R. W. Albers. 1966a. Sodium-potassium-activated adenosine triphosphatase of *Electrophorus* electric organ. II. Effects of *N*-ethylmaleimide and other sulfhydryl reagents. *Journal of Biological Chemistry.* 241:1890–1895.

Fahn, S., G. J. Koval, and R. W. Albers. 1966b. Sodium-potassium-activated adenosine triphosphatase of *Electrophorus* electric organ. I. An associated sodium-activated transphosphorylation. *Journal of Biological Chemistry.* 241:1882–1889.

Fendler, K., G. Nagel, E. Grell, and E. Bamberg. 1988. Electrogenic properties of the Na^+K^+-ATPase. *In* Na,K-ATPase: Cellular Aspects. J. C. Skou, M. Esmann, J. G. Nørby, and A. B. Maunsbach, editors. Alan R. Liss, Inc., New York, NY. In press.

Forbush, B., III. 1985. Rapid ion movements in a single turnover of the Na^+ pump. *In* The Sodium Pump. I. M. Glynn and J. C. Ellory, editors. Company of Biologists, Cambridge, England. 599–611.

Garay, R. P., and P. J. Garrahan. 1973. The interaction of sodium and potassium with the sodium pump in red cells. *Journal of Physiology.* 231:297–325.

Garrahan, P. J., and I. M. Glynn. 1967a. The behaviour of the sodium pump in red cells in the absence of external potassium. *Journal of Physiology.* 192:159–174.

Garrahan, P. J., and I. M. Glynn. 1967b. The sensitivity of the sodium pump to external sodium. *Journal of Physiology.* 192:175–188.

Garrahan, P. J., and I. M. Glynn. 1967c. Factors affecting the relative magnitudes of the sodium:potassium and sodium:sodium exchanges catalysed by the sodium pump. *Journal of Physiology.* 192:189–216.

Garrahan, P. J., and I. M. Glynn. 1967d. The stoichiometry of the sodium pump. *Journal of Physiology.* 192:217–235.

Garrahan, P. J., and I. M. Glynn. 1967e. The incorporation of inorganic phosphate into adenosine triphosphate by reversal of the sodium pump. *Journal of Physiology.* 192:237–257.

Glynn, I. M. 1985. The Na$^+$,K$^+$-transporting adenosine triphosphatase. *In* The Enzymes of Biological Membranes. 2nd ed. A. M. Martonosi, editor. Plenum Publishing Corp., New York, NY. 3:35–114.

Glynn, I. M. 1988. The coupling of enzymatic steps to the translocation of sodium and potassium. *In* Na,K-ATPase: Molecular Aspects. J. C. Skou, M. Esmann, J. G. Nørby, and A. B. Maunsbach, editors. Alan R. Liss, Inc., New York, NY. In press.

Glynn, I. M., Y. Hara, and D. E. Richards. 1984. The occlusion of sodium ions within the mammalian sodium-potassium pump: its role in sodium transport. *Journal of Physiology.* 351:531–547.

Glynn, I. M., and J. F. Hoffman. 1971. Nucleotide requirements for sodium-sodium exchange catalysed by the sodium pump in human red cells. *Journal of Physiology.* 218:239–256.

Glynn, I. M., J. L. Howland, and D. E. Richards. 1985. Evidence for the ordered release of rubidium ions occluded within the Na,K-ATPase of mammalian kidney. *Journal of Physiology.* 368:453–469.

Glynn, I. M., D. E. Richards, and Y. Hara. 1985*b*. The properties and role of occluded ion forms of the Na,K-ATPase. *In* The Sodium Pump. I. M. Glynn and J. C. Ellory, editors. Company of Biologists, Cambridge, England. 589–598.

Glynn, I. M., and S. J. D. Karlish. 1976. ATP hydrolysis associated with an uncoupled sodium flux through the sodium pump: evidence for allosteric effects of intracellular ATP and extracellular sodium. *Journal of Physiology.* 256:465–496.

Glynn, I. M., and V. L. Lew. 1970. Synthesis of adenosine triphosphate at the expense of downhill cation movements in intact human red cells. *Journal of Physiology.* 207:393–402.

Glynn, I. M., V. L. Lew, and U. Lüthi. 1970. Reversal of the potassium entry mechanism in red cells, and with and without reversal of the entire pump cycle. *Journal of Physiology.* 207:371–391.

Glynn, I. M., and D. E. Richards. 1982. Occlusion of rubidium ions by the sodium-potassium pump: its implications for the mechanism of potassium transport. *Journal of Physiology.* 330:17–43.

Hara, Y., and M. Nakao. 1986. ATP-dependent proton uptake by proteoliposomes reconstituted with purified Na$^+$,K$^+$-ATPase. *Journal of Biological Chemistry.* 261:12655–12658.

Hebert, H., P. L. Jørgensen, E. Skriver, and A. B. Maunsbach. 1982. Crystallization patterns of membrane-bound (Na$^+$ + K$^+$)-ATPase. *Biochimica et Biophysica Acta.* 689:571–574.

Hegyvary, C., and R. L. Post. 1971. Binding of adenosine triphosphate to sodium and potassium ion-stimulated adenosine triphosphatase. *Journal of Biological Chemistry.* 246:5234–5240.

Jørgensen, P. L. 1974. Purification and characterization of (Na$^+$ + K$^+$)-ATPase. III. Purification from the outer medulla of mammalian kidney after selective removal of membrane components by SDS. *Biochimica et Biophysica Acta.* 356:36–52.

Jørgensen, P. L. 1975. Purification and characterization of (Na$^+$ + K$^+$)-ATPase. V. Conformational changes in the enzyme. Transitions between the Na-form and the K-form studied with tryptic digestion as a tool. *Biochimica et Biophysica Acta.* 401:399–415.

Jørgensen, P. L., E. Skriver, H. Herbert, and A. B. Maunsbach. 1982. Structure of the Na,K-pump. Crystallization of pure membrane-bound Na,K-ATPase and identification of functional domains of the alpha subunit. *Annals of the New York Academy of Sciences.* 402:207–224.

Karlish, S. J. D., and I. M. Glynn. 1974. An uncoupled efflux of sodium ions from human red cells probably associated with Na-dependent ATPase activity. *Annals of the New York Academy of Sciences.* 242:461–470.

Karlish, S. J. D., and U. Pick. 1981. Sidedness of the effects of sodium and potassium ions on the conformational state of the sodium-potassium pump. *Journal of Physiology.* 312:505–529.

Karlish, S. J. D., D. W. Yates, and I. M. Glynn. 1978. Elementary steps of the $(Na^+ + K^+)$-ATPase mechanism studied with formycin nucleotides. *Biochimica et Biophysica Acta.* 525:230–251.

Kawakami, K., S. Noguchi, M. Noda, H. Takahashi, T. Ohta, M. Kawamura, H. Nojima, K. Nagano, T. Hirose, S. Inayama, H. Hayashida, T. Miyata, and S. Numa. 1985. Primary structure of the alpha-subunit of *Torpedo californica* $(Na^+ + K^+)$ATPase deduced from cDNA sequence. *Nature.* 316:733–736.

Kyte, J., and R. F. Doolittle. 1982. A simple method for displaying the hydropathic character of a protein. *Journal of Molecular Biology.* 157:105–132.

Lee, K. H., and R. Blostein. 1980. Red cell sodium fluxes catalysed by the sodium pump in the absence of K^+ and ADP. *Nature.* 285:338–339.

Lew, V. L., M. A. Hardy, and J. C. Ellory. 1973. The uncoupled extrusion of Na^+ through the Na^+ pump. *Biochimica et Biophysica Acta.* 323:251–266.

Marin, R., and J. F. Hoffman. 1988. Two different types of anion coupled Na transport mediated by the red cell Na/K pump. *In* Na,K-ATPase: Cellular Aspects. J. C. Skou, M. Esmann, J. G. Norby, and A. B. Maunsbach, editors. Alan R. Liss, Inc., New York, NY. In press.

Nørby, J. G., and J. Jensen. 1971. Binding of ATP to brain microsomal ATPase. Determination of the ATP-binding capacity and the dissociation constant of the enzyme-ATP complex as a function of K^+ concentration. *Biochimica et Biophysica Acta.* 233:104–116.

Post, R. L., C. Hegyvary, and S. Kume. 1972. Activation by adenosine triphosphate in the phosphorylation kinetics of sodium and potassium ion transport adenosine triphosphatase. *Journal of Biological Chemistry.* 247:6530–6540.

Post, R. L., and S. Kume. 1973. Evidence for an aspartyl phosphate residue at the active site of sodium and potassium ion transport adenosine triphosphatase. *Journal of Biological Chemistry.* 248:6993–7000.

Post, R. L., S. Kume, T. Tobin, B. Orcutt, and A. K. Sen. 1969. Flexibility of an active center in sodium-plus-potassium adenosine triphosphatase. *Journal of General Physiology.* 54:306*s*–326*s*.

Post, R. L., A. K. Sen, and A. S. Rosenthal. 1965. A phosphorylated intermediate in adenosine triphosphate-dependent sodium and potassium transport across kidney membranes. *Journal of Biological Chemistry.* 240:1437–1445.

Rephaeli, A., D. E. Richards, and S. J. D. Karlish. 1986. Electrical potential accelerates the $E_1P(Na) \rightarrow E_2P$ conformational transition of (Na,K)-ATPase in reconstituted vesicles. *Journal of Biological Chemistry.* 261:12437–12440.

Sachs, J. R. 1970. Sodium movements in the human red blood cell. *Journal of General Physiology.* 56:322–341.

Sachs, J. R. 1981. Mechanistic implications of the potassium-potassium exchange carried out by the sodium-potassium pump. *Journal of Physiology.* 316:263–277.

Sachs, J. R. 1986. Potassium-potassium exchange as part of the overall reaction mechanism of the sodium pump of the human red blood cell. *Journal of Physiology*. 374:221–244.

Shull, G. E., L. K. Lane, and J. B. Lingrell. 1986. Amino-acid sequence of the beta subunit of the (Na$^+$ + K$^+$)ATPase deduced from a cDNA. *Nature*. 321:429–431.

Shull, G. E., A. Schwartz, and J. B. Lingrell. 1985. Amino-acid sequence of the catalytic subunit of the (Na$^+$ + K$^+$)ATPase deduced from a complementary DNA. *Nature*. 316:691–695.

Siegel, G. J., and R. W. Albers. 1967. Sodium-potassium activated adenosine triphosphatase of *Electrophorus* electric organ. IV. Modification of responses to sodium and potassium by arsenite plus 2,3-dimercaptopropanol. *Journal of Biological Chemistry*. 242:4972–4975.

Simons, T. J. B. 1974. Potassium:potassium exchange catalysed by the sodium pump in human red cells. *Journal of Physiology*. 237:123–155.

Simons, T. J. B. 1975. The interaction of ATP-analogues possessing a blocked gamma-phosphate group with the sodium pump in human red cells. *Journal of Physiology*. 244:731–739.

Hematopoiesis: the Ontogeny and Regulation of Red Cells, White Cells, and Platelets

Chapter 2

Introduction to the Regulation of Hematopoiesis

David G. Nathan

*Division of Hematology and Oncology, The Children's Hospital;
the Dana-Farber Cancer Institute; and the Department of
Pediatrics, Harvard Medical School, Boston, Massachusetts*

Cell Physiology of Blood © 1988 by The Rockefeller University Press

In this chapter, we will attempt to review the rapidly growing and changing field of hematopoiesis (Lipton and Nathan, 1987; Sieff, 1987). This area has been the beneficiary of advances in molecular biology that have permitted the cloning of hematopoietic growth hormone genes and hence the production of large amounts of purified hormones with important biological activities. Before we launch into some detailed presentations, it would be useful to provide you with a brief overview because the subject is not necessarily one that is considered in detail by physiologists. This is particularly the case because the transmembrane signaling processes that occupy the attention of physiologists are, to date, entirely unknown in this field.

The bone marrow is a heterogeneous mixture of precursor cells, all of which will die by differentiation in the act of producing the cells of the circulation. As medical students, we painfully learn to identify the morphologic stages of precursor development, recognizing the amplification of cell production that occurs as cells mature and divide from their recognizable blast cell stages to the familiar cells of the peripheral blood. This process is not a self-renewing system. It represents progressive maturation associated with division. Unless the blast cells are renewed from yet another population, the marrow would become aplastic.

The existence of such a renewing population, generally referred to as the progenitor cell cohort, became recognized during the past 30 years (Metcalf and Moore, 1971). The classic studies of Alpen and Cranmore (1959) in bled and normal dogs proved that an erythroid progenitor cell, invisible in the marrow, must give rise to erythroblasts. A few years later, Metcalf and Moore (1971) in Melbourne and Sachs and co-workers (1987) in Israel demonstrated that such progenitor cells could give rise to colonies of differentiated precursor cells under in vitro conditions. But it was quickly noted that such colony formation, representing the differentiation of progenitor cells, required the presence of certain conditioned media. These were therefore said to contain colony-stimulating activities. Some conditioned media contained activities that would stimulate the formation of macrophage colonies; such media were said to contain macrophage colony-stimulating activity (M-CSA). Others produced granulocyte colonies; these were said to contain granulocyte colony-stimulating activity (G-CSA). Some induced formation of colonies of both granulocytes and macrophages, and such media were said to contain granulocyte/macrophage colony-stimulating activity (GM-CSA). Finally, certain media, particularly those derived from T cells, could induce erythroid colonies if erythropoietin was also present in the medium. Since erythroid colonies are often multi-segmented, such media were said to contain erythroid burst-promoting activity (BPA). All of these activities have now been defined as the various colony-stimulating factors. These cloned factors are listed in Table I.

The progenitors that give rise to differentiated hematopoietic colonies are also not capable of indefinite self-renewal. Although they amplify in response to hematopoietic hormones, their destiny is to give rise to the differentiated hematopoietic blast cells, which go on to amplify and differentiate further to form the cells of the peripheral blood. Therefore, such committed progenitors would themselves be incapable of sustaining marrow cellularity. They, too, must be fed from a pool of progenitors, a special class of stem cells, capable of self-renewal as well as differentiation into the various compartments of committed progenitors.

The studies of Till and McCullough (1961) revealed the presence of the so-called spleen colony-forming unit (CFU-S), a progenitor cell with self-renewal capacity.

They injected bone marrow into the circulation of irradiated mice and found discrete colonies of precursor cells in the spleens of these recipients. Transfer of a single colony into a second irradiated mouse gave rise to spleen colonies in the second animal, demonstrating that spleen colony-forming cells remained present in the initial colonies. This is firm evidence for self-renewal. Studies with retroviruses or chromosome markers have demonstrated that each of these colonies is derived from a single spleen colony-forming cell (Lemischka et al., 1986).

Thus, hematopoiesis is actually driven by an extremely small population of cells, the progenitor cell population, some of which are stem cells and most of which are committed progenitors of one of the recognizable precursor lineages. When irradiated

TABLE I
Purified Human Colony-stimulating Factors

Factor	Source*	Molecular weight[‡]	Gene clone	Biological activity	
				Progenitors	Cell lines
		$\times 10^{-3}$			
Multi-CSF or IL-3	MLA-144 Gibbon T cell line	20–26	cDNA: 133aa Mature protein Chromosome 5q	CFU-GGmm BFU-E CFU GM/G/M CFU Meg CFU-Eo	
GM-CSF	MO cell line	18–22	cDNA: 127aa mature protein Genomic clone: 2.5 kb, 3 introns, chromosome 5q21–5q32	CFU-GM CFU-GEMM BFU-E CFU-Meg CFU-Eo	KG-1 HL-60 Differentiation
M-CSF	Human urine	45 dimer	cDNA: 224aa/189aa proteins (26 kD), chromosome 5q	CFU-M	Monocyte Myelo-mono-cytic lines only
G-CSF	Human bladder carcinoma 5637	19.4	cDNA: 177aa mature protein (18.7 kD)	CFU-G	HL-60 Differentiation WEHI3B (D)[‡] Differentiation
Erythropoietin	Human urine	34–39	cDNA: 166aa mature protein (18.4 kD) Genomic clone: 5 exons, single gene, chromosome 7q11–q22	BFU-E CFU-E	—

*Source of purified factor.
[‡]Molecular weight of glycosylated protein.

mice are transplanted with colony-forming cells from a single spleen colony, they become clonal mice (Williams et al., 1984). All of their precursor cells are derived from a single cell. How does a stem progenitor cell make the decision to renew itself or differentiate to a committed progenitor cell? That decision appears to be stochastic and, until recently, it has not been affected by external manipulation. In this symposium, we will hear the first evidence derived by Dr. Makio Ogawa that certain cytokines may influence stem cell decisions to a limited degree.

Thus, if increased numbers of erythroid progenitors are required to meet the needs of bleeding or hemolysis, the number of erythroid progenitors increases as a result of the stimulation of such progenitors by the hormone erythropoietin (Udupa and Reissman, 1979). However, the number of stem cells remains constant unless the cells are exposed to interleukin 3 (IL-3), a stem cell mitogen (Kindler et al., 1976). The stem cell compartment "pays no attention" to action at or near the precursor level.

In summary, as shown in Fig. 1, hematopoiesis is driven initially by pluripotent stem cells with the capacity for both self-renewal and differentiation to committed progenitors, which in turn amplify in response to certain broadly active cytokines, such as IL-3 or GM-CSF. The more mature progenitors, such as CFU-E, CFU-G, and CFU-M, respond to their respective trophic hormones erythropoietin, G-CSF, and M-CSF. IL-3 and GM-CSF also induce the differentiation of the mature progenitors. The relationships of these hormones to their progenitors and cellular sites of origin will be discussed by Dr. Colin Sieff.

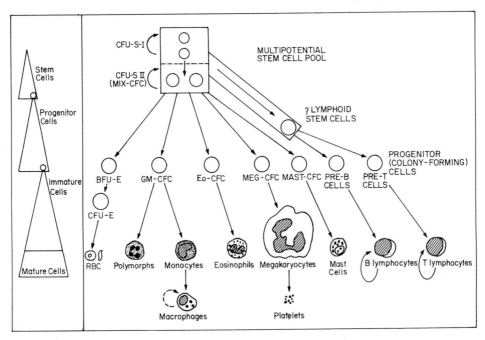

Figure 1. A schematic outline of the progenitor basis of hematopoiesis.

Although smears of bone marrows and lineage charts such as Fig. 1 are instructive, they do not reflect the actual conditions of the bone marrow. Fig. 2 represents a bone marrow section prepared by freeze fracture. Note that the precursor and progenitor cells are packed into fronds surrounded by fibroblasts and endothelial cells. The completed precursor cells must find their way through the endothelial cell gaps and enter the blood via a sinusoid that surrounds the frond. Michael Dexter has provided us with a great deal of information concerning the nature of the stroma in which the precursor and progenitor cells reside. His paper is a discussion of stem cells and the stroma in which they find a hospitable niche.

In the latter half of this section, John Adamson discusses the interaction of erythropoietin with its progenitors. Ronald Hoffman's paper deals with the most arcane of all of the hematopoietic cell systems, the megakaryocyte. Finally, Arthur

Nienhuis's paper discusses our long-sought goal, that of inserting engineered retroviruses into progenitor cells to alter gene expression. That is certainly our hope for the future.

For the present, however, the availability of the newly cloned cytokines that are trophic hormones in hematopoiesis is already stimulating exciting new explorations in clinical investigation. In their studies in simians, Donahue et al. (1976) and Nienhuis et al. (1987) have already demonstrated marked effects of human recombinant GM-CSF on hematopoiesis both in the normal monkey, in the monkey undergoing autotransplantation, and in at least one monkey with retroviral-induced cytopenia. Groopman and his colleagues (1987) have administered GM-CSF to patients with AIDS, who have developed leukocytosis in response to such infusions. Studies of

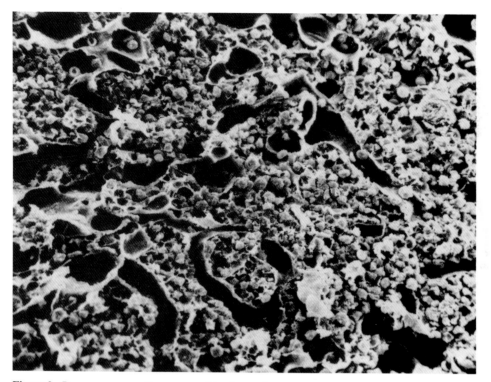

Figure 2. Bone marrow section prepared by freeze fracture.

patients with various forms of marrow failure are now planned and will soon be reported.

These are exciting times in the study of hematopoiesis. It is my hope that the papers collected here will introduce you to this new era and stimulate you to add the strengths of your technology to this burgeoning field.

References

Alpen, E. L., and D. Cranmore. 1959. Cellular kinetics and iron utilization in bone marrow as absorbed by Fe[59] autoradiography. *In* The Kinetics of Cellular Proliferation. F. Stohlman, Jr., editor. Grune & Stratton, New York, NY. 290–300.

Donahue, R. E., E. A. Wang, D. K. Stone, R. Kamen, G. G. Wong, P. K. Sehgal, D. G. Nathan, and S. C. Clark. 1986. Stimulation of haematopoiesis in primates by continuous infusion of recombinant human GM-CSF. *Nature*. 321:872–875.

Groopman, J. E., R. T. Mitsuyasu, M. J. DeLeo, D. H. Oette, and D. W. Golde. 1987. Effect of recombinant human granulocyte-macrophage colony-stimulating factor on myelopoiesis in the acquired immunodeficiency syndrome. *New England Journal of Medicine*. 317:593–588.

Kindler, V., B. Thorens, S. de Kossodo, B. Allet, J. F. Eliason, D. Thatcher, N. Farber, and P. Vassalli. 1986. Stimulation of hematopoiesis in vivo by recombinant bacterial murine interleukin-3. *Proceedings of the National Academy of Sciences*. 83:1001–1005.

Lemischka, I. R., D. H. Raulet, and R. C. Mulligan. 1986. Developmental and potential and dynamic behavior of hematopoietic stem cells. *Cell*. 45:917–927.

Lipton, J. M., and D. G. Nathan. 1987. The anatomy and physiology of hematopoiesis. *In* Hematology of Infancy and Childhood. 3rd ed. D. G. Nathan and F. A. Oski, editors. W. B. Saunders Co., Philadelphia, PA. 128–158.

Metcalf, D., and M. A. S. Moore. 1971. Haematopoietic Cells. Frontiers of Biology. A. Neuberger and E. L. Tatum, editors. Elsevier/North-Holland Amsterdam. 24:366–367.

Nienhuis, A. W., R. E. Donahue, S. Karlsson, S. C. Clark, B. Agricola, N. Antinoff, J. E. Pierce, P. Turner, W. F. Anderson, and D. G. Nathan. 1987. Recombinant human granulocyte-macrophage colony stimulating factor (GM-CSF) shortens the period of neutropenia following autologous bone marrow transplantation in a primate model. *Journal of Clinical Investigation*. 80:573–577.

Sachs, L. 1987. The molecular control of blood cell development (Review). *Science*. 238:1374–1379.

Sieff, C. A. 1987. Hematopoietic growth factors. *Journal of Clinical Investigation*. 79:1549–1557.

Till, J. E., and E. A. McCulloch. 1961. A direct measurement of the radiation sensitivity of normal mouse bone marrow cells. *Radiation Research*. 14:213.

Udupa, K. B., and D. R. Reissman. 1979. In vivo erythropoietin requirements of regenerating erythroid progenitors (BFU-e, CFU-e) in bone marrow of mice. *Blood*. 53:1164–1171.

Williams, D. A., I. R. Lemischka, D. G. Nathan, and R. C. Mulligan. 1984. Introduction of new genetic material into pluripotent haematopoietic stem cells of the mouse. *Nature*. 310:476–480.

Chapter 3

Growth and Differentiation of Hematopoietic Stem Cells

T. M. Dexter, I. L. O. Ponting, R. A. Roberts, E. Spooncer, C. Heyworth, and J. T. Gallagher

Departments of Experimental Haematology and Medical Oncology, Paterson Institute for Cancer Research, Christie Hospital and Holt Radium Institute, Withington, Manchester, England

Cell Physiology of Blood © 1988 by The Rockefeller University Press

Introduction and Historical Perspective

The majority of mature blood cells have only a short life span and must be continuously produced from precursor cells present in the bone marrow (Cronkite and Feinendegen, 1976). Such precursor cells are uni- or bipotential and are capable of undergoing proliferation, differentiation, and development into only one or two of the mature cell types. For example, granulocyte/macrophage colony-forming cells (GM-CFC) proliferate and develop into mature neutrophils and macrophages. Primitive erythroid progenitor cells (erythroid burst-forming units; BFU-E) proliferate and develop into more mature precursor cells (erythroid colony-forming units; CFU-E), which then undergo further growth and hemoglobinization to produce mature erythrocytes. Similarly, "lineage-restricted" progenitor cells have been described that give rise to eosinophils (Eos-CFC), basophils/mast cells (Bas-CFC), megakaryocytes (Meg-CFC), B lymphocytes (BL-CFC), and T lymphocytes (TL-CFC) (Abramson et al., 1977; Metcalf, 1977; Metcalf et al., 1979). But like the mature cells they produce, these lineage-restricted progenitor cells are also "short-lived" in that as they proliferate, they concomitantly undergo development and therefore "lose" their proliferative potential. Because of this, they are often termed "transit cells." Obviously, to fulfill the demands for mature cells that are being imposed upon the system, these progenitor cells must in turn be continuously generated from more primitive cells that have a capacity to persist throughout the life span of the animal. These are termed "stem" cells.

The earliest quantitative assay for the detection of stem cells was developed by Till and McCulloch in 1961. In a series of elegant experiments, they showed that if bone marrow cells are injected into mice that had received a potentially lethal dose of radiation (to ablate endogenous hematopoiesis), some of the injected cells settle in the spleen. There, they undergo clonal proliferation to produce macroscopically visible spleen nodules 8–12 d later. Upon examination, a proportion of these nodules can be shown to contain cells of the various myeloid lineages. In other words, at least some of the spleen colony-formation cells (CFU-S) are multipotential (Till and McCulloch, 1961; Curry and Trentin, 1967; Fowler et al., 1967). Furthermore, it has also been demonstrated that the spleen colonies contain precursor cells committed to only one or two cell lineages (e.g., GM-CFC, BFU-E, Meg-CFC), thus formally establishing a parent-progeny relationship between the multipotent stem cells and the lineage-restricted precursor cells. Other work has clearly shown that some CFU-S are able to give rise to B and T lymphocytes, thus establishing these cells as potential founder cells of the total hematopoietic system (Schrader and Schrader, 1978; Jones-Villeneuve and Phillips, 1980). However, it is clear that not all CFU-S are multipotent stem cells, since the spleen colony assay also detects primitive cells in a relatively advanced stage of differentiation, able to give rise to, e.g., only erythroid cells (Bleiberg et al., 1967; Magli et al., 1982). In addition, CFU-S have been described that can reconsitute only the restricted compartments of the hematopoietic system, producing various mixtures of myeloid cells, B lymphoyctes, and T lymphoyctes (Abramson et al., 1977; Quesenberry and Levitt, 1979; Lemischka et al., 1986). In other words, there appears to be a heterogeneity in the stem cell system. However, a word of caution: to what extent this heterogeneity is a reflection of the assay system employed in looking for changes in developmental potential, as opposed to an intrinsic (and genetically determined) heterogeneity, is still open to question.

Nonetheless, these assays all support the existence of multipotential cells. Using various markers (either chromosomal markers or the more recent retroviral insertion tagging techniques [Abramson et al., 1977; Dick et al., 1985; Lemischka et al., 1986]) to follow the behavior of cells, it has also been established that the hematopoietic system of animals can be effectively "rescued" (following marrow transplantation) by the progeny of a single cell (Abramson et al., 1977; Lemischka et al., 1986). Such reconstituted animals live a normal life span and do not seem to be compromised in their ability to produce mature myeloid cells or lymphocytes. It follows, therefore (in view of the arguments presented previously on the short-lived nature of the precursor cells), that the reconstituting cell is capable of extensive self-renewal, undergoing proliferation to produce more stem cells. This quality for self-renewal is the fundamental characteristic of the hematopoietic stem cell.

Self-Renewal and Differentiation

Since the very persistence of the mature cells of the hematopoietic system requires continuous output from the stem cell reservoir, this reservoir must be maintained at a sufficient level to take into account the proliferative demands of a normal life span. In actual fact, work has shown that the stem cell reserve is probably sufficient for several life spans, even under conditions of extreme hematopoietic stress (Boggs et al., 1982). Put another way, the loss from the stem cell compartment (by differentiation into the lineage-restricted progenitor cells) must be balanced by replenishment (as a consequence of self-renewal of stem cells). This can be formally stated as a probability p for differentiation. Prolonged (albeit minor) alterations in p would have dramatic biological consequences. If too many stem cells underwent differentiation, the stem cell reserve would rapidly become exhausted: if too many stem cells underwent self-renewal rather than differentiation, the production of mature cells would rapidly fail. Clearly, the mechanisms underlying the control of self-renewal and differentiation have implications for both normal and for tumor cell biology.

The Role of Growth Factors in Self-Renewal and Differentiation

Much attention has recently been focused upon the role of soluble mediators, growth promoters, and growth inhibitors in stem cell growth and development. The majority of these agents were initially described and characterized for their effects on committed precursor cells (Bradley and Metcalf, 1966; Ichikawa et al., 1966; Metcalf, 1984), but, perhaps paradoxically, at least some of these have direct or indirect effects upon multipotent stem cells (Burgess et al., 1977; Iscove et al., 1982; Stanley et al., 1986) (Table I). Interleukin 3 (IL-3) is a multilineage stimulating factor that facilitates the growth and development in vitro of many clonogenic precursor cells (GM-CFC, BFU-E, Eos-CFC, Meg-CFC) and mast cells (Iscove et al., 1982; Ihle et al., 1983). Significantly, however, IL-3 also promotes the development of multipotential cells (CFC-Mix) in vitro (Prystowsky et al., 1984) and can act as a powerful inducer of DNA synthesis in CFU-S (Garland and Crompton, 1983). Granulocyte/macrophage colony-stimulating factor (GM-CSF), initially thought to only stimulate proliferation and development of the GM-CFC, is now known to have a wider spectrum of target cells and can facilitate development of Eos-CFC and Meg-CFC (at relatively high factor concentrations). Furthermore, GM-CSF will also stimulate the proliferation of at least some BFU-E and CFC-Mix. In other words, it shows a remarkable overlap to

IL-3. Granulocyte colony-stimulating factor (G-CSF) and macrophage colony-stimulating factor (M-CSF), on the other hand, appear to be lineage-restricted molecules, influencing the growth and development of granulocytes and macrophages, respectively, from their precursor cells.

So far as is known, GM-CSF, G-CSF, and M-CSF have no growth-promoting effects on lymphoid cells. The situation with IL-3 is less clear, and a role in B cell and T cell proliferation has also been claimed (Ihle et al., 1982; Palacios et al., 1984). Furthermore, other growth factors that show activity on both myeloid and lymphoid cells have recently been described (O'Garra et al., 1988; Oppenheim et al., 1986; Stanley et al., 1986; Perschel et al., 1987), but the full range of their biological targets, including effects on multipotent stem cells, is yet to be determined. Therefore, a number of purified, characterized, and molecularly cloned growth-promoting agents are known that can influence hemopoietic cell development, at least in vitro, and that

Table I
Growth Factors in Hematopoiesis

Growth factor	Target Cells									
	CFU-S, CFC-Mix	B cell, pre-B cell	T cell, pre-T cell	GM-, M-, G-CFC	BFU-E	CFU-E	Meg-CFC	Mast cells	Eos-CFC	Mature cells
IL-1	+	+	+	?(+)	?	?	?	?	?	+
IL-2	?	+	+	−	−	−	−	−	−	+
IL-3	+	±	±	+	+	+	+	+	+	+
IL-4	+	+	+	+	+	+	+	+	?	?
IL-5	−	+	−	−	−	−	−	−	+	+
IL-6	?	+	?	?	?	?	?	?	?	?
HLGF-1	+	+	?	?	?	?	?	?	?	?
CFU-S inhibitor	+	?	?	−	−	−	−	−	−	−
CFU-S stimulator	+	?	?	−	−	−	−	−	−	−
GM-CSF	±	−	−	+	±	−	+	−	+	+
CSF-1	−	−	−	+	−	−	−	−	−	+
G-CSF	−	−	−	+	−	−	−	−	−	+
Meg-CSF	−	−	−	−	−	−	+	−	−	+
Erythropoietin	−	−	−	−	−	+	±	−	−	+

+, direct stimulation; −, no stimulation; ?, not determined. CFU-S are stem cells which form nodules in the spleens of irradiated mice. CFC-Mix are cells that produce multilineage colonies when cultured in vitro.

show varying efficacy at recruiting stem cells or precursor cells to undergo proliferation and development. However, superimposed upon the growth-promoting agents are other biological response modifiers that are not, by themselves, acting as growth factors.

For example, factors have been described that can modify the cell cycle status of CFU-S, acting either as an inhibitor or as a stimulator of DNA synthesis (Lord et al., 1976, 1977). A DNA synthesis inhibitor has also been described for BFU-E (Axelrad et al., 1981). Presumably, these molecules act to facilitate or to antagonize the response of cells to growth factors.

Perhaps of more immediate interest is the effect of interleukin 1 (IL-1). Several years ago, a "factor" was described that synergized with other known growth factors to promote clonal expansion of hematopoietic target cells (Bradley and Hodgson, 1979). This was later partially purified and termed hematopoietin 1 (H-1) and was shown by Stanley and colleagues to facilitate the response of primitive cells to lineage-restricted growth factors such as M-CSF (Bartelmez and Stanley, 1985; Jubinsky and Stanley, 1985; Stanley et al., 1986). More recently, however, it has become apparent that H-1 is probably the same molecule as IL-1, and that the latter has all of the biological effects attributed to H-1 (Moore et al., 1987; Mochizuki et al., 1987) and can recruit multipotent cells to an M-CSF–responsive state. This is obviously of profound biological significance since it indicates the possibility of modulating stem cell differentiation.

However, the experiments described can be faulted in the sense that impure target cell populations were used, and it was unclear whether the IL-1 was having a direct effect upon the multipotent stem cells, or an indirect effect upon these cells by stimulating the release of other growth factors by accessory cells present in the target cell population. Certainly, it is known that many different types of cells, including macrophages, lymphocytes, and connective tissue cells, have receptors for and can respond to IL-1.

To further investigate this possibility, we have used fluorescence-activated cell-sorted (FACS) marrow cells that have been highly enriched for day-12 CFU-S according to our published procedure (Lord and Spooncer, 1986). This population is made up of 50–100% "pure" multipotent stem cells and is effectively free of accessory cells that may interfere with the assay. These cells were then plated in soft-gel culture systems, in the presence of various growth factors that are more or less restricted in their target cells. The results are shown in Fig 1. Several points are worthy of comment. First, between 5 and 15% of the cells plated will produce mixed myeloid colonies in response to IL-3. Thus, at least some of the multipotential stem cells present in the enriched population contain receptors for and can respond to IL-3. (It should be stressed here that the relatively low colony-forming efficiency may not be a true reflection of the lack of the ability of the remaining cells to respond to IL-3, but may simply reflect suboptimal growth conditions. In other words, the potential IL-3–inducible colony-forming cells may be much higher.) Significantly, no survival or growth was seen in the presence of IL-1 alone, a finding in agreement with previously published data (Bartelmez and Stanley, 1985). When IL-3 and IL-1 were used in combination, although the effects were variable, only a slight effect on colony formation was usually seen. At least in this system, therefore, IL-1 had little ability to recruit more multipotent cells to an IL-3 responsive state in terms of colony formation. (This is in contrast to other findings, demonstrating a clear synergy when IL-3 and IL-1 are used in combination using 5-fluorouracil [5-FU]-treated but otherwise unfractionated target cells [Stanley et al., 1986].) The effects with M-CSF, however, were dramatic. By itself, M-CSF was found to be a poor stimulus for growth and differentiation of the multipotent cells, usually producing relatively few colonies when compared with IL-3–treated cultures. In some experiments, no colony growth was observed. In combination with IL-1, however, between 5 and 10% of the cells could be induced to produce colonies. Significantly, these contained only macrophages: other lineages were conspicuously absent. Also, co-treatment with IL-3 (i.e., a combination of IL-3, M-CSF, and IL-1) had no appreciable additive effect over and above that seen with M-CSF and IL-1. Thus, the combination of M-CSF and IL-1 is almost certainly recruiting most of the cells that respond to IL-3 alone; i.e., the IL-1 is promoting the stem cells to a state where they can respond to M-CSF and undergo proliferation and development to produce mature macrophages. The underlying mechanisms responsible for this transition are unclear, but presumably involve changes at the level of the M-CSF receptor. Furthermore, possible synergy with other growth factors such as G-CSF and GM-CSF still remain to be elucidated. Nonetheless, these experiments demonstrate an important tool in research aimed at understanding how differentiation of stem cells is controlled.

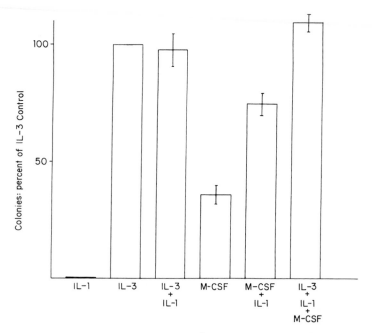

Figure 1. The FACS-sorted cells were used at 10^3 cells/plate and were incorporated in a single agar layer, which also contained 20% fetal calf serum, 10% bovine serum albumin, and Iscove's modified Dulbecco's medium supplemented with L-glutamine, sodium bicarbonate, and antibiotics. The results are shown as percentages of the IL-3 control, which produced an average of 68 colonies/plate after 10 d. All growth factors were recombinant except M-CSF, which was partially purified from L cell–conditioned medium. Plateau concentrations of growth factors were used.

Role of Stromal Cells in Hematopoiesis

The results described above suggest a simple model for the control of hematopoiesis, namely that homeostasis (and the balance between self-renewal and differentiation of stem cells) is determined by growth factors (lineage-indifferent and lineage-restricted) and that the response of cells to these factors is in turn modulated by agents such as IL-1. But this model may be too reductionist. Hematopoiesis in vivo, for example, occurs in association within a complex network of marrow stromal cells. The primary importance of these stromal cells in influencing the growth and development of hematopoietic stem cells and precursor cells was formally demonstrated using long-term bone marrow cultures.

In these cultures, in vitro bone marrow–derived stromal cells are allowed to adhere to, and to grow on, a suitable substratum. Within 2–3 wk, the adherent cells produce a complex multilayer of cells representative of the stromal cell types present in vivo (Allen and Dexter, 1983). For example, using various histochemical and immuno- logical techniques, adherent cells can be detected that are analogous with marrow adipocytes, reticular cells, macrophages, endothelial cells, and fibroblasts (Allen and Dexter, 1976, 1984; Dexter et al., 1977). When these stromal cell cultures are seeded with hematopoietic cells, the stem cells that are present in the inoculum first attach to the stromal cells and subsequently migrate within the three-dimensional matrix provided by the stromal cell multilayer. Within this environment, the stem cells then undergo proliferation, self-renewal, and differentiation to produce the lineage- restricted precursor cells. The latter, under appropriate conditions, then go on to produce mature progeny (Fig. 2). Significantly, self-renewal and differentiation in these conditions can be maintained for months (Dexter et al., 1984*b*). This should be contrasted with the short-term clonogenic assay systems, where, in the presence of the

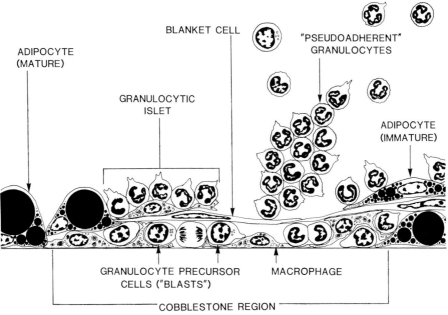

Figure 2. Schematic representation of the interactions between hemopoietic and stromal cells in long-term cultures.

growth factors(s) described previously, proliferation and development readily occur, but little or no self-renewal is found (Bradley and Metcalf, 1966; Pluznik and Sachs, 1966; Metcalf and Burgess, 1982). In other words, the growth factor–supplemented cultures favor the process of differentiation and development rather than self-renewal. In the long-term marrow cultures, on the other hand, in the presence of stromal cells and in the absence of added growth factors, homeostasis is maintained. Obviously, the stromal cell cultures are more strictly comparable with hematopoiesis in vivo and the data imply that the marrow stromal cells can produce the growth factors and other stimuli that are necessary for hematopoiesis. Because of this, much emphasis has been placed upon investigating the interaction between stromal cells and hematopoietic stem cells.

Interactions of Stromal Cells with Hematopoietic Stem Cells

In attempts to reconcile the observations of growth factor vs. stromal cell hematopoiesis, one possibility explored was that hematopoiesis was being driven by diffusible mediators released from stromal cells. These studies, however, have been somewhat disappointing in that the major species of growth factor, released from marrow stromal cells, was found to be M-CSF, much of which was biologically inactive (Shadduck et al., 1983). More recently, biological and molecular studies have also shown that the stromal cell environment can constitutively or inducibly secrete GM-CSF, G-CSF, Meg-CSF, IL-1, IL-4, and the CFU-S cell cycle modulators described previously. However, these factors by themselves are unable to maintain self-renewal of stem cells: when stem cells are cultured in medium conditioned by marrow stromal cells, they rapidly die. Significantly, no IL-3 is produced by, or present in, the stromal cells (using both bioassays and molecular probes) (Spooncer et al., 1986), although, as we will discuss later, an IL-3–like activity is obviously present. Thus, release of growth factors does not appear to be the driving force behind self-renewal of stem cells.

These observations gave credence to the possibility of stromal cell–bound mediators, having their effect by direct cell:cell contact, producing a local permissive milieu for stem cell self-renewal and differentiation (Dexter et al., 1984*a*). Certainly, early studies established the importance of cell contact: if the stem cells are separated from the stromal cells by interposing a thin agar layer or a millipore filter, the stem cells rapidly die (Bentley, 1981). Physical contact is important.

However, the nature of this contact has been a difficult problem to approach. First, the heterogeneity of the stromal cell environment (which stromal cells provide the necessary environment for stem cell self-renewal?) has precluded many studies at the mechanistic level. Second, the low incidence of multipotent stem cells (capable of self-renewal) in normal bone marrow (Dexter et al., 1984*a*) has made it difficult to examine directly the interaction of stem cells with stroma.

The latter problem, however, has been largely overcome by the development of continuously growing multipotent stem cell lines (Spooncer et al., 1984). Using appropriate selective conditions (in this case, selection of stem cells on marrow stromal cells transformed by a virus containing the *src* oncogene), multipotent stem cells can be isolated from long-term marrow cultures, which grow continuously in liquid culture in the presence of IL-3. These cells (FDCP-Mix) have a diploid karyotype, are non-leukemic, and can be induced to undergo multilineage differentiation when inoculated onto normal marrow stromal cells (Spooncer et al., 1984). Significantly, although their continuous growth in liquid culture requires IL-3 (in its absence, they die), their ability to survive and grow on marrow stromal cells occurs in the absence of added IL-3. Since the stromal cells themselves are not producing IL-3 (Spooncer et al., 1986), some other molecule(s) is obviously replacing the normal growth factor requirement. The advantage of such cell lines, of course, is that they can readily be produced in unlimited numbers. Nonetheless, the marrow stromal cell heterogeneity still imposes significant problems in further mechanistic studies.

Because of this, we screened for the ability of the FDCP-Mix cells to attach to, survive, and proliferate on a variety of cloned stromal cell lines (Roberts et al., 1987). Somewhat to our surprise, we found that the FDCP-Mix cells could be maintained most efficiently on the primitive mesenchymal cell line, 3T3. The reasons for this are unclear, but presumably represent the phenotypic diversity of embryonic cells.

Furthermore, FACS-enriched, normal-marrow, day-12 CFU-S also showed an ability to attach, survive, grow, and develop on 3T3 cells with kinetics similar to the FDCP-Mix cells (Figs. 3 and 4). Significantly, however, with both the FACS–CFU-S and the FDCP-Mix cells, little or no self-renewal occurs under these conditions (Roberts et al., 1987). Thus, although part of the "marrow microenvironment" is reflected by the 3T3 cells, one essential component (the ability to promote self-renewal) is obviously lacking. The system does, however, provide a model for examining cellular interactions, differentiation, and development.

In this system, the importance of cell contact is again emphasized by our finding that medium conditioned by 3T3 cells does not support the survival or development of either FACS–CFU-S or the FDCP-Mix cells. Furthermore, when the target hematopoietic cells are separated from the stroma (using a thin layer of agar), the cells die. Additionally, we have produced a series of other IL-3–dependent (but non-multipotent) cell lines that lack an ability to attach to stromal cells and cannot survive or grow on 3T3 cells (or on marrow stromal cells) in the absence of IL-3. These data suggest that physical proximity to the stromal cells is not, by itself, a sufficient stimulus for survival and growth, and that direct physical binding is of fundamental importance. The importance of this is further stressed by the observation that the growth and development of the FDCP-Mix cells does not require metabolically active stromal cells. For example, if the 3T3 cells are lightly fixed in glutaraldehyde (which effectively kills the cells but maintains the integrity of many membrane components), the FDCP-Mix cells can still bind to the fixed cells. Furthermore, they then undergo limited proliferation and develop into mature cells (Roberts et al., 1987). Obviously, production of growth factors by the stromal cells cannot be important under these conditions, and other possibilities need to be considered.

The Extracellular Matrix and Hematopoiesis

The potential importance of extracellular matrix molecules, fibronectin, collagens, and proteoglycans in cellular differentiation and development has been suggested from studies in other systems (Grobstein, 1975; Meier and Hay, 1975; Toole et al., 1977). Also, our own studies using β-D-xylosides (agents that disrupt the synthesis of proteoglycans) suggest an important role of such molecules in hematopoiesis (Spooncer et al., 1983), although whether their effect is direct or indirect has been difficult to determine. Recently, however, a series of intriguing experiments indicates that proteoglycans can bind hematopoietic growth factors and present them to hematopoietic cells (Gordon et al., 1987). This in turn raises the possibility that stromal cell–associated hematopoiesis is facilitated by such a mechanism, i.e., by membrane-bound growth factors that are intimately associated with extracellular matrix molecules and can exert their effects only when intimate physical contact takes place. Since our own results (using variant IL-3–dependent cell lines that cannot bind to stroma; Roberts et al., 1987) suggest that physical proximity is not by itself a sufficient stimulus, this in turn indicates that the stromal cell–bound growth factor exists in a cryptic form and is only available once stem cell binding has occurred. The nature of this binding is still a puzzle but probably reflects specific molecules (receptors?) present on the surface of the stromal cells and "recognition" molecules present on stem cells. Presumably, such stem cell "recognition" molecules are developmentally regulated and our recent studies, which show developmentally related

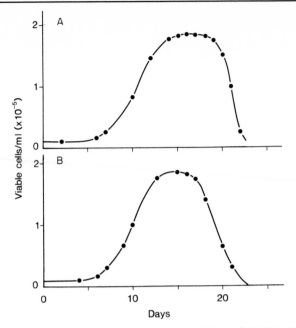

Figure 3. The number of viable cells after the inoculation of (*A*) 10^4 FACS–CFU-S and (*B*) 10^6 FDCP-Mix cells onto normal 3T3 monolayers.

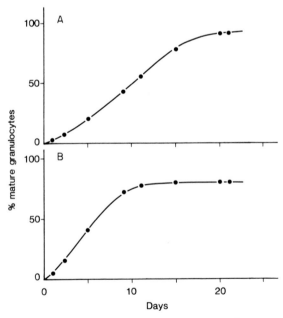

Figure 4. Kinetics of differentiation after the inoculation of (*A*) 10^4 FACS–CFU-S and (*B*) 10^6 FDCP-Mix cells onto normal 3T3 monolayers.

changes in surface membrane glycopeptides of hematopoietic cells (and major differences between stem cells, lineage-restricted precursor cells [Gallagher et al., 1987] and mature granulocytes), would tend to support this model, shown below.

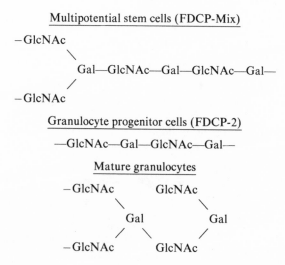

Multipotential stem cells (FDCP-Mix)

Granulocyte progenitor cells (FDCP-2)

—GlcNAc—Gal—GlcNAc—Gal—

Mature granulocytes

Thus, although the control of stem cell self-renewal and differentiation remains an enigma, the developments discussed have laid the foundation for an experimental approach at the mechanistic level.

References

Abramson, S., R. G. Miller, and R. A. Phillips. 1977. The identification in adult bone marrow of pluripotent and restricted stem cells of the myeloid and lymphoid systems. *Journal of Experimental Medicine*. 145:1567–1579.

Allen, T. D., and T. M. Dexter. 1976. Cellular inter-relationships during in vitro granulopoiesis. *Differentiation*. 6:191–194.

Allen, T. D., and T. M. Dexter. 1983. Long-term bone marrow cultures: an ultrastructural review. *In* Scanning Electron Microscopy. 1983. O. Johani, editor. SEM, Inc., Chicago, IL. 1851–1866.

Allen, T. D., and T. M. Dexter. 1984. The essential cells of the haemopoietic microenvironment. *Experimental Haematology*. 12:517–521.

Axelrad, A. A., H. Croizat, and D. Eskinazi. 1981. A washable macromolecule from Fv2^{rr} marrow negatively regulates DNA synthesis in erythropoietic progenitor cells BFU-E. *Cell*. 26:233–244.

Bartelmez, S. H., and E. R. Stanley. 1985. Synergism between haemopoietic growth factors (HGFs) detected by their effects on cells bearing receptors for a lineage specific HGF: assay of haemopoietin-1. *Journal of Cellular Physiology*. 122:370–378.

Bentley, S. A. 1981. Close range cell:cell interaction required for stem cell maintenance in long term bone marrow cultures. *Experimental Haematology*. 9:308–312.

Bleiberg, I., M. Liron, and M. Feldman. 1967. Studies on the regulation of haemopoietic spleen colonies. *Blood*. 29:469–480.

Boggs, D. R., S. S. Boggs, D. F. Saxe, L. A. Gress, and D. R. Canfield. 1982. Hematopoietic stem cells with high proliferative potential: assay of their concentration in marrow by their frequency and duration of cure of W/W^v mice. *Journal of Clinical Investigation.* 70:242–253.

Bradley, T. R., and G. S. Hodgson. 1979. Detection of primitive macrophage progenitor cells in mouse bone marrow. *Blood.* 54:1446–1450.

Bradley, T. R., and D. Metcalf. 1966. The growth of mouse bone marrow cells in vitro. *Australian Journal of Experimental Biology and Medical Science.* 44:287–300.

Burgess, A. W., J. Camakaris, and D. Metcalf. 1977. Purification and properties of colony-stimulating factor from mouse lung conditioned medium. *Journal of Biological Chemistry.* 252:1998–2003.

Cronkite, E. P., and L. E. Feinendegen. 1976. Notions about human stem cells. *Blood Cells.* 2:263–284.

Curry, J. L., and J. J. Trentin. 1967. Hemopoietic spleen colony studies. IV. Phytohemagglutinin and hemopoietic regeneration. *Journal of Experimental Medicine.* 126:819–832.

Dexter, T. M., T. D. Allen, and L. G. Lajtha. 1977. Conditions controlling the proliferation of haemopoietic stem cells in vitro. *Journal of Cellular Physiology.* 91:335–344.

Dexter, T. M., E. Spooncer, R. Schofield, B. I. Lord, and P. Simmons. 1984a. Haemopoietic stem cells and the problem of self-renewal. *Blood Cells.* 10:315–339.

Dexter, T. M., E. Spooncer, P. Simmons, and T. D. Allen. 1984b. Long-term marrow culture: an overview of techniques and experience. *In* Long-Term Bone Marrow Culture. D. G. Wright and J. S. Greenberger, editors. Alan R. Liss, Inc., New York, NY. 57–96.

Dick, J. E., M. C. Magli, D. Huszar, R. A. Phillips, and A. Bernstein. 1985. Introduction of a selectable gene into primitive stem cells capable of long-term reconstitution of the hemopoietic system of W/W^v mice. *Cell.* 42:71–79.

Fowler, J. H., A. M. Wu, J. E. Till, E. A. McCulloch, and L. Siminovitch. 1967. The cellular composition of hemopoietic spleen colonies. *Journal of Cellular Physiology.* 69:65–71.

Gallagher, J. T., A. J. Morris, and T. M. Dexter. 1987. Developmentally-related changes in surface membrane glycopeptides of murine haemopoietic cells. *Biochemical Journal.* 242:857–865.

Garland, J. M., and S. Crompton. 1983. A preliminary report. Preparation containing interleukin 3 (IL-3) promotes proliferation of multipotential stem cells (CFU-S) in the mouse. *Experimental Haematology.* 11:757–761.

Gordon, M. Y., G. P. Riley, S. M. Watt, and M. F. Greaves. 1987. Compartmentalization of a haemopoietic growth factor (GM-CSF) by glycosaminoglycans in the bone marrow environment. *Nature.* 326:403–405.

Grobstein, C. 1975. *In* Extracellular Matrix Influences on Gene Expression. H. C. Slavkin and R. C. Greulich, editors. Academic Press, Inc., New York, NY. 9–16.

Ichikawa, Y., D. H. Pluznik, and L. Sachs. 1966. In vitro control of the development of macrophage and granulocyte colonies. *Proceedings of the National Academy of Sciences.* 56:488–495.

Ihle, J. N., J. Keller, S. Oroszlan, L. E. Henderson, T. D. Copeland, F. Fitch, M. B. Prystowsky, E. Goldwasser, J. W. Schrader, E. Palaszynski, K. M. Dy, and B. Lebel. 1983. Biologic properties of homogeneous interleukin 3. I. Demonstration of WEHI-3 growth factor activity, mast cell growth factor activity, P cell-stimulating factor activity, colony stimulating factor activity and histamine-producing cell-stimulating factor activity. *Journal of Immunology.* 131:282–287.

Ihle, J. N., L. Rebar, J. Keller, J. C. Lee, and A. J. Hopel. 1982. Interleukin 3: possible roles in the regulation of lymphocyte differentiation and growth. *Immunological Reviews*. 63:5–32.

Iscove, N. N., C. A. Roitsch, N. Williams, and L. J. Guilbert. 1982. Molecules stimulating early red cell, granulocyte macrophage, megakaryocyte precursors in culture: similarity in size, hydrophobicity, and charge. *Journal of Cellular Physiology*. 1(Suppl.):65–78.

Jones-Villeneuve, E., and R. A. Phillips. 1980. Potentials for lymphoid differentiation by cells from long-term cultures of bone marrow. *Experimental Haematology*. 8:65–76.

Jubinsky, P. T., and E. R. Stanley. 1985. Purification of haemopoietin 1: a multilineage hemopoietic growth factor. *Proceedings of the National Academy of Sciences*. 82:2764–2768.

Lemischka, I. R., D. H. Raulet, and R. C. Mulligan. 1986. Developmental potential and dynamic behaviour of hematopoietic stem cells. *Cell*. 45:917–927.

Lord, B. I., K. J. Mori, and E. G. Wright. 1977. A stimulator of stem cell proliferation in regenerating bone marrow. *Biomedicine*. 27:223–226.

Lord, B. I., K. J. Mori, E. J. Wright, and L. J. Lajtha. 1976. An inhibitor of stem cell proliferation in normal bone marrow. *British Journal of Haematology*. 34:441–445.

Lord, B. I., and E. Spooncer. 1986. Isolation of haemopoietic spleen colony forming cells. *Lymphokine Research*. 5:59–72.

Magli, M. C., N. N. Iscove, and N. Odartchenko. 1982. Transient nature of early haemato-poietic spleen colonies. *Nature*. 295:527–529.

Meier, S., and E. D. Hay. 1975. Stimulation of corneal differentiation by interaction between cell surface and extracellular matrix. I. Morphometric analysis of transfilter "induction." *Journal of Cell Biology*. 66:275–291.

Metcalf, D. 1977. Haemopoietic colonies in vitro: cloning of normal and leukemic cells. *Recent Results in Cancer Research*. 61:1–227.

Metcalf, D. 1984. The Hemopoietic Colony Stimulating Factors. Elsevier/North Holland, Amsterdam. 494 pp.

Metcalf, D., and A. W. Burgess. 1982. Clonal analysis of progenitor cell commitment of granulocyte or macrophage production. *Journal of Cellular Physiology*. 111:275–283.

Metcalf, D., G. R. Johnson, and T. E. Mandel. 1979. Colony formation in agar by multipotential haemopoietic cells. *Journal of Cellular Physiology*. 98:401–420.

Mochizuki, D. Y., R. J. Tushinski, J. A. Eisenman, P. J. Conlon, and A. Larsen. 1987. Characterization of a human hematopoietic potentiating activity. *Blood*. 68(Suppl. 1):78a. (Abstr.)

Moore, M. A. S., D. Warren, and L. Souza. 1987. In vivo and in vitro action of G-CSF and IL-1 in myelosuppressed mice. *Journal of Cellular Biochemistry*. In press.

O'Garra, A., S. Umland, T. De France, and J. Christiansen. 1988. B-cell factors are pleiotropic. *Immunology Today*. 9:45–54.

Oppenheim, J. J., E. J. Kovacs, K. Maksushima and S. K. Durum. 1986. There is more than one interleukin 1. *Immunology Today*. 7:45–56.

Palacios, R., G. Henson, M. Steinmetz, and J. P. McKearn. 1984. Interleukin-3 supports growth of mouse pre-B-cell clones in vitro. *Nature*. 309:126–131.

Perschel, C., W. E. Paul, J. Ohara, and I. Green 1987. Effects of B-cell stimulating factor-1/interleukin 4 on hematopoietic progenitor cells. *Blood*. 70:254–263.

Pluznik, P. H., and L. Sachs. 1966. The induction of colonies of normal "mast" cells by a substance in conditioned medium. Experimental Cell Research. 43:553–563.

Prystowsky, M. B., G. Otten, M. F. Naujokas, J. Vardiman, J. H. Ihle, E. Goldwasser, and F. W. Fitch. 1984. Multiple hemopoietic lineages are found after stimulation of mouse bone marrow precursor cells with interleukin 3. *American Journal of Pathology.* 117:171–179.

Quesenberry, P., and L. Levitt. 1979. Hematopoietic stem cells. I. *New England Journal of Medicine* 301:755–769.

Roberts, R. A., E. Spooncer, E. K. Parkinson, B. I. Lord, T. D. Allen, and T. M. Dexter. 1987. Metabolically inactive 3T3 cells can substitute for marrow stromal cells to promote the proliferation and development of multipotent haemopoietic stem cells. *Journal of Cellular Physiology.* 132:203–214.

Schrader, J. W., and S. Schrader. 1978. In vitro studies on lymphocyte differentiation. I. Long-term in vitro culture of cells giving rise to functional lymphocytes in irradiated mice. *Journal of Experimental Medicine.* 148:823–828.

Shadduck, R. K., A. Waheed, J. S. Greenberger, and T. M. Dexter. 1983. Production of colony stimulating factor in long-term bone marrow cultures. *Journal of Cellular Physiology.* 114:88–92.

Spooncer, E., D. Boettiger, and T. M. Dexter. 1984. Continuous in vitro generation of multipotential stem cell clones from src-infected cultures. *Nature.* 310:228–230.

Spooncer, E., J. T. Gallagher, F. Krizsa, and T. M. Dexter. 1983. Regulation of haemopoiesis in long-term bone marrow cultures. IV. Glycosaminoglycan synthesis and the stimulation of haemopoiesis by β-D-xylosides. *Journal of Cell Biology.* 96:510–514.

Spooncer, E., C. M. Heyworth, A. Dunn, and T. M. Dexter. 1986. Self-renewal and differentiation of interleukin-3-dependent multipotent stem cells are modified by stromal cells and serum factors. *Differentiation.* 31:111–118.

Stanley, E. R., A. Bartocci, D. Patinkin, M. Rosendaal, and T. R. Bradley. 1986. Regulation of very primitive multipotent, hemopoietic cells by hemopoietin-1. *Cell.* 45:667–674.

Till, J. E., and E. A. McCulloch. 1961. A direct measurement of the radiation sensitivity of normal mouse bone marrow cells. *Radiation Research.* 14:213–222.

Toole, B. P., M. Okayama, R. W. Orkin, M. Yoshimura, M. Muto, and A. Kaji. 1977. Developmental roles of hyaluronic and chondroitin sulfate proteoglycans. *In* Cell and Tissue Interactions. J. W. Lash and M. M. Bareer, editors. Raven Press, New York, NY. 139–154.

Chapter 4

Stem Cell Functions Assessed in Clonal Culture

Makio Ogawa

*Veterans Administration Medical Center, and Department of
Medicine, Medical University of South Carolina, Charleston,
South Carolina*

Cell Physiology of Blood © 1988 by The Rockefeller University Press

Introduction

The central feature of the hematopoietic system is its life-long, continuous cell turnover. This cell renewal process is supported by stem cells, which possess the ability to self-renew and the ability to produce progenitors that are committed to differentiation in single lineages. During the last three decades, studies using clonal cell culture assays for progenitors at various stages of development (Golde, 1984) have resulted in significant elucidation of the mechanisms of hematopoiesis. In this review, I will summarize our recent studies of stem cell proliferation and differentiation in methylcellulose culture.

Background

It is generally held that, in the steady state, most hematopoietic stem cells are not engaged in active proliferation and reside in the so-called G_0 state. Experimental evidence for this concept has been provided by investigators in several laboratories (Becker et al., 1965; Hara and Ogawa, 1978; Hodgson and Bradley, 1979). In our laboratory, sequential observations of the growth of individual multipotential progenitors from spleen cells of 5-fluorouracil–treated mice suggested that most primitive multipotential progenitors are dormant in cell cycling (Suda et al., 1983a). The concept of a true resting state was originally proposed by Lajtha (1963), who envisioned that this state is metabolically distinct from the other phases of cell cycle, and coined the term G_0 state. He proposed that hematopoietic stem cells are in G_0 since cell cycle dormancy confers on the stem cells time to repair DNA damage and thus allows maintenance of the genetic integrity of the stem cell population (Lajtha, 1979). The mechanisms that might initiate cell division of the stem cells remained elusive until only recently. As will be discussed later, there is evidence that interleukin 1 (IL-1) and interleukin 6 (IL-6) appear to shorten the G_0 period and may induce the stem cells to begin active proliferation. I will summarize the role that growth factors play in stem cell proliferation.

Role of Growth Factors

IL-3 appears to be an important factor for continued proliferation of early hematopoietic progenitors. Investigators in several laboratories reported that purified murine IL-3 supports the growth of various types of colonies, including multilineage colonies (Ihle et al., 1983; Rennick et al., 1985; Suda et al., 1985). We have shown that it also supports the formation of multipotential blast cell colonies (Koike et al., 1986; Leary et al., 1987b). We also observed that delayed addition of murine IL-3 to cultures 7 d after cell plating decreases the number of multipotential blast cell colonies to one-half the number in cultures with Il-3 added on day 0 (Suda et al., 1985). It did not, however, alter the proliferative and differentiative characteristics of late-emerging, multipotential blast cell colonies (Suda et al., 1985). On the basis of these observations, we proposed that IL-3 does not trigger the dormant stem cells into active proliferation, but that IL-3 is required for the proliferation of early multipotential progenitors.

Very recently, several factors have been identified that appear to induce proliferation in hematopoietic stem cells. Stanley et al. (1986) reported that hemopoietin 1

(H-1) (Jubinsky and Stanley, 1985), which was purified from human bladder carcinoma cell line 5637, acts synergistically with IL-3 in support of proliferation of murine hematopoietic stem cells. Recent reports by Mochizuki et al. (1987) and Moore and Warren (1987) suggested that IL-1, which is abundant in the supernatant of this cell line, may account for the H-1 activity. In our laboratory, we have observed that IL-6 (also called interferon $\beta 2$ [Zilberstein et al., 1986] and B cell stimulatory factor 2 [Hirano et al., 1986]) is synergistic with IL-3 in support of the active proliferation of murine hematopoietic stem cells (Ikebuchi et al., 1987). In the presence of IL-3, multipotential blast cell colonies emerged after varying lag times from spleen cells of 5-fluorouracil–treated mice, as we reported previously (Suda et al., 1985). When IL-6 was also present, the emergence of multipotential blast cell colonies was significantly hastened, although the rate of the growth of the blast cell colonies was not affected (Ikebuchi et al., 1987). These data suggested that part of the H-1 effect of IL-6 is an apparent shortening of the G_0 period of hematopoietic stem cells. Human IL-6 also works synergistically with human IL-3 in stimulation of the proliferation of early human progenitors (Leary et al., 1987a). In murine studies, IL-6 appears to possess stronger synergistic effects than IL-1 (Ikebuchi et al., 1987). Since IL-1 can stimulate production of IL-6 in a variety of cell types (Yang et al., 1988; Yasukawa et al., 1987), it is possible that part of the H-1 effect of IL-1 may be mediated by IL-6.

Differentiation

Once the stem cells leave G_0 and begin active proliferation, the process of differentiation concomitantly takes place. Several models for stem cell differentiation have been proposed (Ogawa et al., 1983). The hematopoietic inductive microenvironment (HIM) model of Trentin and his associates (Curry and Trentin, 1967; Trentin, 1970) features small anatomical niches that direct differentiation of individual stem cells. The so-called "stem cell competition" model envisions control of stem cell differentiation by lineage-specific humoral factors. For example, VanZant and Goldwasser (Goldwasser, 1975; VanZant and Goldwasser, 1977) proposed that levels of erythropoietin and colony-stimulating factors determine the commitment of stem cells. In addition, a model of predetermined, sequential loss of lineage potentials has been proposed (Johnson, 1981). As discussed in my earlier review (Ogawa et al., 1983), specific criticism can be raised against the data supporting these models. Most importantly, these models are based on studies of populations of progenitors rather than single progenitors. In our laboratory, identification of murine blast cell colonies with high replating efficiencies (Nakahata and Ogawa, 1982) provided us with an opportunity to micromanipulate single progenitors and study their differentiation potentials. When we carried out cytological analysis of multilineage colonies that were derived from single progenitor cells and were cultured under identical conditions, we observed varying lineage combinations expressed in individual colonies (Suda et al., 1983b). Subsequently, studies of paired progenitors revealed heterologous pairs of single-lineage and multilineage colonies expressing diverse lineage combinations (Suda et al., 1984b). These observations were consistent with the notion that differentiation of stem cells is featured by stochastic (random) loss of lineage potentials. Examples of pairs of colonies expressing disparate lineage combinations are presented in Scheme I.

$$
\begin{array}{lll}
1. \left[\begin{array}{l} m \\ nmmastEM \end{array}\right] &
4. \left[\begin{array}{l} nmmastM \\ nmmastEM \end{array}\right] &
7. \left[\begin{array}{l} m \\ mM \end{array}\right] \\[3em]
2. \left[\begin{array}{l} M \\ nmmastEM \end{array}\right] &
5. \left[\begin{array}{l} M \\ nmM \end{array}\right] &
8. \left[\begin{array}{l} E \\ nm \end{array}\right] \qquad \text{(Scheme I)} \\[3em]
3. \left[\begin{array}{l} nm \\ nmmastEM \end{array}\right] &
6. \left[\begin{array}{l} nm \\ mM \end{array}\right] &
9. \left[\begin{array}{l} M \\ m \end{array}\right]
\end{array}
$$

(Abbreviations: m, monocyte/macrophage; n, neutrophil; mast, mast cell; E, erythrocyte; M, megakaryocyte.)

Some pairs, such as pairs 1 and 2, consisted of a single-lineage colony and a multilineage colony. In others, both colonies were multilineage colonies sharing one or more constituent lineages (e.g., pairs 3, 4, 6, and 7). Some pairs shared no lineages (e.g., pairs 8 and 9). Pairs like 1 and 2 indicate that a multipotential progenitor can give rise to a monopotential progenitor directly. Other types of pairs suggest that an alternative mode of commitment is progressive, random loss in lineage potential. Sequential manipulation of paired progenitors and analysis of the differentials of colonies derived from these progenitors indicated that the apparent random commitment in lineage potentials takes place sequentially during stem cell differentiation (Suda et al., 1984*a*). Analyses of human single and paired progenitors also revealed similar results and were consistent with the notion that the principle of human stem cell differentiation is also a random restriction in the lineage potentials (Leary et al., 1984, 1985).

In the study of human multilineage colonies of single cell origin, we observed that individual lineages were represented by a variable number of cells. For example, a mixed colony consisted of 1,340 erythrocytes and 4 eosinophilic leukocytes (Leary et al., 1984). This observation suggested that the number of times that a progenitor divides during the maturation process (the process after commitment to a single lineage) is variable. This observation is also in agreement with earlier observations on single-lineage colonies. Wu (1983) reported that the cumulative frequency distribution of the numbers of secondary T lymphocyte colonies per primary T lymphocyte colony could be approximated by a γ distribution. Analysis of the size of secondary mast cell colonies and computer stimulation of a modification of the "birth and death" model of Till et al. (1964) suggested that mast cell proliferation in culture is also a stochastic process (Pharr et al., 1985). These data were consistent with the notion that both differentiation and proliferation of hematopoietic progenitors are separate stochastic processes. In this model, hematopoietic factors do not influence the differentiation process. Rather, the functions of factors are to support survival and/or proliferation of progenitors.

ACKNOWLEDGMENTS

The author wishes to thank Dr. Pamela N. Pharr, Mrs. Anne G. Leary, and Mrs. Linda S. Vann for their assistance in the preparation of this manuscript.

This work was supported by National Institutes of Health grant AM-32294 and by the Veterans Administration. Dr. Ogawa is a VA Medical Investigator.

References

Becker, A. J., E. A. McCulloch, L. Siminovitch, and J. E. Till. 1965. The effect of differing demands for blood cell production on DNA synthesis by hemopoietic colony forming cells of mice. *Blood.* 26:296–308.

Curry, J. L., and J. J. Trentin. 1967. Hemopoietic spleen colony studies. I. Growth and differentiation. *Developmental Biology.* 15:395–413.

Golde, D. W. 1984. Methods in Hematology: Hematopoiesis. Churchill Livingstone, New York, NY. 73–149.

Goldwasser, E. 1975. Erythropoietin and differentiation of red blood cells. *Federation Proceedings.* 34:2285–2292.

Hara, H., and M. Ogawa. 1978. Murine hemopoietic colonies in culture containing normoblasts, macrophages, and megakaryoctyes. *American Journal of Hematology.* 4:23–34.

Hirano, T., K. Yasukawa, H. Harada, T. Taga, Y. Watanabe, T. Matsuda, S. Kashiwamura, K. Nakajima, K. Koyama, A. Iwamatsu, S. Tsunasawa, F. Sakiyama, H. Matsui, Y. Takahara, T. Taniguchi, and T. Kishimoto. 1986. Complementary DNA for a novel human interleukin (BSF-2) that induces B lymphocytes to produce immunoglobulin. *Nature.* 324:73–77.

Hodgson, G. S., and T. R. Bradley. 1979. Properties of haematopoietic stem cells surviving 5-fluorouracil treatment: evidence for a pre-CFU-S cell? *Nature.* 281:381–382.

Ihle, J. N., J. Keller, S. Oroszlan, L. E. Henderson, T. D. Copeland, F. Fitch, M. B. Prystowsky, E. Goldwasser, J. W. Schrader, E. Palaszynski, M. Dy, and B. Lebel. 1983. Biologic properties of homogeneous interleukin 3. I. Demonstration of WEHI-3 growth factor activity, mast cell growth factor activity, P cell-stimulating factor activity, colony-stimulating factor activity and histamine-producing cell-stimulating factor activity. *Journal of Immunology.* 131:282–287.

Ikebuchi, K., G. G. Wong, S. C. Clark, J. N. Ihle, Y. Hirai, and M. Ogawa. 1987. Interleukin-6 enhancement of interleukin-3-dependent proliferation of multipotential hemopoietic progenitors. *Proceedings of the National Academy of Sciences.* 84:9035–9039.

Johnson, G. R. 1981. Is erythropoiesis an obligatory step in the commitment of multipotential hematopoietic stem cells? *In* Experimental Hematology Today. S. J. Baum, G. D. Ledney, and A. Kahn, editors. Springer-Verlag, New York, NY. 13–20.

Jubinsky, P. T., and E. R. Stanley. 1985. Purification of hemopoietin 1: a multilineage hemopoietic growth factor. *Proceedings of the National Academy of Sciences.* 82:2764–2768.

Koike, K., J. N. Ihle, and M. Ogawa. 1986. Declining sensitivity to interleukin 3 of murine multipotential hemopoietic progenitors during their development. Application to a culture system that favors blast cell colony formation. *Journal of Clinical Investigation.* 77:894–899.

Lajtha, L. G. 1963. On the concept of the cell cycle. *Journal of Cellular and Comparative Physiology.* 62(Suppl. 1):143–144.

Lajtha, L. G. 1979. Stem cell concepts. *Differentiation.* 14:23–34.

Leary, A. G., K. Ikebuchi, G. G. Wong, Y.-C. Yang, S. C. Clark, and M. Ogawa. 1987*a*. Interleukin-6 is a synergistic factor for human hemopoietic stem cells. *Blood.* 70(Suppl. 1):176*a*. (Abstr.)

Leary, A. G., Y.-C. Yang, S. C. Clark, J. C. Gasson, D. W. Golde, and M. Ogawa. 1987*b*. Recombinant gibbon interleukin-3 (IL-3) supports formation of human multilineage colonies and blast cell colonies in culture: comparison with recombinant human granulocyte-macrophage colony-stimulating factor (GM-CSF). *Blood.* 70:1343–1348.

Leary, A. G., M. Ogawa, L. C. Strauss, and C. I. Civin. 1984. Single cell origin of multilineage colonies in culture: evidence that differentiation of multipotent progenitors and restriction of proliferative potential of monopotent progenitors are stochastic processes. *Journal of Clinical Investigation.* 74:2193–2197.

Leary, A. G., L. C. Strauss, C. I. Civin, and M. Ogawa. 1985. Disparate differentiation in hemopoietic colonies derived from human paired progenitors. *Blood.* 66:327–332.

Mochizuki, D. Y., J. A. Eisenman, P. J. Conlon, A. D. Larsen, and R. J. Tushinski. 1987. Interleukin 1 regulates hematopoietic activity, a role previously ascribed to hemopoietin 1. *Proceedings of the National Academy of Sciences.* 84:5267–5271.

Moore, M. A. S., and D. J. Warren. 1987. Synergy of interleukin-1 and granulocyte colony-stimulating factor: *in vivo* stimulation of stem-cell recovery and hematopoietic regeneration following 5-fluorouracil treatment of mice. *Proceedings of the National Academy of Sciences.* 84:7134–7138.

Nakahata, T., and M. Ogawa. 1982. Identification in culture of a class of hemopoietic colony-forming units with extensive capability to self-renew and generate multipotential colonies. *Proceedings of the National Academy of Sciences.* 79:3843–3847.

Ogawa, M., P. N. Porter, and T. Nakahata. 1983. Renewal and commitment to differentiation of hemopoietic stem cells: an interpretive review. *Blood.* 61:823–829.

Pharr, P. N., J. Nedelman, H. P. Downs, M. Ogawa, and A. J. Gross. 1985. A stochastic model for mast cell proliferation in culture. *Journal of Cellular Physiology.* 125:379–386.

Rennick, D. M., F. D. Lee, T. Yokota, K. Arai, H. Cantor, and G. J. Nabel. 1985. A cloned MCGF cDNA encodes a multilineage hematopoietic growth factor: multiple activities of interleukin 3. *Journal of Immunology.* 134:910–914.

Stanley, E. R., A. Bartocci, D. Patinkin, M. Rosendaal, and T. R. Bradley. 1986. Regulation of very primitive, multipotent, hemopoietic cells by hemopoietin-1. *Cell.* 45:667–674.

Suda, T., J. Suda, and M. Ogawa. 1983*a*. Proliferative kinetics and differentiation of murine blast cell colonies in culture: evidence for variable G_0 periods and constant doubling rates of early pluripotent hemopoietic progenitors. *Journal of Cellular Physiology.* 117:308–318.

Suda, T., J. Suda, and M. Ogawa. 1983*b*. Single-cell origin of mouse hemopoietic colonies expressing multiple lineages in variable combinations. *Proceedings of the National Academy of Sciences.* 80:6689–6693.

Suda, J., T. Suda, and M. Ogawa. 1984*a*. Analysis of differentiation of mouse hemopoietic stem cells in culture by sequential replating of paired progenitors. *Blood.* 64:393–399.

Suda, T., J. Suda, and M. Ogawa. 1984*b*. Disparate differentiation in mouse hemopoietic colonies derived from paired progenitors. *Proceedings of the National Academy of Sciences.* 81:2520–2524.

Suda, T., J. Suda, M. Ogawa, and J. N. Ihle. 1985. Permissive role of interleukin 3 (IL-3) in proliferation and differentiation of multipotential hemopoietic progenitors in culture. *Journal of Cellular Physiology.* 124:182–190.

Till, J. E., E. A. McCulloch, and L. Siminovitch. 1964. A stochastic model of stem cell proliferation, based on the growth of spleen colony forming cells. *Proceedings of the National Academy of Sciences.* 51:29–36.

Trentin, J. J., 1970. Influence of hematopoietic organ stroma (hematopoietic inductive microenvironments) on stem cell differentiation. *In* Regulation of Hematopoiesis. A. S. Gordon, editor. Appleton-Century-Crofts, New York, NY. 161–186.

VanZant, G., and E. Goldwasser. 1977. Simultaneous effects of erythropoietin and colony-stimulating factor on bone marrow cells. *Science.* 198:733–735.

Wu, A. M. 1983. Regulation of self-renewal of human T lymphocyte colony-forming units (TL-CFUs). *Journal of Cellular Physiology.* 117:101–108.

Yang, Y.-C., S. Tsai, G. G. Wong, and S. C. Clark. 1988. Interleukin-1 regulation of hematopoietic growth factor production by human stromal fibroblasts. *Journal of Cellular Physiology.* 134:292–296.

Yasukawa, K., T. Hirano, Y. Watanabe, K. Muratani, T. Matsuda, S. Nakai, and T. Kishimoto. 1987. Structure and gene expression of human B cell stimulatory factor-2 (BSF-2/IL-6) gene. *European Molecular Biology Organization Journal.* 6:2939–2945.

Zilberstein, A., R. Ruggieri, J. H. Korn, and M. Revel. 1986. Structure and expression of cDNA and genes for human interferon-beta-2, a distinct species inducible by growth stimulatory cytokines. *European Molecular Biology Organization Journal.* 5:2529–2537.

Chapter 5

Human Colony-stimulating Factors and Stromal Cell Function

Colin A. Sieff, Charlotte M. Niemeyer, and Douglas V. Faller

Division of Hematology and Pediatric Oncology, The Children's Hospital and Dana-Farber Cancer Institute, and Department of Pediatrics, Harvard Medical School, Boston, Massachusetts

Cell Physiology of Blood © 1988 by The Rockefeller University Press

Introduction

During the past three years, the genes for several human hematopoietic growth factors or colony-stimulating factors (CSFs) have been cloned and recombinant proteins have been produced and purified. These include multi-CSF, or interleukin 3 (IL-3) (Yang et al., 1986), granulocyte-macrophage CSF (GM-CSF) (Wong et al., 1985), granulocyte CSF (G-CSF) (Nagata et al., 1986), and macrophage CSF (M-CSF) (Kawasaki et al., 1985). Although the amino acid sequences of these four proteins are distinct, there is considerable overlap in their biological activities, and the genes for three of them, IL-3, GM-CSF, and M-CSF, together with the M-CSF receptor, have been localized to the long arm of chromosome 5 (Le Beau et al., 1987a, b; Pettenati et al., 1987).

Clonal assays for hematopoietic progenitor cells have clearly demonstrated that the CSFs are essential for in vitro progenitor survival, proliferation, and differentiation. It is interesting that they also affect mature cell function. This property, together with the observation that cells common to every organ can synthesize the CSFs, suggests that, in addition to their role in blood cell production, the CSFs may also have an important local role at a site of infection or inflammation.

Normal adult hematopoiesis occurs only in the microenvironment of the bone marrow, where macrophages, endothelial cells, and fibroblast-derived reticular cells and fat cells are the major constituents of the stromal cellular network (Dexter et al., 1977; Bentley, 1982). Since the production of the CSFs is not restricted to bone marrow "stromal" cells, it is clear that these cells must function in other ways to specifically nurture developing hematopoietic stem and progenitor cells.

In this report, we summarize and compare some of the functional properties of IL-3 and GM-CSF (two CSFs that act on the earliest progenitor cells), discuss the cellular origin of the CSFs, and indicate other potentially important functional roles that can be attributed to certain stromal cells.

Results and Discussion

Comparison of the Biological Activities of GM-CSF and IL-3

Many of the "accessory" cells now known to synthesize specific CSFs (see below) are present in unfractionated Ficoll-Hypaque–separated bone marrow or fetal liver cells. To assess the biological activities of highly purified recombinant proteins, it is therefore necessary to deplete the bone marrow of these factor-producing cells. We used plastic adherence to deplete monocytes, followed by incubation of the nonadherent cells with a panel of monoclonal antibodies that selectively label maturing erythroid, myeloid, and lymphoid cells (Sieff et al., 1986). The antibody-labeled cells were then removed by a panning method using plastic petri dishes that had previously been coated with rabbit anti-mouse immunoglobulins. The nonadherent antibody-negative cells were enriched for progenitor cells and did not contain significant numbers of factor-producing cells; they depended entirely on an exogenous source of CSF for colony formation in methylcellulose semisolid cultures. Previous experiments with purified recombinant GM-CSF showed that, despite its misleading nomenclature, this factor can act as a stimulus for early erythroid burst-forming units (BFU-E) and mixed colony-forming units (CFU-Mix), provided that erythropoietin (Ep) is added to the cultures (Sieff et al., 1985). Two subsets of BFU-E can be distinguished, one that responds to Ep and GM-CSF, and one that responds to Ep alone (Sieff et al., 1986). Ep is thus required for the terminal maturation of all erythroid progenitors, but if its

TABLE I
Response of Bone Marrow Cells to Multipoietins

Stimulus	Concentration	Colonies per 10^5 enriched bone marrow cells				
		BFU-E*	CFU-Mix*	CFU-G	CFU-M	CFU-Eo
	pM					
GM-CSF[‡]	500	420	60	240	20	20
IL-3[§]	500	939	60	60	40	20

*Erythropoietin (2 U/ml) was added on day 3 of culture.
[‡]Purified human recombinant GM-CSF (expressed in Chinese Hamster Ovary cells).
[§]Purified human recombinant IL-3 (expressed in *E. coli*).

addition is delayed for 3 d, the "background" of Ep-responsive BFU-E is virtually eliminated (Sieff et al., 1986). Table I shows the BFU-E–derived and CFU-Mix–derived colonies observed in a representative experiment using enriched bone marrow progenitor cells with GM-CSF added on day 0 and Ep on day 3. The table also shows that GM-CSF stimulates the formation of colonies derived from granulocyte, macrophage, and eosinophil progenitors as well (CFU-G, CFU-M, CFU-GM, CFU-Eo); megakaryocyte progenitors are also responsive (not shown).

Recently, the gene for the primate homologue of murine IL-3 was cloned from the gibbon T cell line MLA 144 (Yang et al., 1986). The human gene was subsequently isolated and expessed in *Escherichia coli*. Table I shows the results of a bioassay using highly purified human *E. coli* IL-3. It is clear that this protein induces the formation of colonies derived from a similar wide range of progenitor cells, as does GM-CSF. When IL-3 and GM-CSF were compared for their capacity to stimulate different progenitor classes, it was found that IL-3 induced more BFU-E–derived colonies than did GM-CSF (Fig. 1; Sieff et al., 1987*a*). The combination of both factors induced no

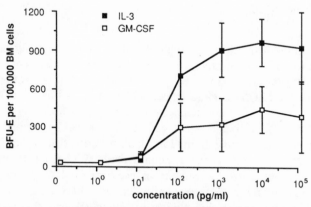

Figure 1. IL-3 stimulates more BFU-E than does GM-CSF. Enriched bone marrow progenitors were incubated with increasing concentrations of either highly purified recombinant human IL-3 or GM-CSF, and erythropoietin (2 U/ml) was added on day 3. While the half-maximal and maximal concentrations for the two CSFs were similar (10–100 pg/ml and 100–1,000 pg/ml, respectively), more BFU-E–derived colonies were observed with IL-3 than with GM-CSF. Each point represents the means of triplicate cultures ± 1 standard deviation from one representative experiment.

greater number of BFU-E–derived colonies than IL-3 alone, which suggests that GM-CSF stimulates a subset of the BFU-E that respond to IL-3. It was also found that IL-3 in combination with G-CSF induced the formation of more granulocyte colonies than either factor alone. Furthermore, the colonies that formed in the presence of both factors were larger than those that formed in either factor alone (Table II).

These data suggest that IL-3 and G-CSF may act on distinct but possibly overlapping subsets of granulocytic progenitor cells. GM-CSF also acts on mature neutrophils to induce a rapid increase in cell surface expression of a family of leukocyte adhesion molecules; this is associated with an increase in neutrophil aggregability (Arnaout et al., 1986). These effects occur at concentrations (picomolar) of GM-CSF similar to those that induce colony formation. It is interesting that while G-CSF has properties similar to those of GM-CSF, IL-3 does not have these actions on mature granulocytes (Arnaout, M. A., and C. A. Sieff, unpublished data). Binding studies with ^{125}I-labeled GM-CSF suggest that these effects are directly mediated, since neutrophils express low numbers (100–500 sites/cell) of high-affinity receptors for GM-CSF (kD $\simeq$ 150 pM). The expression of GM-CSF receptors on normal progenitor

TABLE II
IL-3 and G-CSF Synergize on CFU-G

Stimulus	Percent	Colonies per 10^5 enriched bone marrow		
		BFU-E	CFU-G	CFU-M
IL-3*	0.01	440	140 (400)[‡]	180
G-CSF[§]	0.03	0	140 (560)[‡]	20
IL-3 + G-CSF	0.1 + 0.03	400	440 (2,800)[‡]	160

*Conditioned medium from monkey COS-1 cells transfected with gibbon IL-3 cDNA.
[‡]Mean colony cell number per 6–12 sequentially picked CFU-G–derived colonies.
[§]Conditioned medium from monkey COS-1 cells transfected with human G-CSF cDNA.

cells has not been verified because of the difficulty in obtaining sufficient numbers of primitive cells for binding studies. However, limiting dilution assays of colony-forming cells in the presence of either GM-CSF or IL-3 indicate that the action of these two growth factors is direct (C.A. Sieff, unpublished data).

Cellular Origin of the CSFs

One of the major limitations to defining the cellular origin of the CSFs is the inability to obtain a full representative spectrum of normal cell types from different organs. A second limitation is the lack of specificity of the bioassays. Our approach to these problems has been to use cDNA probes to identify the specific pattern of CSF messenger RNA content in cell populations that are of the same type as the marrow microenvironmental vascular and stromal cells. Primary umbilical vein endothelial cells (HUVE) release a low level of detectable biological CSF activity into medium in which they are cultured for 48 h (Sieff et al., 1987b). If the cytokines interleukin 1 (IL-1) or tumor necrosis factor (TNF) are added to the medium, a much greater level of biological activity can be detected using the progenitor assay for BFU-E and CFU-GM. Northern analysis of total cellular RNA extracted from these cells shows that the unstimulated cells contain mRNA for M-CSF but no (or barely detectable) GM-CSF or G-CSF mRNA. IL-1/TNF stimulation results in an increase of GM-CSF and G-CSF mRNAs to easily detectable levels but no alteration in the level of M-CSF mRNA (Sieff et al., 1988). In contrast, endothelial cells infected with an amphotrophic pseudotype of the Kirsten sarcoma virus produced mRNAs for all three CSFs constitutively. Primary and infected fibroblasts show a pattern of CSF mRNA content similar to the primary HUVE. Biological activity after induction with IL-1 or TNF can be detected in fibroblasts from a variety of sources, including primary bone marrow and fetal liver fibroblasts, infected synovial fibroblasts, and both primary and infected dermal fibroblasts. At the RNA level, M-CSF transcripts are constitutively produced, but GM-CSF and G-CSF messages are detectable only after induction with IL-1 or TNF. Thus, uninduced, infected endothelial cells, unlike the primary HUVE or primary or infected fibroblasts, contain GM-CSF and G-CSF mRNA. We were interested to see whether these cells constitutively produce IL-1. We therefore extracted RNA from these cell types and hybridized with a ^{32}P-labeled IL-1 cDNA probe. While HUVE and the primary and infected fibroblasts contained no detectable IL-1 mRNA, the infected endothelial cells were strongly positive for IL-1 mRNA.

TABLE III
CSF mRNA Production by Human Cell Type

	M-CSF	GM-CSF	G-CSF	IL-3
Monocytes (LPS)	+	(+)	(+)	−
Fibroblasts (IL-1/TNF)	+	(+)	(+)	−
Endothelial cells (IL-1/TNF)	+	(+)	(+)	−
T lymphocytes (PHA/PMA)		(+)		(+)

+, Constitutive expression; (+), inducible expression with stimulus shown in parentheses.

These results were confirmed at the protein level when conditioned media from the different cells were tested on an IL-1–dependent cell line. Only the infected endothelial cells released IL-1 into the medium.

Circulating blood cells also secrete CSFs. In order to determine the specific pattern of CSF synthesis, we extracted RNA from blood mononuclear cells, mono-cytes, and lymphocytes. Our preliminary results show that monocytes respond to lipopolysaccharide (LPS) induction with increased levels of GM-CSF and G-CSF mRNA. M-CSF mRNA levels are not increased. Blood mononuclear cells and a monocyte-depleted, lymphocyte-enriched fraction of these cells respond to phytohe-magglutinin (PHA) and phorbol myristate acetate (PMA) with an increase in GM-CSF and IL-3 mRNA content.

To summarize, endothelial cells, fibroblasts, and monocytes appear to produce M-CSF constitutively (Table III). Endothelial cells and fibroblasts respond to IL-1 or TNF, and monocytes respond to LPS, by showing increased levels of GM-CSF and G-CSF mRNA. None of these cell types produces IL-3 mRNA. Lymphocyte-enriched cell fractions of blood mononuclear cells contain GM-CSF and IL-3 mRNA after stimulation with PHA or PMA.

Stromal Fibroblastoid Cells and Fibronectin

Since the CSFs are produced by a variety of cell types that are present in almost every organ of the body, an essential unresolved question remains regarding the nature of the specificity of the adult bone marrow or fetal liver microenvironment for hematopoietic stem cells and progenitors. Tsai et al. (1987) showed that a fetal liver fibroblastoid cell strain, cultured as a monolayer, could function as a surface to which erythroid progenitors attached preferentially. A proportion of BFU-E and the more mature CFU-E adhered to the monolayer, but only very few CFU-GM did so. Further experiments showed that purified fibronectin could substitute for the fibroblastoid cells as an adherent substrate. Antibodies to fibronectin, including antibodies specific for its cell-binding domain, inhibited the adhesion of BFU-E and CFU-E to both fibroblastoid cell monolayers and to fibronectin-coated petri dishes (Tsai et al., 1987). Such

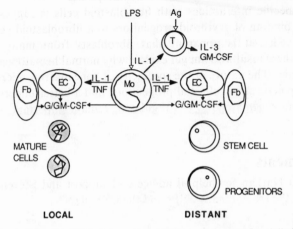

Figure 2. Cytokine-mediated cellular interactions that may occur at local (infection, inflammation) and distant (bone marrow) sites. In response to bacterial LPS, monocytes (Mo) release IL-1/TNF, G-CSF, and GM-CSF. IL-1 and/or TNF induce the release of G-CSF and GM-CSF by local or distant fibroblasts (Fb) and endothelial cells (EC). IL-1, in conjuction with antigen, induces T lymphocytes (T) to release IL-3 and GM-CSF. Locally, these CSFs enhance neutrophil and eosinophil function, while in bone marrow, hematopoiesis and granulopoiesis are increased through direct effects on pluripotent stem cells and lineage-committed progenitor cells.

results are consistent with the earlier mouse erythroleukemia (MEL) cell studies of Patel and Lodish (1984, 1986), who showed that uninduced MEL cells attach tightly and specifically to fibronectin. Dimethylsulfoxide-induced erythroid differentiation of these cells is accompanied by a parallel loss of fibronectin receptors and cell adhesion to fibronectin. Thus, fibronectin may provide an adherent surface to which developing erythroid progenitors and precursors can attach.

Summary and Conclusions

Hematopoiesis and the function of mature blood cells are linked by the CSFs, a class of distinct glycoproteins that have considerable overlap in their actions, tissue sources, and production by cells in response to different inducing agents. These relationships are most evident when the system is stimulated in response to exogenous agents (Fig.

2). Monocytes may lie at the center of this response, as LPS induces the release of GM-CSF and G-CSF, as well as the monokines IL-1 and TNF. GM-CSF and G-CSF act on mature cells at the site of inflammation, increasing their functional capacity by concentration-dependent effects on chemotaxis, increased phagocytosis, enhanced superoxide production, and increased antibody-dependent cellular cytotoxicity (Vadas et al., 1983*a*, *b*; Gasson et al., 1984; Weisbart et al., 1985). The release by monocytes of IL-1 and TNF may amplify this response by inducing local fibroblasts and endothelial cells to produce more GM-CSF and G-CSF. The release of GM-CSF and G-CSF into the circulation could increase bone marrow granulocyte production. Increased circulating levels of IL-1 and/or TNF would also be expected to increase bone marrow stromal cell production of GM-CSF and G-CSF. In conjunction with foreign antigen, IL-1 may also induce T lymphocytes to produce both GM-CSF and IL-3, and these two factors would be expected to increase bone marrow hematopoiesis.

Erythroid-specific interactions with fibroblastoid cells is suggested by in vitro studies showing binding of erythroid progenitors to a fibroblastoid cell strain and to fibronectin. However, it is apparent that fibroblasts from many tissues secrete fibronectin, and these results do not yet explain why normal hematopoiesis is restricted to the bone marrow. The nature of the specificity of bone marrow microenvironmental cells for hematopoietic stem cells and progenitors, and a better understanding of how normal steady state hematopoiesis is regulated, remain important areas for future research.

Acknowledgments

We thank David Nathan for helpful advice and support and Steven Clark and Yu Chung Yang for the recombinant CSFs and the CSF cDNAs.

References

Arnaout, M. A., E. A. Wang, S. C. Clark, and C. A. Sieff. 1986. Human recombinant granulocyte macrophage colony stimulating factor increases cell to cell adhesion and surface expression of adhesion promoting surface glycoproteins on mature granulocytes. *Journal of Clinical Investigation.* 78:597–601.

Bentley, S. A. 1982. Bone marrow connective tissue and the haemopoietic microenvironment. *British Journal of Haematology.* 50:1–6.

Dexter, T. M., T. D. Allen, and L. G. Lajtha. 1977. Conditions controlling the proliferation of hemopoietic stem cells in vitro. *Journal of Cellular Physiology.* 91:337–344.

Gasson, J. C., R. H. Weisbart, S. E. Kaufman, S. C. Clark, R. M. Hewick, G. G. Wong, and D. W. Golde. 1984. Purified human granulocyte-macrophage colony-stimulating factor: direct action on neutrophils. *Science.* 226:1339–1342.

Kawasaki, E. S., M. B. Ladner, A. M. Wang, J. Van Arsdell, M. K. Warren, M. Y. Coyne, V. L. Schweickart, M. T. Lee, K. J. Wilson, A. Boosman, E. R. Stanley, P. Ralph, and D. F. Mark. 1985. Molecular cloning of a complementary DNA encoding human macrophage-specific colony-stimulating factor (CSF-1). *Science.* 230:291–296.

Le Beau, M. M., N. D. Epstein, S. J. O'Brien, A. W. Nienhuis, Y.-C. Yang, S. C. Clark, J. D. Rowley. 1987*a*. The interleukin 3 gene is located on human chromosome 5 and is deleted in myeloid leukemias with a deletion of 5q. *Proceedings of the National Academy of Sciences.* 84:5913–5917.

Le Beau, M. M., C. A. Westbrook, M. O. Diaz, R. A. Larson, J. D. Rowley, J. C. Gasson, D. W. Golde, and C. J. Sherr. 1987*b*. Evidence for the involvement of GM-CSF and FMS in the deletion (5q) in myeloid disorders. *Science.* 231:984–987.

Nagata, S., M. Tsuchiya, S. Asano, P. Kaziro, T. Yamazaki, O. Yamamoto, Y. Hirate, N. Kubota, M. Oheda, H. Nomura, and M. Ono. 1986. Molecular cloning and expression of cDNA for human granulocyte colony-stimulating factor. *Nature.* 319:415–418.

Patel, V. P., and H. F. Lodish. 1984. Loss of adhesion of murine erythroleukemia cells to fibronectin during erythroid differentiation. *Science.* 224:996–998.

Patel, V. P., and H. F. Lodish. 1986. The fibronectin receptor on mammalian erythroid precursor cells: characterization and developmental regulation. *Journal of Cell Biology.* 102:449–456.

Pettenati, M. J., M. M. Le Beau, R. S. Lemons, E. A. Shima, E. S. Kawasaki, R. A. Larson, C. J. Sherr, M. O. Diaz, and J. D. Rowley. 1987. Assignment of CSF-1 to 5q 33.1: evidence for clustering of genes regulating hematopoiesis and for their involvement in the deletion of the long arm of chromosome 5 in myeloid disorders. *Proceedings of the National Academy of Sciences.* 84:2970–2974.

Sieff, C. A., S. G. Emerson, R. E. Donahue, D. G. Nathan, E. A. Wang, G. G. Wong, and S. C. Clark. 1985. Human recombinant granulocyte-macrophage colony stimulating factor: a multi-lineage hematopoietin. *Science.* 230:1171–1173.

Sieff, C. A., S. G. Emerson, A. Mufson, T. G. Gesner, and D. G. Nathan. 1986. Dependence of highly enriched human bone marrow progenitors on hemopoietic growth factors and their response to recombinant erythropoietin. *Journal of Clinical Investigation.* 77:74–81.

Sieff, C. A., C. M. Niemeyer, S. J. Mentzer, and D. V. Faller. 1988. Interleukin-1, tumor necrosis factor, and the production of colony-stimulating factors by cultured mesenchymal cells. *Blood.* 72:In press.

Sieff, C. A., C. M. Niemeyer, D. G. Nathan, S. C. Ekern, F. Bieber, Y. C. Yang, G. Wong, and S. C. Clark. 1987*a*. Stimulation of human hematopoietic colony formation by recombinant gibbon multi-colony-stimulating factor or interleukin 3. *Journal of Clinical Investigation.* 80:818–823.

Sieff, C. A., S. Tsai, and D. V. Faller. 1987*b*. Interleukin 1 induces cultured human endothelial cell production of granulocyte-macrophage colony-stimulating factor. *Journal of Clinical Investigation.* 79:48–51.

Tsai, S., V. Patel, E. Beaumont, H. F. Lodish, D. G. Nathan, and C. A. Sieff. 1987. Differential binding of erythroid and myeloid progenitors to fibroblasts and fibronectin. *Blood.* 69:1587–1594.

Vadas, M. A., N. A. Nicola, and D. Metcalf. 1983*a*. Activation of antibody dependent cell-mediated cytotoxicity of human neutrophils and eosinophils by separate colony-stimulating factors. *Journal of Immunology.* 130:795–799.

Vadas, M. A., G. Varigos, N. Nicola, S. Pincus, A. Dessem, D. Metcalf, and F. L. Battye. 1983*b*. Eosinophil activation by colony-stimulating factor in man: metabolic effects and analysis by flow cytometry. *Blood.* 61:1232–1241.

Weisbart, R. H., D. W. Golde, S. C. Clark, G. G. Wong, and J. C. L. Gasson. 1985. Human granulocyte-macrophage colony-stimulating factor is a neutrophil activator. *Nature.* 314:361–363.

Wong, G. G., J. S. Witek, P. A. Temple, K. M. Wilkens, A. C. Leary, D. P. Luxenberg, S. S. Jones, E. L. Brown, R. M. Kay, E. C. Orr, C. Shoemaker, D. W. Golde, R. J. Kaufman, R. M.

Hewick, E. A. Wang, and S. C. Clark. 1985. Human GM-CSF: molecular cloning of the cDNA and purification of the natural and recombinant proteins. *Science.* 228:810–815.

Yang, Y. C., A. B. Ciarletta, P. A. Temple, M. Chung, S. Kovacic, J. S. Witek-Giannotti, A. C. Leary, R. Kriz, R. E. Donahue, G. G. Wong, and S. C. Clark. 1986. Human interleukin-3 (multi-CSF): identification by expression cloning of a novel hematopoietic growth factor related to murine IL-3. *Cell.* 47:3–10.

Chapter 6

Eythropoietin: In Vitro and In Vivo Studies of the Regulation of Erythropoiesis

John W. Adamson

Department of Medicine, University of Washington, Seattle, Washington

Cell Physiology of Blood © 1988 by The Rockefeller University Press

Introduction

The first report of the cloning and expression of the human erythropoietin (Epo) gene appeared in 1985 (Jacobs et al., 1985; Lin et al., 1985). The gene encoding the mature protein is comprised of four exons and three introns. There is also a sequence that codes for a 27-amino-acid-long secretory leader piece (Fig. 1). The structural gene for human Epo exists in the genome as a single copy, and was initially assigned to chromosome 7 (Powell et al., 1986), and then mapped to the long arm of that chromosome in the region of 7q21 (Law et al., 1986). The mature protein is made up of 165 amino acids and the polypeptide predicts a molecular weight of 18,400. The biologically active hormone exists as a monomer with two internal disulfide bonds linking cysteines at positions 7 and 161, and 29 and 33. The molecular weight of the biologically active hormone is ~34,000. The higher weight is due to glycosylation of the molecule. There are three Asn-linked glycosylation sites at positions 24, 38, and 83, and a single O-linked glycosylation site at the serine at position 126. Glycosylation of the molecule is required for its survival in vivo. If the carbohydrates are removed by treatment of Epo with glycosidases, biological activity in vitro is not lost. Site-directed mutagenesis, which replaces the carbohydrate-linked Asn sites, also results in a molecule with comparable biological activity and protein as determined by radioimmu-

HUMAN EPO GENE

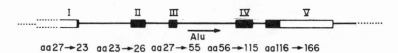

Figure 1. The organization of the structural gene for human erythropoietin. There are four exons and three introns, which account for the coding region for the polypeptide hormone. In addition, the organization of the coding sequences for the secretory leader piece is shown.

noassay (RIA). Detailed in vitro studies of completely nonglycosylated Epo are difficult, however, because of the aggregation of the molecule. The carbohydrate construction of recombinant human Epo and that of native Epo appear to be virtually identical, and both native and recombinant molecules function equally well in vivo and in vitro (Egrie et al., 1985).

Other than the role of the carbohydrate in promoting the in vivo survival of Epo, little is known of the structure/function relationships of the polypeptide. Sytkowski and Donahue (1986) have suggested that amino acids 99–129 may be important in the binding of Epo to its receptor. They produced a series of peptides spanning a large proportion of the Epo molecule. Although these peptides themselves did not have Epo-like activity and did not block the in vitro action of Epo, antibodies prepared to those peptides spanning the 99–129 region interfered with the in vitro action of Epo.

The cDNAs for mouse (McDonald et al., 1986) and monkey (Lin et al., 1986) Epo have also been cloned and sequenced. There is a high degree of homology between all three species and, overall, >90% homology between human and monkey Epo. The tertiary structure of human Epo, as predicted by the Chou and Fasman method, is shown in Fig. 2. In addition, the first introns of the human and mouse Epo genes have a high degree of homology, while homology is not maintained in the other introns. Such

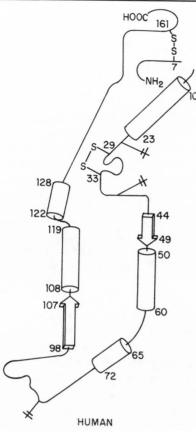

Figure 2. The predicted tertiary structure of human erythropoietin as predicted by the method of Chou and Fasman (1978). (From McDonald et al., 1986, with permission of the authors and the publisher.)

retained homology suggests an important regulatory function for the first intron, but the precise significance of this is unknown.

Regulation of Epo Production

Physiological studies that began nearly 30 years ago clearly implicated the kidney as the important site of Epo production in mammals (Krantz and Jacobson, 1970). Animals subjected to hypoxia will produce high levels of Epo, which can be measured in the plasma by bioassay or RIA. Nephrectomized animals do not produce measurable levels of Epo, whereas animals subjected to ureteral ligation and subsequent hypoxia have an unimpaired Epo response. What has been controversial are the mechanisms and cellular sources resulting in Epo production as a result of hypoxia.

The availability of probes for Epo messenger RNA has allowed further dissection of its cellular sources. Schuster et al. (1987) reported that Epo mRNA could not be detected in preparations of whole cellular RNA from normal rat kidneys; however, when the rats were exposed to severe hypoxia, Epo-specific mRNA appeared within minutes and peaked within a few hours. The appearance of mRNA was followed shortly by the appearance of bioactivity in the plasma of these animals. With cessation of hypoxia, mRNA levels fell and bioactivity declined. The same workers isolated tubular cell preparations and glomeruli by differential sedimentation from rat kidneys.

Only mRNA prepared from isolates rich in tubular cells contained Epo-specific transcripts. The glomeruli preparations had no Epo mRNA. Lacombe et al. (1987) reported similar findings using the technique of in situ hybridization. Rather than the tubular cells themselves, peritubular interstitial cells (presumably endothelial cells) were implicated as the source of Epo. Thus, while the precise cellular source of Epo remains to be determined, the best available evidence would suggest that the tubular or peritubular complex functions as the Epo-generation site.

Plasma Epo

The mature protein circulates in the plasma at a concentration of 15–25 mU/ml, as determined by current-sensitive RIAs (Garcia et al., 1982). Under normal circumstances, the activity determined by RIA parallels that determined by bioassay. Epo production increases with hypoxia or anemia. In a variety of carefully studied human conditions, an inverse log-linear relationship between plasma Epo levels and hemoglobin concentration was found. Thus, with established anemia and a hemoglobin concentration of 5–8 g/dl, plasma Epo levels of 400–1,000 mU/ml would be anticipated.

Although details of Epo pharmacokinetics have not been published, preliminary studies using recombinant human Epo in patients with chronic renal failure indicate a half-life on the average of 6–7 h (range, 5–11) with a single exponential of decay (Egrie et al., 1987). We have carried out a study of the plasma half-life of native human Epo in one patient with chronic renal failure who was subsequently treated with recombinant human Epo. The half-life of the native human Epo was 9.3 h, while that of recombinant human Epo was 9.0 h. Thus, this single study indicates that there are likely to be no significant differences between the half-life of native or recombinant human Epo in man.

Interactions of Epo with Its Target Cell
Studies In Vitro

The cellular anatomy of erythropoiesis in vitro has been established by studies of various populations of progenitor cells, which can be detected only through their growth in culture and their relative dependence on specific growth factors, including Epo. The earliest recognizable erythroid progenitor in culture is the erythroid burst-forming cell (BFU-E). This progenitor has the capacity to form large hemoglobinized colonies that contain as many as 10^4 cells. Epo is required for the hemoglobinization of the cells in these colonies, but the early divisions of the progenitor are dependent upon other growth factors referred to as having burst-promoting activity (BPA). In the absence of BPA, despite the presence of Epo, large erythroid bursts will not form (Yen et al., 1985).

Growth factors that have been isolated, had their genes cloned, and possess BPA include granulocyte/macrophage colony-stimulating factor (GM-CSF) (Kaushansky et al., 1986) and interleukin 3 (IL-3) (Yang et al., 1986). BFU-E, as they mature and differentiate, give rise to erythroid colony-forming cells (CFU-E), which can be assayed in vitro independently. These progenitor cells have an absolute requirement for Epo in order to form colonies in culture and do not appear to respond in a meaningful way to other growth factors.

In Vivo Studies

Studies in experimental animals and man suggest that the primary target of Epo action is the CFU-E or progenitors intermediate between BFU-E and CFU-E (Adamson et al., 1978; Kimura et al., 1986). This conclusion is supported by several lines of evidence. First, if mice are hypertransfused to produce a state of polycythemia, the numbers and cell cycle kinetics of marrow BFU-E are unchanged, while the numbers of CFU-E decline to ~20–25% of normal (Adamson et al., 1978). Second, if rats are made chronically anemic (average hematocrit, 15%) through sustained iron deficiency, and then the numbers and cell cycle kinetics of the various erythroid progenitors are determined, only the CFU-E compartment is affected (Kimura et al., 1986). Such studies demonstrated no alterations in the numbers or cell cycle kinetics of BFU-E, while a three- to fourfold increase in total CFU-E was found (Fig 3). From such studies, the concept has emerged that Epo acts during the terminal differentiation and maturation steps of erythropoiesis, and it is required for the maintenance of the population of CFU-E.

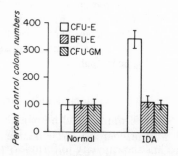

Figure 3. The relative changes in the populations of CFU-GM, BFU-E, and CFU-E in the chronically anemic (hematocrit, 15%) rat. Chronic anemia in the animal was established by a combination of bleeding and an iron-deficient diet. Only the numbers of CFU-E were increased above control values. The numbers of GM-CFC (as control) and BFU-E were unchanged, and there was no alteration in their cell cycle activity, as determined by the tritiated thymidine suicide technique (data not shown). Values for control (normal) animals are expressed as 100% and changes in the animals with iron-deficiency anemia (IDA) are normalized to the control values. (From Kimura et al., 1986, with permission of the authors and the publisher.)

Although Epo is thought generally to be a lineage-specific factor, several workers have demonstrated the stimulation by Epo of acetylcholinesterase generation by murine megakaryocytes (Ishibashi et al., 1987) or the enhancement of megakaryocyte colony formation (McLeod et al., 1976). Thus, at least in vitro, Epo appears to be able to induce or accelerate megakaryocyte cytoplasmic maturation.

Studies of the Epo Receptor

Epo interacts with its target cells through a plasma membrane receptor. The availability of recombinant human Epo has triggered a series of early studies of receptor biology. Sawyer and co-workers (1987), using spleen cells from Friend virus (anemia strain)-infected mice, have demonstrated binding of radioiodinated Epo to cells. Using an enriched progenitor cell population, they reported an average of 300 high-affinity receptors per cell. Initial studies suggested two classes of receptors having different affinities. This has not been confirmed by other workers, however, and most investigators have reported a single class of receptors, as determined by Scatchard analysis.

Problematic in the analysis of these data is the fact that radioiodinated Epo is not biologically active. Nevertheless, displacement analyses by Broudy et al. (1988), as well as binding studies carried out with metabolically labeled recombinant human Epo, confirm the data obtained with the radioiodinated hormone. If the target cells are exposed to radioiodinated Epo and their membranes are then solubilized and subjected to polyacrylamide gel electrophoresis, two bands of radioactivity are found (Sawyer et al., 1987). If the molecular weight of Epo is subtracted, the results indicate a receptor of at least two subunits having molecular weights of 85,000 and 100,000. The structure and assembly of the Epo receptor remain to be determined.

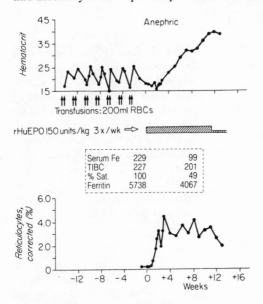

Figure 4. The effect of recombinant human Epo on the transfusion requirements and hematocrit of a patient with severe anemia and end-stage renal disease. With the advent of recombinant human Epo therapy, transfusion dependence ceased and the hematocrit returned to a near-normal range. The serum ferritin values fell, indicating the mobilization of iron from storage sites and its incorporation into the newly synthesized hemoglobin. (From Eschbach et al., 1987, with permission of the authors and the publisher.)

Broudy et al. (1988) have studied a number of murine and human cell lines for Epo binding properties. They found that a Rauscher erythroleukemia cell line, reported to be responsive to Epo, had an average of 1,700 receptors per cell. A single class of receptor was found with an affinity of 440 pM. In contrast, the human erythroleukemia cell line OCIM1 had an average of 3,000 receptors per cell with an affinity of 280 pM. The receptor on these cells had at least two subunits of 105,000 and 95,000 mol wt. Interestingly, receptor numbers could be modulated by exposure of cells either to the tumor promoter TPA or to dimethylsulfoxide (DMSO). TPA-stimulated cells down-regulated receptor numbers to 350 per cell, while the cell line GM 979, exposed for 96 h to DMSO, increased receptor numbers from 1,600 to 8,400 per cell.

Epo as a Therapeutic

The availability of recombinant human Epo has made possible clinical trials with the hormone in patients with predicted Epo-deficiency states. The first trials were conducted in England and the United States. Winearls et al. (1986) and Eschbach et al. (1987) administered recombinant human Epo to severely anemic patients with chronic renal failure and reported a marked improvement in hemoglobin and hematocrit and the abrogation of transfusion requirements (Fig. 4). Erythropoietic inhibitors, suggested by some to be important in the pathogenesis of the anemia of chronic renal failure, were of insufficient magnitude to blunt the response. The recombinant human Epo was effective in a dose-dependent fashion (Fig. 5) and, at this writing, some patients have been maintained at near-normal hematocrits for over two years without evidence of anti-Epo antibody formation or diminished responsiveness. In this condition, therefore, it seems clear that recombinant human Epo will be an effective therapeutic and that improved rehabilitation and prevention of the complications of repeated transfusions can be expected. Encouraging preliminary results have also been reported in patients with anemia associated with rheumatoid arthritis (Means et al., 1987).

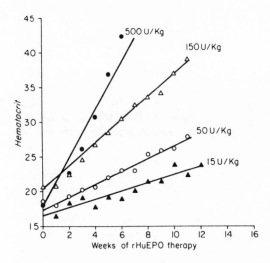

Figure 5. Rate of rise of hematocrit in groups of patients with severe anemia and end-stage renal disease, who were treated with different doses of recombinant human Epo. As can be seen, there was a dose-dependent increase in the rate of rise of hematocrit, and increases of hematocrit by as many as 10 points were observed within 3–4 wk at the highest doses used. (From Eschbach et al., 1987, with permission of the authors and the publisher.)

Summary

The application of recombinant DNA technology to the field of hematology has contributed greatly to our understanding of Epo gene structure and regulation, cellular expression and regulation of hormone production, pharmacokinetics, receptor cell biology and, ultimately, the value of this hormone as a therapeutic. Areas that will undoubtedly prove fruitful for future research include the mechanisms by which hypoxia influences gene expression, the structure/function relationships of the Epo molecule, isolation and cloning of the gene for the Epo receptor, mechanisms of transmembrane signaling and nuclear activation, and the further application of recombinant human Epo in the treatment of other diseases. Epo is but one example of the contribution of modern biology to the understanding of growth factor regulation of hematopoiesis and to the availability of these factors for clinical trial.

Acknowledgments

The assistance of Ms. Lezlie Hall in preparation of the manuscript is gratefully acknowledged. Special thanks are given to AMGen, Thousand Oaks, California, for providing unique reagents for the studies of receptor biology and for supporting the initial clinical trials with recombinant human Epo.

Portions of the work reported were supported by research grants DK-19410, CA-31615, and HL-13823, and Clinical Research Center grant FR-0037 from the National Institutes of Health.

References

Adamson, J. W., B. Torok-Storb, and N. Lin. 1978. Analysis of erythropoiesis by erythroid colony formation in culture. *Blood Cells.* 4:89–103.

Broudy, V. C., N. Lin, J. Egrie, C. de Haën, T. Weiss, T. Papayannopoulou, and J. W. Adamson. 1988. Identification of the receptor for erythropoietin on human and murine erythroleukemia cell lines and modulation by phorbol ester and dimethylsulfoxide. *Proceedings of the National Academy of Sciences.* In press.

Chou, P. Y., and G. D. Fasman. 1978. Empirical predictions of protein conformations. *Annual Review of Biochemistry.* 47:251–276.

Egrie, J. C., J. K. Browne, P. Lai, and F.-K. Lin. 1985. Characterization of recombinant monkey and human erythropoietin. *In* Experimental Approaches for the Study of Hemoglobin Switching. G. Stamatoyannopoulos and A. W. Nienhuis, editors. Alan R. Liss, Inc., New York, NY. 339–350.

Egrie, J. C., J. W. Eschbach, T. McGuire, and J. W. Adamson. 1988. Pharmacokinetics of recombinant human erythropoietin (rHuEpo) administered to hemodialysis (HD) patients. *Kidney International.* 33:262. (Abstr.)

Eschbach, J. W., J. C. Egrie, M. R. Downing, J. K. Browne, and J. W. Adamson. 1987. Correction of the anemia of end-stage renal disease with recombinant human erythropoietin. *New England Journal of Medicine.* 316:73–78.

Garcia, J. F., S. N. Ebbe, L. Hollander, H. O. Cutting, M. E. Miller, and E. P. Cronkite. 1982. Radioimmunoassay of erythropoietin: circulating levels in normal and polycythemic human beings. *Journal of Laboratory and Clinical Medicine.* 99:624–631.

Ishibashi, T., J. A. Koziol, and S. A. Burstein. 1987. Human recombinant erythropoietin promotes differentiation of murine megakaryocytes in vitro. *Journal of Clinical Investigation.* 79:286–289.

Jacobs, K., C. Shoemaker, R. Rudersdorf, S. D. Neill, R. J. Kaufman, A. Mufson, J. Seehra, S. S. Jones, R. Hewick, E. F. Fritsch, M. Kawakita, T. Shimizu, and T. Miyake. 1985. Isolation and characterization of genomic and cDNA clones of human erythropoietin. *Nature.* 313:806–810.

Kaushansky, K., P. J. O'Hara, K. Berkner, G. M. Segal, F. S. Hagen, and J. W. Adamson. 1986. Genomic cloning, characterization, and multilineage growth-promoting activity of human granulocyte-macrophage colony-stimulating factor. *Proceedings of the National Academy of Sciences.* 83:3101-3105.

Kimura, H., C. A. Finch, and J. W. Adamson. 1986. Hematopoiesis in the rat. Quantitation of hematopoietic progenitors and the response to iron deficiency anemia. *Journal of Cellular Physiology.* 126:298–306.

Krantz, S. B., and L. O. Jacobson. 1970. Erythropoietin and the Regulation of Erythropoiesis. The University of Chicago Press, Chicago, IL. 331 pp.

Lacombe, C., P. Bruneval, J. L. Da Silva, J. P. Camilleri, J. Bariety, P. Tambourin, and B. Varet. 1987. Expression of the erythropoietin gene in the hypoxic adult mouse. *Blood.* 70(Suppl. 1):176*a*. (Abstr.)

Law, M. L., G.-Y. Cai, F.-K. Lin, Q. Wei, S.-Z. Haung, J. H. Hartz, H. Morse, C.-H. Lin, C. Jones, and F.-T. Kao. 1986. Chromosomal assignment of the human erythropoietin gene and its DNA polymorphism. *Proceedings of the National Academy of Sciences.* 83:6920–6924.

Lin, F.-K., C.-H. Lin, P.-H. Lai, J. K. Browne, J. C. Egrie, R. Smalling, G. M. Fox, K. K. Chen, M. Castro, and S. Suggs. 1986. Monkey erythropoietin gene: cloning, expression and comparison with the human erythropoietin gene. *Gene.* 44:201–209.

Lin, F.-K., S. Suggs, C.-H. Lin, J. K. Browne, R. Smalling, J. C. Egrie, K. K. Chen, G. M. Fox, F. Martin, Z. Stabinsky, S. M. Badrawi, P.-H. Lai, and E. Goldwasser. 1985. Cloning and expression of the human erythropoietin gene. *Proceedings of the National Academy of Sciences.* 82:7580–7584.

McDonald, J. D., F.-K. Lin, and E. Goldwasser. 1986. Cloning, sequencing, and evolutionary analysis of the mouse erythropoietin gene. *Molecular and Cellular Biology.* 6:842–848.

McLeod, D. L., M. M. Shreeve, and A. A. Axelrad. 1976. Induction of megakaryocyte colonies with platelet formation in vitro. *Nature.* 261:492–494.

Means, R. T., N. J. Olsen, S. B. Krantz, S. E. Graber, E. N. Dessypris, W. J. Stone, T. P. Pincus, and V. O'Neil. 1987. Treatment of the anemia of rheumatoid arthritis and recombinant human erythropoietin: clinical and in vitro results. *Blood.* 70(Suppl. 1):139*a*. (Abstr.)

Powell, J. S., K. L. Berkner, R. V. Lebo, and J. W. Adamson. 1986. Human erythropoietin gene: high level expression in stably transfected mammalian cells and chromosome localization. *Proceedings of the National Academy of Sciences.* 83:6465–6469.

Sawyer, S. T., S. B. Krantz, and J. Luna. 1987. Identification of the receptor for erythropoietin by cross-linking to Friend virus-infected erythroid cells. *Proceedings of the National Academy of Sciences.* 84:3690–3694.

Schuster, S. J., J. H. Wilson, A. J. Erslev, and J. Caro. 1987. Physiologic regulation and tissue localization of renal erythropoietin messenger RNA. *Blood.* 70:316–318.

Sytkowski, A. J., and K. A. Donahue. 1986. Identification of a functional domain of human erythropoietin using site-specific anti-peptide antibodies. *Blood.* 68(Suppl. 1):181*a*. (Abstr.)

Winearls, C. G., D. O. Oliver, M. J. Pippard, C. Reid, M. R. Downing, and P. M. Cotes. 1986. Effect on human erythropoietin derived from recombinant DNA on the anaemia of patients maintained by chronic haemodialysis. *Lancet.* II:1175–1178.

Yang, Y.-C., A. B. Ciarletta, P. A. Temple, M. P. Chung, S. Kovacic, J. S. Witek-Giannotti, A. C. Leary, R. Kirz, R. E. Donahue, G. G. Wong, and S. C. Clark. 1986. Human IL-3 (multi-CSF): identification by expression cloning of a novel hematopoietic growth factor related to murine IL-3. *Cell.* 47:3–10.

Yen, Y. P., P. Zabala, K. Doney, G. Clemons, S. Gillis, J. S. Powell, and J. W. Adamson. 1985. Hematopoietic growth factors in human serum. Erythroid burst-promoting activity in normal subjects and in patients with severe aplastic anemia. *Journal of Laboratory and Clinical Medicine.* 106:384–392.

Chapter 7

The Role of Purified and Recombinant
Hematopoietic Growth Factors in
the Regulation of Various Stages
of Human Megakaryocytopoiesis

Ronald Hoffman, John Straneva, Li Lu, Bruce Roth,
Edward Bruno, and Robert Briddell

Department of Medicine, Hematology/Oncology Section, Indiana
Elks Cancer Research Center, Indiana University School of
Medicine, Indianapolis, Indiana

Cell Physiology of Blood © 1988 by The Rockefeller University Press

Introduction

Platelets are the cellular constituents present in human blood that are important in the maintenance of normal hemostasis (Williams and Levine, 1982). Megakaryocytes, the cell of origin of platelets, reside primarily in the bone marrow and are relatively rare cells comprising between 0.04 and 0.08% of all nucleated cells in the marrow (Berkow et al., 1984; Levine, 1980). The megakaryocyte is unusual when compared with other human cells in that it is normally polyploid, having nuclear ploidy values between 8 and 64 N (Berkow et al., 1984; Levine, 1980). In addition, one of the distinguishing features of megakaryocytes is their large size, with mean diameters of 50.5 and 33.8 μm when measurements are made on fixed and unfixed cellular specimens, respectively (Berkow et al., 1984). Why, from an evolutionary point of view, the process that leads to platelet production in man is so different from the processes that lead to the production of other human hematopoietic cells remains an unresolved issue.

A great deal of data, accumulated over the last three decades, indicate that platelet production in mammals is regulated by a number of humoral growth factors capable of maintaining a constancy of peripheral platelet numbers (Gewirtz, 1986). These factors stimulate the marrow to accelerate platelet production in response to increased demands. Recent applications of protein purification and gene cloning to characterize hematopoietic growth factors have made available for the first time large quantities of purified and recombinant hematopoietic regulatory molecules (Clark and Kamen, 1987). The availability of these molecules has facilitated the understanding of the regulation of human blood cell production. Using these purified or recombinant biomolecules, a number of laboratories have reported that several of these factors can influence a number of stages of megakaryocytopoiesis (Quesenberry et al., 1985; Williams et al., 1985; Peschel et al., 1987; Robinson et al., 1987; Sparrow et al., 1987).

There is a sequence of steps that occurs during the process of platelet formation. Platelets ultimately originate from a pluripotent hematopoietic stem cell capable of differentiation to each of the hematopoietic lineages. The process of commitment from the pluripotent hematopoietic stem cell to the megakaryocyte progenitor cell (colony-forming unit, megakaryocyte [CFU-MK]), the first step in the process of megakaryocyte development, is poorly understood. After this event, the megakaryocyte progenitor cell undergoes proliferation and eventual differentiation. The transition from a proliferating cell to a cell that undergoes endomitosis, resulting in nuclear polyploidization, then occurs. The number of mitoses that each megakaryocyte progenitor cell experiences is thought to determine the number of endomitotic nuclear replications that the progeny undergo (Paulus et al., 1982). This eventually results in a population of mature megakaryocytes heterogeneous with respect to ploidy. Concurrently with or after these events, megakaryocyte membrane and cytoplasmic maturation, associated with membrane morphogenesis and granule assembly, occurs. The final event in thrombopoiesis is actual platelet release. This event is thought by the majority of investigators to occur primarily in the bone marrow (Wright, 1906), while others suggest that the lung might also be an important site of platelet production (Trowbridge et al., 1982). The actual process of platelet release from megakaryocytes has been attributed to either formation of pseudopodia, cellular fragmentation, or specialization of megakaryocyte demarcation membranes (Gewirtz, 1986). Whether all of these cellular events lead eventually to platelet release is currently being explored. This chapter will deal with our evolving understanding of the regulation of platelet production in man.

Regulation at the Progenitor Cell Level

Megakaryocyte progenitor cells present in marrow and peripheral blood are assayed in a variety of semisolid media by their capacity to give rise to colonies composed of megakaryocytes (Mazur et al., 1981; Messner et al., 1982). Studies of crude plasma, serum, and urine obtained from patients with various clinical disorders of thrombopoiesis have led to the detection of a megakaryocyte colony-stimulating activity (MK-CSA) in these body fluids (Enomoto et al., 1980; Hoffman et al., 1981; Kawakita et al., 1981). The detectability of MK-CSA appears to be inversely related to bone marrow megakaryocyte numbers rather than to changes in circulating platelet numbers (Hoffman et al., 1981). These studies have led to attempts to purify a megakaryocyte-specific growth factor, megakaryocyte colony-stimulating factor (MK-CSF), a unique lineage-specific regulator of human megakaryocytopoiesis at the progenitor cell level (Hoffman et al., 1985; Mazur and South, 1985; Shimizu et al., 1987). Our laboratory and others have previously reported the purification from human plasma, urine, or dog sera of an MK-CSF to apparent homogeneity by a variety of physical and biochemical steps (Hoffman et al., 1985; Mazur and South, 1985; Shimizu et al., 1987). The molecular weight of the MK-CSF has been estimated to be between 26,000 and 175,000 (Hoffman et al., 1985; Mazur and South, 1985; Shimizu et al., 1987). Each of the preparations appear to have activity that is restricted to the CFU-MK, and they do not affect the cytoplasmic maturation of more differentiated megakaryocytes (Straneva et al., 1987). The efforts in our own laboratory to purify human MK-CSF have been only partially successful, since amino acid sequencing of our preparation revealed contamination with α_1 acid glycoprotein (unpublished observation). This was not entirely unexpected since, although most previously isolated growth factors have been shown to act at concentrations of 10^{-12}–10^{-13} M (Clark and Kamen, 1987), the MK-CSF preparation was optimally active at 10^{-9} M (Hoffman et al., 1985). The partially purified MK-CSF not only increased colony formation but also increased the number of cells composing individual megakaryocyte colonies (Hoffman et al., 1985). An additional effect of MK-CSF has recently been defined by Arriaga et al. (1987), who have shown that by influencing the mitotic history of the megakaryocyte progenitor cell, differing concentrations of aplastic canine sera, which contain MK-CSA, can actually alter the balance between mitosis and endomitosis, resulting in changes in megakaryocyte ploidy.

Our group had originally envisioned that the process of regulation of the CFU-MK was controlled by a single MK-CSA (Hoffman et al., 1981). The studies of Williams et al. (1985), Quesenberry et al. (1985), Vainchenker et al. (1979), and Robinson et al. (1987), however, have clearly shown this to be incorrect. These groups have conclusively demonstrated that several other hematopoietic growth factors, capable of influencing a number of other hematopoietic lineages, are also able to influence megakaryocyte development. While interleukin 3 (IL-3) and granulocyte/ macrophage colony-stimulating factor (GM-CSF) have been shown to have a major role in this process, the role of erythropoietin remains uncertain. In our studies, a serum-depleted clonal assay for megakaryocyte progenitor cells was recently used to assess the effect of a variety of hematopoietic growth factors on megakaryocyte colony formation. As can be seen in Fig. 1, none of the lineage-specific growth factors such as recombinant erythropoietin (rEpo), recombinant granulocyte colony-stimulating factor (rG-CSF), macrophage colony-stimulating factor (CSF-1), and recombinant interleukin 1α (IL-1α) were capable of augmenting megakaryocyte colony formation. By contrast, rIL-3 and rGM-CSF enhanced cloning efficiency 8-fold and 10-fold, respectively, above baseline values. Each of these molecules was an equivalent stimulator of megakaryocyte colony formation ($P > 0.1$). These data suggest that rGM-CSF and rIL-3 are important regulators of in vitro megakaryocytopoiesis at the progenitor cell level.

Similar data have now been generated using populations of human bone marrow enriched for CFU-MK. Using monoclonal antibodies specific for the My10 and

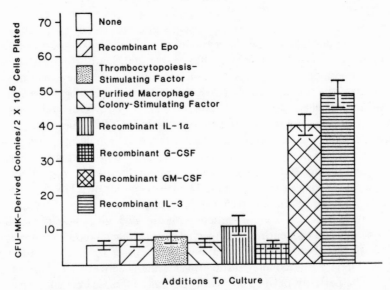

Figure 1. Effect of optimal doses of purified or recombinant hematopoietic growth factors on megakaryocyte colony formation by low-density marrow cells. Each point represents the mean ± SEM of data pooled from duplicate or quadruplicate assays of seven separate, normal marrow donors ($N = 20$).

HLA-DR antigens, we have shown an expression of these antigens by the CFU-MK. By using fluorescence-activated cell sorting and differential antigenic expression, we can localize the CFU-MK to an My10^{+++}DR$^+$ population of cells with ~1.3% of these cells being assayable CFU-MK. This enrichment process, involving macrophage and T cell depletion followed by cell sorting, leads to a 40-fold purification of CFU-MK. As can be seen in Table I, GM-CSF and IL-3 are each capable of stimulating colony formation by these enriched marrow populations. In addition, the effects of these two regulators were found to be additive, since the effect of combinations of rGM-CSF and rIL-3 was greater than that of either molecule alone.

The role for Epo in CFU-MK regulation remains less well established. Epo production is directly related to the severity of anemia in man (Adamson, 1968). Observations in clinical medicine of the association of anemia and thrombocytosis led to the hypothesis that Epo plays a regulatory role in thrombopoiesis. Recently, in fact, McDonald et al. (1987) have presented information that high doses of rEpo stimulate platelet production in vivo, while our own laboratory has reported that Epo can enhance the cloning efficiency of a megakaryocytic cell line (Hoffman et al., 1987). However, in our studies using a serum-depleted assay system, rEpo did not augment megakaryocyte colony formation (Table II). Our findings are in conflict with the findings of several other laboratories (Vainchenker et al., 1979; Dukes et al., 1986). An explanation for these discrepancies might be provided by the recent report of Dessypris et al. (1987*b*), who reported that rEpo was able to stimulate megakaryocyte development only in the presence of suboptimal concentrations of phytohemagglutinin-

TABLE I
Effect of Purified or Recombinant Hematopoietic Growth Factors on
Megakaryocyte Colony Formation by Enriched Fractions
of Human Marrow

Addition to culture	CFU-MK–derived colony formation* $(5 \times 10^3$ cells plated)
None	2.5 ± 0.8[‡]
HPCM[§]	39.7 ± 6.3
rG-CSF (25 U/ml)[∥]	2.6 ± 0.8
CSF-1 (1,000 U/ml)[¶]	1.5 ± 0.5
rEpo (1.0 U/ml)**	1.3 ± 0.5
rGM-CSF (25 U/ml)[‡‡]	9.8 ± 2.8
rGM-CSF (100 U/ml)	11.2 ± 1.3
rIL-3 (50 U/ml)[§§]	9.7 ± 2.4
rIL-3 (200 U/ml)	17.8 ± 3.6
rGM-CSF (25 U/ml)+rIL-3 (50 U/ml)	22.0 ± 1.5

*NALT⁻ normal bone marrow cells were sorted, and a population of cells that expressed high densities of the My10 antigen and low densities of HLA-DR (My10^{+++}DR$^+$) were assayed in a serum-depleted culture system.
[‡]Results are expressed as means ± SEM of duplicate points performed in four separate experiments ($N = 8$).
[§]Serum-free human placenta–conditioned medium constituted 10% of the culture mix.
[∥]Recombinant granulocyte colony-stimulating factor.
[¶]Macrophage colony-stimulating factor.
**Recombinant erythropoietin.
[‡‡]Recombinant granulocyte/macrophage colony-stimulating factor.
[§§]Recombinant human IL-3.

stimulated, leukocyte-conditioned media (PHA-LCM) and human serum. This might explain the lack of stimulation with Epo observed in our studies, since the assay system was devoid of PHA-LCM and was serum depleted. The work of Dessypris et al. (1987*a*) suggests that several hematopoietic factors are capable of interacting and thereby influencing in vitro megakaryocytopoiesis. Along this line, Peschel et al. (1987) have shown a synergistic relationship between B cell stimulatory factor 1 (BSF-1), IL-1α, and rEpo in augmenting megakaryocyte colony formation. Definition of such molecular interactions remains an important focus of clinical investigation.

Vainchenker et al. (1982), Messner et al. (1982), and Solberg et al. (1985) have shown that human plasma is more potent than serum in supporting in vitro megakaryocyte colony formation. Their data suggested that platelet constituents act as chalones and down-regulate progenitor cell proliferation. Dessypris et al. (1987b) have recently isolated a glycoprotein with a molecular weight of 12,000–17,000 that is a platelet release product. This glycoprotein does not affect the early proliferative phase of the CFU-MK but acts on day 6–8 CFU-MK–derived cells by adversely affecting their maturation. In addition, Ishibashi et al. (1987b) have reported that transforming growth factor is inhibitory to megakaryocyte colony formation at picomolar concentrations. Since the glycoprotein isolated by Dessypris et al. (1987b) is distinct from transforming growth factor B, these reports suggest that, in addition to agonists of

TABLE II
Effect of TSF on Cytoplasmic Maturation of Recognizable Megakaryocytes

Days of incubation	Treatment	Maturation stage profile*				Statistical significance[‡]
		I	II	III	IV	
		%	%	%	%	
1	NABS[§]	2	13	47	38	—
	TSF (6.25)[ǀ]	0	13	46.5	42.5	NS
	TSF (9.4)	0	10	45	45	NS
	TSF (12.5)	0	7.5	47.5	45	0.05
	TSF (25)	0	7.5	46.5	46	0.04
2	NABS	0	7.3	49.3	43.5	—
	TSF (6.25)	0	7.5	51.8	40.8	NS
	TSF (9.4)	0	9.3	40.3	50.5	0.04
	TSF (12.5)	0	4.8	4.5	50.3	0.09
	TSF (25)	0	3.7	45	51.3	0.02
3	NABS	0	2	30	68	—
	TSF (6.25)	0	0	18	82	0.001
	TSF (9.4)	0	0	10	90	0.001
	TSF (12.5)	0	0	11	89	0.001
	TSF (25)	0	0	7	93	0.001

*Morphologically recognizable human megakaryocytes isolated by centrifugal elutriation and placed in short-term liquid cultures supplemented with normal human serum with increasing concentrations (6.25–25 μg/ml $\cdot$ 10^5 cells) of TSF (obtained from T.P. McDonald). Maturation was followed by assigning cells ($N = 4{,}000$ MK) to maturation stages as previously described (Straneva et al., 1986).
[‡]Statistical difference from controls (NABS); P values; NS, not significant.
[§]Normal human serum from an AB blood donor.
[ǀ]Final concentration of TSF in micrograms per milliliter.

megakaryocyte progenitor cell development, there are several potentially potent inhibitors of this process. A balance between these opposing biological activities might be important in maintaining the optimal production of platelets in health and disease.

Regulation of Terminal Maturation of Megakaryocytes

The cytoplasmic maturation of megakaryocyte precursor cells into fully developed megakaryocytes has been thought also to be controlled by a humoral factor (Levin and Evatt, 1979). Whole blood concentrations of this factor have been shown to be affected by perturbations in circulating platelet numbers (DeGabriele and Pennington, 1967; Levin and Evatt, 1979). This factor has been isolated from thrombocytopenic plasma and termed thrombopoietin (Evatt et al., 1974, 1979; Hill and Levin, 1986). Such a factor has also been purified from human embryonic kidney cell–conditioned media and termed thrombocytopoiesis stimulating factor (TSF) (McDonald et al., 1975). Recently, Tayrien and Rosenberg (1987) have purified to apparent homogeneity a molecule from media conditioned by human embryonic kidney cells and from thrombocytopenic plasma. This glycoprotein specifically stimulates megakaryocyte cytoplasmic maturation (Greenberg et al., 1987). This so-called megakaryocyte-stimulating factor (MSF) may be akin to TSF. Megakaryocyte cytoplasmic maturation results in the synthesis of constituent proteins and assembly of multiple secretory granules including alpha granules, dense bodies, and lysosomes (Gewirtz, 1986). Since megakaryocytes are relatively rare components of human marrow, the study of these maturational events has proven to be extremely difficult. In order to expedite these investigations, we used a long-term human megakaryocyte leukemia cell culture and attempted to trigger the further development of these cells, using phorbol esters as differentiating agents (Sledge et al., 1986). While two nontransforming phorbols failed to affect these cells, the transforming phorbol myristate acetate (PMA) induced a phenotype with characteristics of more mature megakaryocytes. These phenotypic changes included increased adherence to plastic or glass, further polyploidization, an increase in cell size, and increased expression of membrane glycoproteins and alpha granule constituents. Two-color flow cytometric analysis has allowed simultaneous analysis of cell size (forward-angle light scatter), DNA content, expression of surface membrane components, and alpha granule constituents. These studies have provided evidence that nuclear endomitosis, membrane development, and cytoplasmic maturation occur in parallel during megakaryocyte development (Fig. 2). Although 4 N populations of PMA-treated and untreated cells express factor VIII–related antigen and platelet glycoproteins to the same degree, all other ploidy levels possessed a statistically significant increase in the expression of these antigens as compared with control cells (Fig. 2). The cells that expressed the highest levels of factor VIII–related antigen and platelet glycoproteins after phorbol treatment also demonstrated the highest ploidy levels and represented the largest cells, as measured by forward-angle light scatter (Fig. 2).

In order to study the regulation of terminal maturation events of normal human megakaryocytes, we have attempted to devise an in vitro system more closely related to the in vivo situation. For this purpose, we have isolated, from normal human bone marrow, enriched fractions of megakaryocytes by means of density centrifugation and counterflow centrifugal elutriation (Berkow et al., 1984). The cells are then suspended in culture media for variable periods of time in the presence or absence of crude sera or hematopoietic growth factors. The cytoplasmic maturation of the cells is monitored by a morphological classification system in which megakaryocytes progress from the most immature stage of development (stage I) to the most mature form (stage IV) (Straneva et al., 1986). In the absence of thrombocytopenic plasma, the progression from more immature cells to more differentiated megakaryocytes that release platelets occurs after 5–7 d of incubation. This process has been shown to be accelerated by thrombocytopenic plasma, and the activity responsible for this acceleration of maturation has been termed megakaryocyte maturation factor (MMF). MMF is distinct from MK-CSF since MMF is not neutralizable by an anti–MK-CSF antiserum, and

MK-CSF preparations have not been able to promote cytoplasmic maturation (Straneva et al., 1987). We have also assessed several hematopoietic growth factors for their ability to accelerate megakaryocyte maturation. In Table II, the effect of various concentrations of a preparation of TSF are shown, and in Table III, the effects of Epo are demonstrated. Both TSF and Epo accelerated megakaryocyte cytoplasmic maturation. Several laboratories have also indicated that IL-3 can influence megakaryocyte maturation (Ishibashi and Burstein, 1986). In addition, several molecules have been shown to inhibit the maturation process (Desspyris et al., 1987*b*; Ishibashi et al., 1987*b*). These studies therefore indicate that the precise regulation of cytoplasmic maturation involves not only stimulatory molecules, but also inhibitory substances.

Conclusion

Although much has been learned about the humoral regulation of megakaryocytopoiesis, a great deal still remains unclear. Two alternative hypotheses are currently being entertained. One hypothesis suggests that one molecule with global actions, a so-called

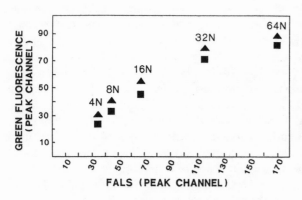

Figure 2. Relationship of cell size, ploidy, and expression of platelet glycoprotein (PGP, ▲) and factor VIII–related antigen (VIII:RAg, ■), as determined by two-color flow cytometric analysis. Cell size is represented by the peak channel of forward-angle light scatter (FALS), and the expression of PGP and VIII:RAg is represented by the peak channel of green fluorescence at each ploidy level. The largest cells have the highest ploidy and also express the highest levels of PGP and VIII:RAg.

thrombopoietin, is responsible for controlling the multiple cellular events occurring as megakaryocytopoiesis progresses from the progenitor cell to the end-stage platelet (Levin and Evatt, 1979). A second hypothesis suggests that two levels of regulation exist: one at the level of megakaryocyte progenitor cells, and the other at the level of the immature megakaryocyte precursor cell (Williams et al., 1982). These two levels of development are thought to be regulated by two distinct sets of growth factors. As can be seen from this discussion, these two hypotheses are not mutually exclusive. There appear to be factors that can influence only progenitor cell proliferation, such as the elusive MK-CSF (Hoffman et al., 1985; Straneva et al., 1987) and GM-CSF (Quesenberry et al., 1985), and others that exclusively alter cytoplasmic maturation such as MSF (TSF) (Greenberg et al., 1987). In addition, IL-3 appears to be unique in that it can alter both stages of megakaryocyte development (Robinson et al., 1987; Ishibashi and Burstein, 1986). Interestingly, while the addition of BSF-1 is needed for Epo to affect megakaryocyte progenitor cell dynamics, (Peschel et al., 1987), Epo alone is capable of accelerating megakaryocyte cytoplasmic maturation (Ishibashi et al., 1987*a*). Megakaryocyte polyploidization and resultant nuclear ploidy is thought to be determined by the previous mitotic history of the progenitor cell (Paulus et al., 1982). This balance between proliferation and endomitosis does not appear to be entirely predetermined, since IL-3, erythropoietin (Ishibashi et al., 1987*a*), thrombocytopenic dog sera (Arriaga et al., 1987), and thrombopoietin (Levin et al., 1982) have been shown to be able to alter nuclear ploidy. The precise role of inhibitors of megakaryocyte progenitor proliferation and terminal maturation is presently being defined (Dessypris et al., 1987*b*; Ishibashi et al., 1987*b*). These inhibitory molecules appear to be an additional mechanism by which the process of megakaryocytopoiesis might be carefully controlled.

TABLE III
Effect of Epo on Cytoplasmic Maturation of Recognizable Megakaryocytes

Days of incubation	Treatment	Maturation stage profiles*				Statistical significance‡
		I	II	III	IV	
		%	%	%	%	
1	NABS§	2.5	26.5	41	30	—
	Epo (1)∥	1	23.5	38	37.5	NS
	Epo (5)	1.5	23.5	36	39	NS
	Epo (10)	1.5	21.5	41.5	35.5	NS
2	NABS	1.5	20.5	45	33	—
	Epo (1)	0	11.5	49.5	39	0.02
	Epo (5)	0	14	40.5	45.5	0.02
	Epo (10)	0	8.2	43.6	48.2	0.001
3	NABS	0	7.7	35.9	56.5	—
	Epo (1)	0	3.2	36.1	60.8	NS
	Epo (5)	0	1.1	37.2	61.7	0.01
	Epo (10)	0	0	34.7	65.3	0.009

*Morphologically recognizable human MK were isolated by centrifugal elutriation and placd in short-term liquid cultures supplemented with normal human serum with increasing concentrations (1–10 U/ml·10^5 cells) of Epo. Maturation was followed by assigning cells ($N = 2,247$ MK) to maturation stages as previously described (Straneva et al., 1986).
‡Statistical difference from controls (NABS); *P* values; NS, not significant.
§Normal human serum from an AB blood donor.
∥Partially purified Epo (57 U/mg protein; Toyobo Co., New York, NY) added.

Acknowledgments

This work was supported in part by funds provided by the National Cancer Institute, National Institutes of Health.

References

Adamson, J. W. 1968. The erythropoietin/hematocrit relationship in normal and polycythemic man: implications of marrow regulation. *Blood.* 32:597–604.

Arriaga, M., K. South, J. L. Cohen, and E. M. Mazur. 1987. Interrelationship between mitosis and endomitosis in cultures of human megakaryocyte progenitor cells. *Blood.* 69:486–492.

Berkow, R. L., J. Straneva, E. Bruno, G. S. Beyer, J. S. Burgess, and R. Hoffman. 1984. Isolation of human megakaryocytes by density centrifugation and counterflow centrifugal elutriation. *Journal of Laboratory and Clinical Medicine*. 103:811–818.

Clark, S. C., and R. Kamen. 1987. The human hematopoietic colony-stimulating factors. *Science*. 236:1229–1236.

DeGabriele, G., and D. G. Pennington. 1967. Physiology of the regulation of platelet production. *British Journal of Haematology*. 13:202–209.

Dukes, P. P., J. C. Egrie, T. W. Strickland, J. K. Broure, and F. K. Lin. 1986. Megakaryocyte colony stimulating activity of recombinant human and monkey erythropoietin. *In* Megakaryocyte Development and Function. R. F. Levine, N. Williams, J. Levin, and B. L. Evatt, editors. Alan R. Liss, Inc., New York, NY. 105–109.

Dessypris, E. N., J. H. Gleaton, and O. L. Armstrong. 1987*a*. Effect of human recombinant erythropoietin on human marrow megakaryocyte colony formation *in vitro*. *British Journal of Haematology*. 65:265–269.

Dessypris, E. N., J. H. Gleaton, S. T. Sawyer, and O. L. Armstrong. 1987*b*. Suppression of maturation of megakaryocyte colony forming unit *in vitro* by a platelet release glycoprotein. *Journal of Cellular Physiology*. 130:361–368.

Enomoto, K., M. Kawakita, S. Kihimoto, N. Katayana, and T. Miyake. 1980. Thrombopoiesis and megakaryocyte colony stimulating factor in the urine of patients with aplastic anemia. *British Journal of Haematology*. 45:551–556.

Evatt, B. L., J. Levin, and K. M. Algazy. 1979. Partial purification of thrombopoietin from plasma of thrombocytopenic rabbits. *Blood*. 54:377–387.

Evatt, B. L., D. P. Shreiner, and J. Levin. 1974. Thrombopoietic activity of fractions of rabbit plasma: studies in rabbits and mice. *Journal of Clinical and Laboratory Medicine*. 83:364–371.

Gewirtz, A. M. 1986. Human megakaryocytopoiesis. *Seminars in Hematology*. 23:27–42.

Greenberg, S. M., D. J. Kuter, and R. D. Rosenberg. 1987. In vitro stimulation of megakaryocyte maturation by megakaryocyte stimulating factor. *Journal of Biological Chemistry*. 262:3269–3276.

Hill, R., and J. Levin. 1986. Partial purification of thrombopoietin using lectin chromatography. *Experimental Hematology*. 14:752–759.

Hoffman, R., E. Mazur, E. Bruno, and V. Floyd. 1981. Assay of an activity in the serum of patients with disorders of thrombopoiesis that stimulates formation of megakaryocytic colonies. *New England Journal of Medicine*. 305:533–538.

Hoffman, R., J. Straneva, H. H. Yang, E. Bruno, and J. Brandt. 1987. New insights into the regulation of human megakaryocytopoiesis. *Blood Cells*. 13:75–86.

Hoffman, R., H. Yang, E. Bruno, and J. Straneva. 1985. Purification and partial characterization of a megakaryocyte colony stimulating factor from human plasma. *Journal of Clinical Investigation*. 75:1174–1182.

Ishibashi, T., and S. Burstein. 1986. Interleukin 3 promotes the differentiation of isolated megakaryocytes. *Blood*. 67:1512–1514.

Ishibashi, T., J. A. Koziol, and S. A. Burstein. 1987*a*. Human recombinant erythropoietin promotes differentiation of murine megakaryocytes in vitro. *Journal of Clinical Investigation*. 79:286–287.

Ishibashi, T., S. L. Miller, and S. Burstein. 1987*b*. Type B transforming growth factor is a potent inhibitor of murine megakaryocytopoiesis in vitro. *Blood.* 69:1737–1741.

Kawakita, M., K. Enomoto, N. Katayama, S. Kishimoto, and T. Miyake. 1981. Thrombopoiesis and megakaryocyte colony stimulating factors in the urine of patients with idiopathic thrombocytopenic purpura. *British Journal of Haematology.* 48:609–615.

Levin, J., and B. L. Evatt. 1979. Humoral control of thrombopoiesis. *Blood Cells.* 5:105–121.

Levin, J., F. Levin, D. F. Hull III, and D. G. Pennington. 1982. The effects of thrombopoietin on megakaryocyte CFU, megakaryocytes and thrombopoiesis with studies of ploidy and platelet size. *Blood.* 80:989–998.

Levine, R. F. 1980. Isolation and characterization of normal human megakaryocytes. *British Journal of Haematology.* 45:487–497.

Mazur, E. M., R. Hoffman, J. Chasis, S. Marchesi, and E. Bruno. 1981. Immunofluorescent identification of human megakaryocyte colonies using an antiplatelet glycoprotein antiserum. *Blood.* 57:277–286.

Mazur, E. M., and K. South. 1985. Human megakaryocyte colony stimulating factor in sera from aplastic dogs: partial purification, characterization and determination of hematopoietic cell lineage specificity. *Experimental Hematology.* 13:1164–1171.

McDonald, T. P., R. Clift, R. D. Lange, C. Holan, J. E. Tribby, and G. H. Barlow. 1975. Thrombopoietin production by human embryonic kidney cells in culture. *Journal of Laboratory and Clinical Medicine.* 85:59–66.

McDonald, T. P., M. D. Cottrell, R. E. Clift, W. C. Cullen, and F. K. Lin. 1987. High doses of recombinant erythropoietin stimulate platelet production in mice. *Experimental Hematology.* 15:719–721.

Messner, H. A., N. Jamal, and C. Izaquire. 1982. The growth of large megakaryocyte colonies from human bone marrow. *Journal of Cellular Physiology.* (Suppl. 1):45–51.

Paulus, J. M., M. Prenant, and J. F. Deschamp. 1982. Polyploid megakaryocytes develop randomly from a multi compartmental system of committed progenitors. *Proceedings of the National Academy of Sciences.* 79:4410–4414.

Peschel, C., W. E. Paul, J. Ohara, and I. Green. 1987. Effects of B cell stimulatory factor-1/interleukin 4 on hematopoietic progenitor cells. *Blood.* 70:254–263.

Quesenberry, P. J., J. N. Ihle, and E. McGrath. 1985. The effect of interleukin-3 and GM-CSA2 on megakaryocyte and myeloid clonal colony formation. *Blood.* 65:214–217.

Robinson, B. E., H. E. McGrath, and P. E. Quesenberry. 1987. Recombinant murine granulocyte macrophage colony stimulating factor has megakaryocyte colony stimulating activity and augments megakaryocyte colony stimulation by interleukin 3. *Journal of Clinical Investigation.* 79:1648–1652.

Shimizu, T., C. Whitacre, T. Kato, H. Mizoguchi, and T. Miyake. 1987. Purification of human sialylated megakaryocyte colony stimulating factor. *Experimental Hematology.* 15:140. (Abstr.)

Sledge, G. W., Jr., M. Glant, J. Jansen, N. A. Heerema, B. J. Roth, M. Goheen, and R. Hoffman. 1986. Establishment in long term culture of megakaryocytic leukemia cells (EST-IU) from the marrow of a patient with leukemia and a mediastinal germ cell neoplasm. *Cancer Research.* 46:2155–2159.

Solberg, L. A., N. Jamal, and H. A. Messner. 1985. Characterization of human megakaryocytic colony formation in human plasma. *Journal of Cellular Physiology.* 124:67–74.

Sparrow, R. L., O. Swee-Huat, and N. Williams. 1987. Haemopoietic growth factors stimulating murine megakaryocytopoiesis: interleukin-3 is immunologically distinct from megakaryocyte potentiator. *Leukemia Research.* 11:31–36.

Straneva, J. E., M. P. Goheen, S. L. Hui, E. Bruno, and R. Hoffman. 1986. Terminal cytoplasmic maturation of human megakaryocytes in vitro. *Experimental Hematology.* 14:919–929.

Straneva, J. E., H. H. Yang, S. L. Hui, E. Bruno, and R. Hoffman. 1987. Effects of megakaryocyte colony stimulating factor on terminal cytoplasmic maturation of human megakaryocytes. *Experimental Hematology.* 15:657–663.

Tayrien, G., and R. D. Rosenberg. 1987. Purification and properties of a megakaryocyte stimulating factor present in the serum free conditioned medium of human embryonic kidney cells and thrombocytopenic plasma. *Journal of Biological Chemistry.* 262:3262–3268.

Trowbridge, E. R. A., J. F. Martin, and D. N. Slater. 1982. Evidence for a theory of physical fragmentation of megakaryocytes implying that all platelets are produced in the pulmonary circulation. *Thrombosis Research.* 28:461–475.

Vainchenker, W., J. Bought, J. Guichaid, and J. Breton-Gorius. 1979. Megakaryocyte colony formation from human bone marrow precursors. *Blood.* 54:940–945.

Vainchenker, W., J. Chapman, J. F. Deschamps, G. Vinci, J. Bouget, M. Titeux, and J. Breton-Gorius. 1982. Normal human serum contains a factor(s) capable of inhibiting megakaryocytic colony formation. *Experimental Hematology.* 10:650–660.

Williams, N., R. R. Eger, H. M. Jackson, and D. J. Nelson. 1982. Two factor requirements for murine megakaryocyte colony formation. *Journal of Cellular Physiology.* 110:101–104.

Williams, N., and R. F. Levine. 1982. The origin, development and regulation of megakaryocytes. *British Journal of Haematology.* 52:173–180.

Williams, N., R. Sparrow, K. Gill, D. Yasmeen, and I. McNeice. 1985. Murine megakaryocyte colony stimulating factor: its relationship to interleukin 3. *Leukemia Research.* 9:1487–1496.

Wright, J. H. 1906. The origin and nature of blood platelets. *Boston Medical and Surgical Journal.* 154:643–645.

Chapter 8

Transfer of Genes in Hematopoietic Cells with Retroviral Vectors

Arthur Nienhuis, David Bodine, Timothy Browder, Siu-Wah Chung, Cynthia Dunbar, Stefan Karlsson, and Peter Wong

Clinical Hematology Branch, National Heart, Lung, and Blood Institute, National Institutes of Health, Bethesda, Maryland

Cell Physiology of Blood © 1988 by The Rockefeller University Press

Introduction

The ability to transfer genes into hematopoietic stem cells in a form that allows lineage-specific expression is a potentially valuable experimental tool. One application of this strategy is the deliberate modification of the active genetic repertoire of a stem cell to determine how a specific gene product affects progenitor and precursor cell proliferation and differentiation. Such studies may be useful in defining oncogenic mechanisms. Another application is the treatment of genetic diseases in which the primary defect is expressed in bone marrow–derived cells, e.g., thalassemia, sickle cell anemia, or adenosine deaminase deficiency.

Although several techniques are useful for introducing genetic information into tissue culture cells (Graham and van der Eb, 1973; Schaffner, 1980; Chu et al., 1987), gene transfer into hematopoietic stem cells has proved more challenging. The only successful technique involves the use of retroviral vectors. This strategy uses retroviral particles containing a recombinant genome to infect marrow cells (Williams et al., 1984; Miller et al., 1984). Animals are monitored for reconstitution and gene transfer by direct detection of the transferred gene, by analysis of its RNA product, or by detection of the encoded protein directly (Williams et al., 1984, 1986; Miller et al., 1984; Eglitis et al., 1985; Keller et al., 1985; Dick et al., 1985; Lemischka et al., 1986; Kantoff et al., 1987; Lim et al., 1987).

Retroviral vectors have been developed for transfer and expression of the human beta-globin gene (Karlsson et al., 1987; Cone et al., 1987). The ultimate goal of this investigation is to devise strategies for genetic therapy for sickle cell anemia and thalassemia. Appropriately designed retroviral vectors (Wong et al., 1987) have also been used to transfer growth factor genes into hematopoietic progenitor and stem cells. The consequences of endogenous production of the growth factor have been defined both in vitro and in vivo. A myeloproliferative syndrome has been created in mice by virtue of endogenous production of hematopoietic growth factors.

General Strategy for Creation of Recombinant Retroviral Vectors

RNA viruses infect cells by interaction of the virally encoded envelope protein with specific cellular surface molecules that function as receptors. Surface binding is followed by membrane fusion or endocytosis, leading to release of the retroviral contents into the cytoplasm (White et al., 1983). Only in the case of the HIV virus has the receptor molecule been identified; the HIV envelope protein interacts with the surface molecule defined by antibodies with CD4 specificity (Dagleish et al., 1984; Maddon et al., 1986; Stein et al., 1987). Genomic RNA is converted into DNA by reverse transcriptase and transferred into the nucleus, where it integrates into a host cell chromosome. The structure of the integrated proviral form of the retrovirus is defined and highly reproducible; the genes for the virally encoded proteins are flanked by long terminal repeats (LTR) that contain transcription control signals (Varmus, 1982). However, integration into host cell chromosomes appears to be at random sites.

In a productive retroviral infection, the integrated provirus is transcribed to yield both full-length genomic RNA and spliced RNA species. The spliced species encode the various viral proteins, including the gag products that form the core of the retroviral particle, reverse transcriptase, and the envelope proteins. The full-length

RNA species are paackaged to serve as the genetic information contained in newly formed retroviral particles. Inclusion of an RNA molecule into a retroviral particle requires interaction of the gag proteins with a "packaging signal" that is found near the 5' end of full-length proviral transcripts (Varmus, 1982).

Production of replication-incompetent, helper-free recombinant retroviruses for the purposes of gene transfer requires separation of the two functions of the proviral genome in a "packaging" or "helper cell" (Mann et al., 1983; Cone and Mulligan, 1984; Miller et al., 1985; Miller and Buttimore, 1986). Separate proviral genomes are transcribed to yield either RNA molecules that can be packaged into new retroviral particles or RNA molecules that are translated into the proteins required for particle

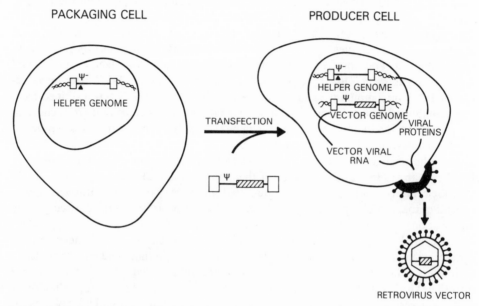

Figure 1. Packaging of retroviral vectors with helper cells. Helper or packaging cell lines are created by introduction of a defective proviral genome using conventional DNA transfer techniques. This proviral genome encodes the information necessary for synthesis of proteins needed to form a retroviral particle. A recombinant retroviral genome is introduced into the packaging cell line to derive producer clones. The vector viral RNA is packaged in viral proteins encoded by the helper virus genes, allowing formation of recombinant retroviral particles. Such particles can then be used to infect target cell populations, resulting in transfer of genetic information. (Figure provided by Martin Eglitis.)

formation (Fig. 1). Creation of a "packaging" or "helper" cell line requires introduction of a defective retroviral genome. The molecularly cloned proviral form is deliberately mutated to remove the DNA sequences that encode the packaging signal. Further modifications such as removal of the 3' LTR reduces the possibility that RNA molecules transcribed from this mutated helper proviral genome will recombine with vector proviral RNA species to form molecules that encode all of the functions of an infectious retrovirus (Miller and Buttimore, 1986). Such recombination leads to the emergence of replication-competent wild-type virus in recombinant retroviral stocks.

The mutated proviral genome is introduced into mouse 3T3 cells by conventional DNA transfection techniques. Clones are selected, and those that contain the desired proviral genome and produce the proteins needed for particle formation are identified and used in subsequent experiments. Currently useful packaging cell lines include the $\psi2$ (Mann et al., 1983) and PA317 (Miller and Buttimore, 1986) cell lines. Retroviral particles generated in the $\psi2$ packaging cell line have ecotropic host range and infect mouse cells preferentially. The envelope protein sythesized by these PA317 cells has an amphotropic host range and therefore recombinant particles infect cells of a wide range of species including man.

Cells lines producing recombinant retroviruses are obtained when a second proviral genome is introduced into a packaging cell line. The gene of interest is inserted into a cloned provirus by recombinant techniques using plasmid vectors in *Escherichia coli*. The coding sequences for the protein product can be inserted into the proviral genome in a fashion that takes advantage of the transcriptional signals of the provirus. Alternatively, the intact transcriptional unit can be inserted with its own promoter and splicing signals. Examples of both types of constructions are described in more detail below. Once the appropriate recombinant plasmid is obtained, its DNA is introduced into the packaging cell line by conventional techniques. Within 2–3 d after introduction of DNA, recombinant retroviral particles are produced. RNA transcribed from the recombinant proviral genomes can be packaged by components that are encoded in the packaging cell line by the "helper" genome. A useful strategy has been developed in which these initial viral particles are used to infect a second packaging line of opposite host range (e.g., ecotropic into amphotropic or amphotropic into ecotropic) (Miller et al., 1985). The advantage of this approach is that producer clones can be identified that contain only a single integrated intact copy of the proviral genome, so that all of the particles produced should be capable of transferring the desired genetic information.

The final step in obtaining a useful cell line for gene transfer into bone marrow is the screening of individual producer clones to identify those that yield viral particles in high titers. The titer reflects the capacity of individual producer clones to release infectious viral particles into culture medium over time. Highly efficient production of virus is necessary for successful infection of bone marrow cells. The capacity of individual clones to produce infectious recombinant particles is a stable genetic characteristic, so that once a useful clone is identified, it becomes a permanent valuable reagent for subsequent gene transfer experiments.

Gene Transfer into Mouse Hematopoietic Cells

Gene transfer experiments in vivo have been performed in the context of bone marrow transplantation. Bone marrow or fetal liver cells are removed from donor animals and infected with recombinant retroviral particles. Two strategies have been used, "supernatant" and "co-culture." For the supernatant method, fresh culture medium is added to semiconfluent producer cells. After 24 h the medium, now containing a high concentration of retroviral particles, is removed. Bone marrow cells are incubated in such supernatants for periods ranging from two to several hours, during which the initial steps of infection occur. A potentially more efficient method of virus infection involves the intimate contact of hematopoietic and producer cells by co-culture. The producer cells adhere to the tissue culture flask, whereas the hematopoietic cells can be removed with gentle washing. After 48–72 h of co-culture, the hematopoietic cells can be recovered for reinjection into recipient animals.

Two types of animals can be used as recipients. Lethally irradiated animals can be repopulated with hematopoietic stem cells (Williams et al., 1984). Alternatively, the genetically defective W/W^v strain can be repopulated with minimal or no irradiation because normal stem cells have a competitive advantage (Dick et al., 1985). It is important to have a means to monitor repopulation, either by using syngenic strains with differing hemoglobin types or by transplantation of female cells into male animals. Disappearance of Y chromosomal DNA from hematopoietic cells, in the latter case, can be used as an index of repopulation. The efficiency of gene transfer can be evaluated only in animals in which significant repopulation of the hematopoietic tissue with donor cells has been accomplished.

Retroviruses have the potential of infecting all dividing cells obtained from fetal liver or bone marrow (Fig. 2). Infection of differentiating precursors or hematopoietic progenitors that lack self-renewal potential is of no consequence with respect to the ultimate goal of stable gene transfer. The progeny of such cells disappear quickly in

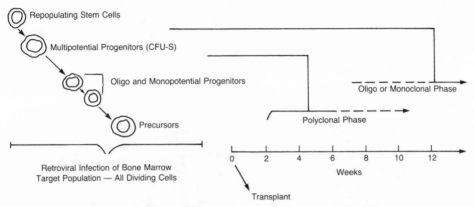

Figure 2. Gene transfer into mouse hematopoietic cells. Bone marrow or fetal liver cells provide targets for retroviral infection. An active state of cell proliferation is required for transfer and integration of retroviral information. Cells at various stages of hematopoietic differentiation are infected. Reconstitution is characterized by a polyclonal phase in which circulating blood cells are derived from multipotential progenitors. Long-term reconstitution requires repopulating stem cells and is usually oligo- or monoclonal.

recipient animals. The desired target for retroviral infection includes the multipotential progenitors that in mice are capable of forming spleen colonies, and stem cells that ultimately repopulate the hematopoietic tissues of recipient animals. Gene transfer is monitored by examining individual spleen colonies 10–14 d after injection of infected hematopoietic cells. Alternatively, one can monitor the circulating blood elements between 2 and 8 wk, when the cells derived from these multipotential progenitors are circulating. By 12–16 wk, reconstitution by repopulating stem cells is accomplished in animals in which transplantation has been successful.

Various strategies have been devised to enhance the probability that repopulating stem cells and multipotential progenitors are infected with recombinant retroviruses (Dick et al., 1985). Injection of 5-fluorouracil into donor animals effectively destroys dividing hematopoietic cells. This cell cycle–specific agent kills later-stage progenitor and precursor cells that are in an active state of division. A consequence of this abrupt destruction of late cells in the hematopoietic pathway is activation of earlier cells into the cell cycle as marrow regeneration begins. Enrichment in early cells is obtained and such cells are rendered more susceptible to infection and integration of retrovirally encoded genetic information.

Most retroviral particles used to date include a selectable marker. This allows preselection of infected cells before marrow engraftment is attempted. The neo^R gene encodes a phosphotransferase that inactivates the neomycin analogue, G418. Successfully infected marrow can be preselected by incubation in G418 for several days (Dick

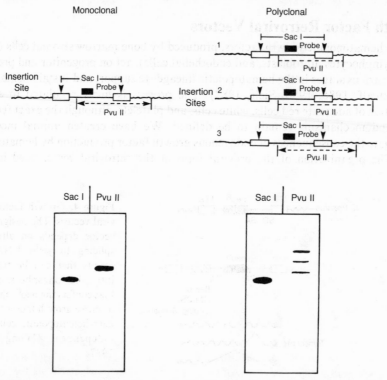

Figure 3. Retroviral insertion as a marker of clonality. The retroviral genome is integrated as a defined structure, although the insertion site in host cell chromosomal DNA is random. Restriction endonuclease analysis can be used to "count" the number of insertion sites in a population of cells. Because the vast majority of infected cells contain only a single integrated proviral genome, the number of insertion sites provides an estimate of the number of "clones" represented. Enzymes that cut twice in the integrated proviral genome define its structure. Enzymes that cut once in the proviral genome yield DNA fragments that are of unique length for each insertion site, since the relative distance between the insertion site and the next site for that enzyme is random and highly variable. Examples of data obtained on analysis of monoclonal or polyclonal populations are shown.

et al., 1985). Alternatively, infection of hematopoietic cells with vectors containing a methotrexate-resistant form of dihydrofolate reductase allows selection in methotrexate before engraftment (Miller et al., 1985).

A remarkable feature of hematopoietic reconstitution in transplant recipients has emerged from experiments in which retroviral-mediated gene transfer is successful. Hematopoiesis can be sustained for long periods by the progeny of a single stem cell (Dick et al., 1985; Lemischka et al., 1986). Restriction endonuclease analysis, as illustrated in Fig. 3, can be used to estimate the number of integration sites represented in the cells of hematopoietic tissues. The vast majority of successfully infected cells have a single integration site, so that the number of integration sites provides an estimate of the number of "clones" of stem cell progeny that are contributing to hematopoiesis at any given time in the animal. Clonal succession has been observed in that hematopoiesis can be sustained for long periods by one stem cell with gradual replacement by the progeny of a second or even a third stem cell.

Growth Factor Retroviral Vectors

Several hematopoietic growth factors, produced by bone marrow stromal cells (monocytes, lymphocytes, fibroblasts, and endothelial cells), act on progenitor and precursor cells at various stages in the hematopoietic lineages to support and sustain hematopoiesis (Metcalf, 1985, 1986; Sieff, 1987). The process is finely regulated to achieve production of adequate red cells, white cells, and platelets, although the exact feedback loops and mechanisms remain to be defined. We have created animal models to investigate the consequences of endogenous growth factor production by hematopoietic cells. The organization of the proviral form of the retroviral vector used in these

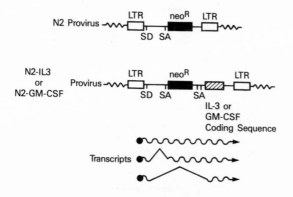

Figure 4. Growth factor retroviral vectors. The design of this vector depends on alternative splicing to yield RNA transcripts that can be translated into the phosphotransferase that confers the neo^R phenotype or into a growth factor that renders hematopoietic cells factor independent (Wong et al., 1987).

experiments is shown in Fig. 4 (Wong et al., 1987). The N2 vector, designed in the laboratory of E. Gilboa (Hwang et al., 1984), contains the gene that confers resistance to G418. Most of the sequences that encode the normal retroviral products (gag, pol, and envelope) have been removed, although a segment of the 5' end is preserved. The N2 vector consistently allows identification of highly efficient producer clones. The N2 vector was modified by insertion of the gene for interleukin 3 (IL-3) or granulocyte/monocyte colony-stimulating factor (GM-CSF) along with RNA splicing signals (Wong et al., 1987). The transcripts generated from the integrated form of this retroviral vector are shown. Alternative splicing yields mRNA species that can be translated into the phosphotransferase (neo^R) or the growth factor.

Infection of growth factor–dependent established hematopoietic cell lines with N2–IL-3 renders such cells factor dependent (Wong et al., 1987). Endogenous growth factor synthesis establishes an autocrine mechanism whereby the growth factor and receptor apparently interact within the Golgi and eliminate the need for growth factor within the culture medium (Browder, T.M., J.S. Abrams, P.M.C. Wong, and A.W. Nienhuis, manuscript in preparation). Characterization of infected fetal liver or bone marrow cells in clonal hematopoietic culture indicates that normal progenitors also acquire the property of growth factor independence when successfully infected with this virus.

Animals reconstituted with bone marrow cells infected with the N2–IL-3 retrovirus vector exhibit a myeloproliferative syndrome (Wong, P.M.C., S.-W. Chung, C.E. Dunbar, D.M. Bodine, S. Ruscetti, and A.W. Nienhuis, manuscript in preparation). This occurs in ~20–25% of recipients; there is marked elevation in the white blood cell count (200,000–900,000/mm^3) predominantly with neutrophils and late-stage myeloid cells (bands). Enlargement of the liver and spleen in such animals is due to infiltration by mature hematopoietic cells. Bone marrow hyperplasia is striking and involves predominantly the myeloid series. Despite these striking distortions of hematopoiesis, the animals survive for several weeks. Clinically, the disorder resembles chronic myelogenous leukemia.

One definition of leukemia is the evolution of a clonal population of autonomously proliferating cells independently of normal regulatory mechanisms. The genetic basis for such disordered growth is thought to be mutations in cellular proto-oncogenes. In our experiments, the IL-3 gene appears to act in an oncogenic fashion in sustaining unregulated hematopoiesis. Usually, this is clonal, as shown by analysis of proviral integration sites (Fig. 3). Remarkably, a single "mutation" may be sufficient to establish this syndrome. We speculate that endogenous production of IL-3 in this experimental model subserves the same functional consequences as the bcr-abl rearranged gene product (Mes-Masson et al., 1986) encoded on the Philadelphia chromosome in human chronic myelogenous leukemia.

Globin Retroviral Vectors

In beta-thalassemia, there is severe deficiency of beta-globin synthesis. Individuals with sickle cell anemia have an abnormal hemoglobin that undergoes polymerization upon deoxygenation. Introduction of a normal globin gene into the bone marrow cells of both types of patients could have remarkable therapeutic benefit. A high level of expression and continued production of red cells that express the newly introduced gene are required.

To achieve tissue-specific expression, we have introduced an intact globin gene into the N2 vector (Karlsson et al., 1987). The transcriptional control sequences that are necessary for tissue-specific expression, as defined in transgenic mice, are wholly contained within the globin genes and the immediate 5' flanking sequences. In transgenic animals, the human beta-globin gene is preferentially expressed in the definitive erythroid lineage (Costantini et al., 1985; Townes et al., 1985; Kollias et al., 1986). Expression is often low and many animals exhibit no expression at all. Other transcriptional control elements at greater distance from the beta-globin gene are thought to be required to achieve consistent and high-level expression (Grosveld et al., 1987). New vectors are being made that incorporate these sequences in an effort to improve the outcome of globin gene transfer into the bone marrow cells of mice.

The vector outlined in Fig. 5 has been used to insert globin genes into mouse erythroleukemia cells (Karlsson et al., 1987). These continuously proliferating cells can be induced with various chemical substances to undergo normal erythroid maturation in culture (Marks et al., 1985). Expression of the newly introduced gene increased in a normal fashion with erythroid maturation. A normal-sized, correctly spliced globin mRNA was produced from the inserted proviral genome. Human beta-globin could be detected by immunocytochemical techniques in most of the cells.

The same vector has been used for in vivo gene transfer (Karlsson, S., D. Bodine, Th. Pappayammopoulou, L. Perry, and A.W. Nienhuis, manuscript in preparation). The vast majority of mice have red cells containing human beta-globin chains up to 8 wk after transplantation. This production of red cells containing human hemoglobin apparently derives from the multipotential progenitor differentiation and maturation in the spleen. Once hematopoiesis becomes oligo- or monoclonal at 12–16 wk, globin gene expression usually disappears. We believe that this reflects the difficulty of gene transfer into stem cells.

Two barriers must be overcome for successful gene transfer for therapeutic purposes in hemoglobin disorders. The level of expression is relatively low, only 1–10% of the level of expression of the mouse globin genes. This problem can be overcome by inclusion of additional *cis*-active sequence elements that serve to modulate globin gene expression (Grosveld et al., 1987). The second problem is the apparent infrequency of stem cell infection. Efforts are being directed toward stem cell purification and

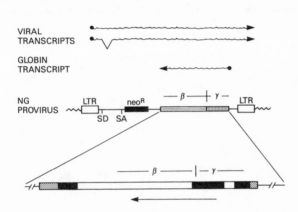

Figure 5. Globin retroviral vector. This vector is designed so that the globin gene is inserted in the reverse transcriptional orientation from the RNA generated from the LTR. Globin transcriptional control signals and introns are preserved during propagation and passage of the recombinant retrovirus. The integrated proviral genome in the target cell generates two types of transcripts in opposite transcriptional orientation (Karlsson et al., 1987).

manipulation of cells to make them more susceptible to infection by induction of proliferation and by design of vectors that can target specifically to hematopoietic stem cells. The newly available hematopoietic growth factors and increasing knowledge regarding hematopoietic mechanisms may allow definition of infection conditions that lead to more efficient gene transfer into stem cells.

References

Chu, G., H. Hayakawa, and P. Berg. 1987. Electroporation for the efficient transfection of mammalian cells with DNA. *Nucleic Acids Research.* 15:1311–1326.

Cone, R. D., and R. C. Mulligan. 1984. High-efficiency gene transfer into mammalian cells: generation of helper-free recombinant retrovirus with broad mammalian host range. *Proceedings of the National Academy of Sciences.* 81:6349–6353.

Cone, R. D., A. Weber-Benarous, D. Baorto, and R. C. Mulligan. 1987. Regulated expression of a complete human beta-globin gene encoded by a transmissible retrovirus vector. *Molecular and Cellular Biology.* 7:887–897.

Costantini, F., G. Radice, J. Magram, G. Stamatoyannopoulos, T. Papayannopoulou, and K. Chada. 1985. Developmental regulation of human globin genes in transgenic mice. *Cold Spring Harbor Symposium of Quantitive Biology.* 50:361–370.

Dalgleish, A. G., P. C. L. Beverley, P. R. Clapham, D. H. Crawford, M. F. Greaves, and R. A. Weiss. 1984. The CD4 (T4) antigen is an essential component of the receptor for the AIDS retrovirus. *Nature.* 312:763–768.

Dick, J. E., M. C. Magli, D. Huszar, R. A. Phillips, and A. Bernstein. 1985. Introduction of a selectable gene into primitive stem cells capable of long-term reconstitution of the hemopoietic system of W/Wᵛ mice. *Cell.* 42:71–79.

Eglitis, M. A., P. Kantoff, E. Gilboa, and W. F. Anderson. 1985. Gene expression in mice after high efficiency retroviral-mediated gene transfer. *Science.* 230:1395–1398.

Graham, F. L., and A. J. van der Eb. 1973. A new technique for the assay of infectivity of human adenovirus 5 DNA. *Virology.* 52:456–467.

Grosveld, F., M. Blom van Assendelft, D. R. Greaves, and G. Kollias. 1987. Position independent high level expression of human beta-globin gene in transgenic mice. *Cell.* 51: 975–985.

Hwang, L. S., J. Park, and E. Gilboa. 1984. Role of intron-contained sequences in formation of Moloney murine leukemia virus env mRNA. *Molecular and Cellular Biology.* 4:2289–2297.

Kantoff, P. W., A. Gillio, L. R. McLachlin, C. Bordignon, M. A. Eglitis, N. A. Kernan, R. C. Moen, D. B. Kohn, S.-F. Yu, E. Karson, S. Karlsson, J. A. Zweibel, E. Gilboa, M. Blaese, A. W. Nienhuis, R. J. O'Reilly, and W. F. Anderson. 1987. Expression of human adenosine deaminase in non-human primates after retroviral-mediated gene transfer. *Journal of Experimental Medicine.* 166:219–234.

Karlsson, S., T. Papayannopoulou, S. G. Schweiger, G. Stamatoyannopoulos, and A. W. Nienhuis. 1987. Retroviral-mediated transfer of genomic globin genes leads to regulated production of RNA and protein. *Proceedings of the National Academy of Sciences.* 84:2411–2415.

Keller, G., C. Paige, E. Gilboa, and E. F. Wagner. 1985. Expression of a foreign gene in myeloid and lymphoid cells derived from multipotent haematopoietic precursors. *Nature.* 318:149–154.

Kollias, G., N. Wrighton, J. Hurst, and F. Grosveld. 1986. Regulated expression of human A gamma-, beta-, and hybrid gamma beta-globin genes in transgenic mice: manipulation of the developmental expression patterns. *Cell.* 46:89–94.

Lemischka, I. R., D. H. Raulet, and R. C. Mulligan. 1986. Developmental potential and dynamic behavior of hematopoietic stem cells. *Cell.* 45:917–927.

Lim, B., D. A. Williams, and S. H. Orkin. 1987. Retrovirus-mediated gene transfer of human adenosine deaminase: expression of functional enzyme in murine hematopoietic stem cells in vivo. *Molecular and Cellular Biology.* 7:3459–3465.

Maddon, P. J., A. G. Dalgleish, J. S. McDougal, P. R. Clapham, R. A. Weiss, and R. Axel. 1986. The T4 gene encodes the AIDS virus receptor and is expressed in the immune system and the brain. *Cell.* 47:333–348.

Mann, R., R. C. Mulligan, and D. Baltimore. 1983. Construction of a retrovirus packaging mutant and its use to produce helper-free defective retrovirus. *Cell.* 33:153–159.

Marks, P. A., M. Sheffery, and R. A. Rifkind. 1985. Modulation of gene expression during terminal cell differentiation. *Progress in Clinical and Biological Research.* 191:185–203.

Mes-Masson, A. M., J. McLaughlin, G. Q. Daley, M. Paskind, and O. N. Witte. 1986. Overlapping cDNA clones define the complete coding region for the P210c-abl gene product associated with chronic myelogenous leukemia cells containing the Philadelphia chromosome. *Proceedings of the National Academy of Sciences.* 83:8768–8772.

Metcalf, D. 1985. The granulocyte-macrophage colony stimulating factors. *Cell.* 43:5–6.

Metcalf, D. 1986. The molecular biology and functions of the granulocyte-macrophage colony-stimulating factors. *Blood.* 67:257–267.

Miller, A. D., and C. Buttimore. 1986. Redesign of retrovirus packaging cell lines to avoid recombination leading to helper virus production. *Molecular and Cellular Biology.* 6:2895–2902.

Miller, A. D., R. J. Eckner, D. J. Jolly, T. Friedmann, and I. M. Verma. 1984. Expression of a retrovirus encoding human HPRT in mice. *Science.* 225:630–632.

Miller, A. D., M. F. Law, and I. M. Verma. 1985. Generation of helper-free amphotropic retroviruses that transduce a dominant-acting, methotrexate-resistant dihydrofolate reductase gene. *Molecular and Cellular Biology.* 5:431–437.

Schaffner, W. 1980. Direct transfer of cloned genes from bacteria to mammalian cells. *Proceedings of the National Academy of Sciences.* 77:2163–2167.

Sieff, C. A. 1987. Hematopoietic growth factors. *Journal of Clinical Investigation.* 79:1549–1557.

Stein, B. S., S. D. Gowda, J. D. Lifson, R. C. Penhallow, K. G. Bensch, and E. G. Engleman. 1987. pH-independent HIV entry into CD4-positive T cells via virus envelope fusion to the plasma membrane. *Cell.* 49:659–668.

Townes, T. M., J. B. Lingrel, H. Y. Chen, R. L. Brinster, and R. D. Palmiter. 1985. Erythroid-specific expression of human beta-globin genes in transgenic mice. *EMBO Journal.* 4:1715–1723.

Varmus, H. E. 1982. Form and function of retroviral proviruses. *Science.* 216:812–820.

White, J., M. Kielian, and A. Helenius. 1983. Membrane fusion proteins of enveloped animal viruses. *Quarterly Review of Biophysics.* 16:151–195.

Williams, D. A., I. R. Lamischka, D. G. Nathan, and R. C. Mulligan. 1984. Introduction of new genetic material into pluripotent haematopoietic stem cells of the mouse. *Nature.* 310:476–480.

Williams, D. A., S. H. Orkin, and R. C. Mulligan. 1986. Retrovirus-mediated transfer of human adenosine deaminase gene sequences into cells in culture and into murine hematopoietic cells in vivo. *Proceedings of the National Academy of Sciences.* 83:2566–2570.

Wong, P. M. C., S.-W. Chung, and A. W. Nienhuis. 1987. Retroviral transfer and expression of the interleukin-3 gene in hematopoietic cells. *Genes and Development.* 1:358–365.

Organization and
Function of Membrane
and Skeletal Proteins

Chapter 9

The Red Cell Membrane Skeleton: A Model with General Biological Relevance but Pathological Significance for Blood

Peter Agre, Barbara L. Smith, Ali M. Saboori, and Andrew Asimos

Departments of Medicine and Cell Biology, The Johns Hopkins University School of Medicine, Baltimore, Maryland

Cell Physiology of Blood © 1988 by The Rockefeller University Press

Introduction

The erythrocyte membrane has been more thoroughly investigated than any other biological membrane system. While new discoveries continue to be made, tremendous progress has been achieved over the past 15 years, and it is certain that many of the most important principles of erythrocyte membrane physiology have now been identified. The erythrocyte is a highly specialized cell, but the similarities between erythroid and non-erythroid plasma membranes are striking. The principles identified in studies of erythrocytes have provided an approach to the membranes of non-erythroid cells, most of which are exceedingly complex. Another result of these basic investigations has been the identification of the fundamental defects in a group of common congenital anemias, spherocytosis and elliptocytosis. Observations made on these naturally occurring mutant erythrocytes have supplied additional information about the normal function of several erythrocyte membrane components. The pace of discovery in this area has led to major advances in several new areas, and it is very likely that research into erythrocyte membranes will continue to provide information of both general and tissue-specific importance. Since the erythrocyte membrane skeleton has been the subject of several recent reviews, the goal of this chapter will be to summarize features of general relevance, as well as specific points of pathological significance in erythrocytes.

The Erythrocyte Membrane Skeleton

The shape and reversible deformability characteristic of the normal erythrocyte are determined by the membrane skeleton, a highly organized network of proteins that lies immediately beneath the lipid bilayer (reviewed in Cohen, 1983; Bennett, 1985; Marchesi, 1985). Despite being intimately associated with the lipid bilayer, the membrane skeleton retains the shape of the original erythrocyte after extraction of the lipid bilayer with non-ionic detergents (Steck, 1974). Unlike the complex cytoskeleton of other cells, the membrane skeleton is composed only of proteins that reside adjacent to the plasma membrane and does not include long filaments or tubules that span nucleated cells diametrically.

The membrane lipid bilayer is composed of an asymmetric assembly of phospholipids, with the majority of the outer leaflet composed of phosphatidylcholine and sphingomyelin, while the inner leaflet is composed of phosphatidylserine and phosphatidylethanolamine. This asymmetric distribution of phospholipid is restrained by the protein components of the membrane. The proteins that penetrate the lipid bilayer are referred to as "integral" membrane proteins and include the anion transporter (also known as "band 3," based upon electrophoretic migration in sodium dodecyl sulfate polyacrylamide gels; Steck, 1974) and the glycophorins (the site of certain blood groups), as well as certain less abundant proteins, some of which are thought to be other transporters.

Proteins that are associated with the inner leaflet of the membrane bilayer but do not penetrate the bilayer are referred to as "peripheral" membrane proteins. The largest and most abundant peripheral protein is spectrin. If erythrocyte membrane components are described in architectural terms, spectrin should be likened to the beams and girders of a building's skeleton. Spectrin is composed of two distinct subunits, alpha and beta chains, which are synthesized independently. While individually unstable, the subunits exist naturally as an alpha-beta heterodimer. At concentrations known to exist near the bilayer, these heterodimers associate in a head-to-head manner, forming tetramers and larger oligomers. The association of spectrin with the membrane bilayer results from a high-affinity association of the beta spectrin chain with ankyrin, a large protein with another domain that associates with a site on the cytoplasmic domain of the anion transporter.

A second site of multiple protein associations exists at the tail ends of the spectrin oligomers. The associations at this site are a complex of several proteins, including the tails of five to eight spectrin tetramers, a short protofilament composed of ~14 actin monomers, a globular protein referred to as 4.1 (also based upon electrophoretic migration), tropomyosin, adducin (Bennett, 1988), and probably other proteins that remain to be identified. Protein 4.1 has several distinct functions, and recent

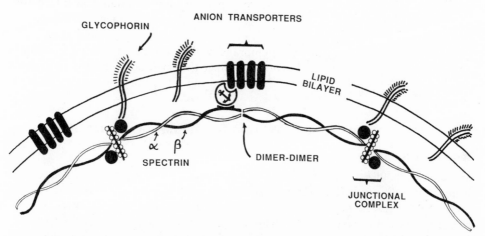

Figure 1. Diagram of the cross-section of the erythrocyte membrane representing selected components. Integral membrane proteins span the lipid bilayer and include glycophorins (representing all species) and anion transporters. A subset of each population is linked to the underlying membrane skeleton. The membrane skeleton is assembled from peripheral membrane proteins. Spectrin consists of nonidentical alpha and beta subunits, which form a heterodimer. Two heterodimers associate at the point labeled "dimer-dimer" to form a tetramer. Junctional complexes consist of protein 4.1 (dark circles), short actin protofilaments (small open circles, each representing an actin monomer), and tropomyosin (dark filament) lying in the groove between two actin protofilaments. Other points of assembly include: ankyrin (anchor), associating with beta spectrin and anion transporter, spectrin tails, associating with junctional complexes, and protein 4.1, associating with glycophorins.

investigations indicate that this protein is heterogeneous, with variations in subpopulations resulting from alternative splicing at the RNA level, protein phosphorylations, and glycosylation with an unusual cytoplasmic form of carbohydrate. It is known that a subpopulation of 4.1 interacts directly with sites on cytoplasmic domains of certain integral membrane proteins in the bilayer. A high-affinity class of sites resides on cytoplasmic domains of the glycophorins (Mueller and Morrison, 1981; Anderson and Lovrien, 1984), and the interaction with these sites is regulated by phosphatidylinositol-4,5-bisphosphate (Anderson and Marchesi, 1985). An additional association of lower affinity has also been identified between 4.1 and a site on the cytoplasmic domain of the anion transporter (Pasternack et al., 1985).

These complex associations result in a beautiful assembly notable for surprising symmetry. Using negative-staining electron microscopy, investigators have demonstrated a repetitive hexagonal arrangement of the membrane skeleton components that resembles the geodesic dome (Byers and Branton, 1985; Shen et al., 1986; Liu et al., 1987). The arms of the skeleton are the spectrin tetramers, and the points of intersection are the junctional complexes composed of an average of six spectrins radiating from a nucleus of actin, 4.1, and tropomyosin. The numerology of the copy number for each of the proteins is in close agreement with this geometry. There are ~100,000 copies of ankyrin per cell, 100,000 copies of spectrin tetramer, and 30,000 of the tail-end spectrin-4.1-actin complexes. This complex assembly provides a branch point where the one-dimensional spectrin tetramers can join with multiple others forming a two-dimensional lattice.

Recent work from several laboratories has demonstrated that other proteins related to spectrin, ankyrin, protein 4.1, and other membrane proteins exist in other cell types (reviewed in Bennett, 1985). Indeed, the spectrin-based membrane skeleton is likely to be the basis of membrane stabilization in most tissues, as well as a mechanism for restriction of a variety of receptors, transporters, and antigens at locations on the cell surface determined by the membrane skeleton within the cell.

Clinical Pathology of the Erythrocyte Membrane Skeleton

Hereditary Spherocytosis

Despite long-standing interest on the part of hematologists, the group of congenital anemias, including spherocytosis and elliptocytosis, remained an enigma until the development of current methods in membrane biology. Numerous lines of investigation had previously been pursued, with identification of numerous abnormalities in these erythrocytes. In all cases, however, the abnormalities were subsequently decided to be secondary to a more fundamental defect (reviewed in Lux, 1983; Palek and Lux, 1983). As their names imply, the erythrocytes in these disorders are spherical or elliptical rather than having the normal biconcave disk shape, and primary defects within the membrane skeleton were suspected. Despite being relatively common, of mild to moderate clinical severity, and inherited in autosomal dominant fashions, larger degrees of heterogeneity were observed, with some patients appearing to have inherited their anemias in nondominant manners and with greater clinical severities.

Several different mouse mutants were discovered, with the homozygotes expressing marked spherocytic anemias with erythrocytes strikingly deficient in spectrin (Greenquist et al., 1978; Lux et al., 1979). Subsequent evaluations demonstrated that while all mutants were deficient in spectrin, the underlying basis of the spectrin deficiencies resulted from (*a*) reduced synthesis of alpha spectrin, (*b*) reduced stability of alpha spectrin, (*c*) reduced synthesis of beta spectrin, and (*d*) reduced stability of ankyrin (Bodine et al., 1984). These important studies established the principle that a multitude of defects in at least three different genes can produce a common phenotypic feature: spherocytic erythrocytes deficient in spectrin.

Human spherocytosis soon proved susceptible to similar forms of biochemical investigation. A group of patients were identified with unusually marked spherocytic anemia and erythrocyte membranes notably deficient in spectrin (Agre et al., 1982). The range of the spectrin deficiency varied between 30 and 74% of the normal level, and parameters reflecting the clinical severity of the anemia varied similarly (Agre et al., 1985). Recent evaluations of patients with the more common dominantly inherited spherocytosis were found to have membranes with slighter deficiencies of spectrin, ranging from 63 to 81% of normal (Agre et al., 1986). Biophysical analyses of the erythrocytes from these patients confirmed the suspicion that spectrin deficiency leads to reduced membrane elasticity, measured in single cells (Waugh and Agre, 1988), and membrane deformability, measured in the entire population of cells in the circulation of an individual (Chasis et al., 1988). Other membrane proteins appear to be quantitatively reduced secondary to the spectrin deficiency, but the magnitude of the secondary reductions is invariably smaller than the deficiency of spectrin. The underlying basis of the spectrin deficiency in most families with the nondominant (recessive) form of spherocytosis appears to be an abnormality in synthesis or assembly of a variant of alpha spectrin. Most affected individuals appear to be homozygotes for a polymorphism in the alpha II domain, a region in the interior of the molecule that can be studied by two-dimensional analysis of limited tryptic digests of spectrin (Winkelmann et al., 1986a). The primary defect in most cases of dominantly inherited spherocytosis remains to be identified, but alpha spectrin has been excluded by studies of large kindreds with probes to alpha spectrin cDNA (Winkelmann et al., 1986b). A subset of the dominantly affected patients appears to have a defect in a region of spectrin where the association with 4.1 occurs, and three such families have been reported (Goodman et al., 1982; Wolfe et al., 1982), although a quantitative reduction in spectrin has subsequently been identified in these patients (Becker and Lux, 1985). The variety of clinical severities that correlate with the overall erythrocyte spectrin content demonstrates the fundamental importance of spectrin in supporting the membrane, for even very minor deficiencies with spectrin levels of 80% of normal result in significant degrees of membrane instability and notable clinical hemolysis. It warrants emphasis that the genetic defects resulting in spectrin deficiency are likely to include primary defects in alpha and beta spectrin as well as defects in other membrane skeleton proteins that are required for assembly of spectrin into the membrane skeleton, and the search for the primary genetic defects continues.

While the common feature of spherocytic erythrocytes appears to be spectrin deficiency with a secondary loss of membrane surface area. Precisely how the surface membrane bilayer is lost remains to be defined. In the most severely deficient erythrocytes, there appears to be spontaneous shedding of sections of membrane that are presumably not well supported from beneath by a membrane skeleton that is grossly deficient in spectrin. This event is no doubt accelerated by the spleen; however, it may occur to some degree in splenectomized patients expressing the most severe forms of the disease.

Hereditary Elliptocytosis and Pyropoikilocytosis

This group of anemias is common among blacks and not infrequent among Mediterraneans, although the most severely affected patients, the pyropoikilocytics, are seen least frequently. Although usually mild, the level of clinical severity is variable, and can be quite marked among the pyropoikilocytic patients, who have notable erythrocyte fragmentation (resembling that seen in severe thermal burns) with significant degrees of anemia (reviewed in Lux, 1983; Palek and Lux, 1983). Many of the elliptocytic patients and nearly all of the pyropoikilocytic patients have inherited abnormalities in the alpha I domain of spectrin (near the NH_2-terminal dimer-dimer association site). The common feature of these patients is a reduction of the dimer-dimer association, which reduces the stability and elasticity of the membrane skeleton. Interestingly, there appear to be at least three distinct forms of alpha I defects, based upon two-dimensional electrophoretic analysis of limited tryptic spectrin digests. When the abnormal spectrins were sequenced, it was discovered that multiple different mutations exist (Marchesi et al., 1987). While several different amino acid substitutions were identified in the different families, replacement of different residues with proline was seen in three different families, which suggests that the pathological consequence of the mutations led to alteration in the alpha helix by substitution of proline at crucial locations in the portion of spectrin where the dimer-dimer association occurs.

Evaluations of families with elliptocytosis failed to identify abnormalities in many families, and the search for other defects continued. Protein 4.1 plays an important role in the assembly of the membrane skeleton, and a family with parental consanguinity and deficiencies of 4.1 was one of the first membrane skeleton defects reported (Tchernia et al., 1981). The initial 4.1-deficient family expressed the mutation as an autosomal dominant trait. The heterozygotes were missing approximately half of their 4.1, but were only mildly affected clinically, while the homozygotes with total deficiency of 4.1 suffered with a severe anemia. Additional cases of partial 4.1 deficiency have been identified in Europe (Alloisio et al., 1985), and a single American patient was found with total 4.1 deficiency (Mueller and Morrison, 1981).

The use of immunoblots with anti-4.1 and a 4.1 radioimmunoassay has recently led to the identification of three large American kindreds that express clinically mild elliptocytosis of dominant inheritance in linkage with (*a*) a high-molecular-weight 4.1 variant, (*b*) a low-molecular-weight 4.1 variant, and (*c*) a simple, partial 4.1 deficiency in the different families (McGuire and Agre, 1987). Elliptocytosis in these kindreds is also linked with Rh blood type, an association no doubt resulting from the close chromosomal location of the genes for 4.1 and Rh on the short arm of chromosome 1 (Conboy et al., 1986). These patients with the 4.1-associated form of elliptocytosis are generally less severely affected than those with defects in spectrin dimer-dimer association. Recent evaluation of the abnormal 4.1 proteins from these families has indicated that the sites of insertion (in the high-molecular-weight form) and deletion (in the low-molecular-weight form) reside at the site of RNA splicing near the portion of 4.1, where association with spectrin is known to occur (Letsinger et al., 1986).

Protein 4.1 is known to be synthesized relatively late in the development of the erythrocyte, and it is probable that 4.1 may function to enhance the overall stability of the membrane skeleton, which has already been assembled. It is interesting to note that clinical defects relating to deficient 4.1 are of much less clinical severity than corresponding deficiencies of spectrin. While the significance of 4.1 cannot be questioned, the relative importance has been likened to the importance of the lug nuts on an automobile wheel. The machine looks nearly normal and can be driven at moderate speeds, with each wheel secured by only half of the nuts. However, if the mechanic forgets all of the lug nuts, severe consequences will quickly be noted.

Despite success in identification of molecular defects in many of the kindreds with spherocytosis and elliptocytosis, it must be considered that primary defects are likely to occur in other identified and yet-to-be-identified membrane skeleton proteins, and it is hoped that such patients will provide mutant forms of these proteins that may provide probes for identification of functions of the native proteins.

Newly Identified Integral Membrane Proteins Linked to the Membrane Skeleton

The Rh Antigen and a 28,000-M_r Protein

With the exception of the Rh blood group system, the major blood groups have been worked out in biochemical detail (reviewed in Rosse, 1988). Much confusion has arisen concerning the biochemical nature of the Rh(D) antigen, and the complexity of the Rh-associated antigens (D, C/c, and E/e) remains poorly understood. The Rh(D) antigen is in several ways distinct from the other blood groups, and recent work indicates that it is a constituent of the erythrocyte membrane skeleton. Unlike the other blood groups, which exist in millions of copies per cell, there are only ~30,000 copies of Rh(D) per erythrocyte, and the antigens appear fixed at intervals on the membrane. There is evidence that the Rh(D) antigen is of some structural importance to the erythrocyte, since rare individuals lacking D, C/c, and E/e (the Rh null syndrome) have erythrocytes that are fragile, misshapen, and leaky to cations. A 30,000–32,000-M_r integral membrane protein was recently identified on erythrocytes bearing the Rh(D) antigen based upon surface ^{125}I-labeling and precipitation with Rh(D)-immune globulin (Gahmberg, 1982; Moore et al., 1982). The curious distribution on the erythrocyte membrane was explained by the observation that the 30,000-M_r Rh-associated integral membrane protein (Rh-IMP) is associated with the membrane skeleton (Gahmberg and Karhi, 1984; Ridgwell et al., 1984) but can be extracted with special conditions. This suggests that the Rh-IMP is indirectly associated with the membrane skeleton by linkage to another integral membrane protein, which is itself directly linked to the membrane skeleton (Bloy et al., 1987).

Our laboratory has recently pursued the 30,000-M_r Rh-IMP as a biochemical approach to the Rh antigen system (Agre et al., 1987). Methods were developed to quantitatively immunoprecipitate Rh-IMP from erythrocytes. Employing extensive proteolytic digestion of inside-out membrane vesicles, no evidence for a cytoplasmic domain was identified. Despite extreme immunogenicity and an apparent extracellular domain containing a tyrosine, extensive proteolytic digestion of intact erythrocytes failed to degrade the Rh-IMP on the external surface of the membrane, which suggests that there is minimal extracellular mass. The Rh-IMP was purified using traditional chromatographic methods, and it is now feasible to conduct biochemical studies on the Rh-IMP, which can be produced in hundreds of micrograms in pure form. Amino acid composition of the pure Rh-IMP demonstrates a large proportion of hydrophobic amino acids (12% leucine), and even the pure Rh-IMP appears to oligomerize even in the presence of sodium dodecyl sulfate. Back-calculation from the purified Rh-IMP demonstrated that ~60,000 Rh-IMPs exist per cell, which is twice the number of Rh(D) antigens known to occur. Together, these findings are consistent with the Rh-IMP existing as an oligomer while in the membrane. Recent work has demonstrated that the Rh-IMPs purified from specific Rh phenotypes bear characteristic protein polymorphisms (Saboori et al., 1987). While it is likely that additional cofactors are needed to reconstitute the Rh antigen, it seems very likely that the genetic basis of the antigen is involved in the amino acid sequence of the Rh-IMP.

Early purifications of the Rh-IMP were hampered by the copurification of a 28,000-M_r integral membrane protein, which is also linked to the membrane skeleton. This 28,000-M_r protein is somehow associated with the Rh antigen, since it can be partially immunoprecipitated with Rh(D)-immune globulin. Unlike the Rh-IMP, there is no evidence that the 28,000-M_r protein has an extracellular domain. However, the 28,000-M_r protein clearly has cytoplasmic mass and can be degraded to a 20,000-M_r fragment that remains imbedded in the bilayer. Polyclonal antibodies that react with the 28,000-M_r protein on immunoblots were affinity-purified and used to screen for analogues to this protein in other tissues. No such proteins were found in any tissue except in kidney, where a large degree of immunoreactive material of similar molecular weight was noted. Immunocytochemical staining of human kidney preparations demonstrated restriction of staining to the proximal convoluted tubules, where intense staining indicates an abundance of the protein. Purification of the kidney analogue has not yet been accomplished, but despite appearing identical to the erythrocyte form, immunoblots with antibodies to several other erythrocyte proteins demonstrated that there was no significant contamination of the kidney membrane preparations with erythrocytes. No direct role had been identified for this 28,000-M_r integral protein. It is possible that it plays some enzymatic function as a transporter. Alternatively, it may be simply a structural protein with importance as a mechanism to restrict locations of other integral membrane proteins to sites determined from below by the membrane skeleton.

Conclusions

The erythrocyte membrane skeleton has been thoroughly explored, but new components continue to be identified. This membrane system has proven to be a valuable model for the approach to membranes of more complex cells. Naturally occurring mutants have been identified, and the study of these erythrocytes has provided answers to questions in clinical hematology and further illustrates important functions of the membrane components.

Acknowledgments

This work was made possible by the enthusiastic participation of our patients and their families. Support was provided by grants from the National Institutes of Health, the March of Dimes (Basil O'Connor Award), and the American Heart Association (Established Investigator Award).

References

Agre, P., A. Asimos, J. F. Casella, and C. McMillan. 1986. Inheritance pattern and clinical response to splenectomy as a reflection of erythrocyte spectrin deficiency in hereditary spherocytosis. *New England Journal of Medicine.* 315:1579–1583.

Agre, P., J. F. Casella, W. H. Zinkham, C. McMillan, and V. Bennett. 1985. Partial deficiency of erythrocyte spectrin in hereditary spherocytosis. *Nature.* 314:380–383.

Agre, P., E. P. Orringer, and V. Bennett. 1982. Deficient red cell spectrin in severe, recessively inherited spherocytosis. *New England Journal of Medicine.* 301:1155–1161.

Agre, P., A. M. Saboori, A. Asimos, and B. L. Smith. 1987. Purification and partial characterization of the M_r 30,000 integral membrane protein associated with the erythrocyte Rh(D) antigen. *Journal of Biological Chemistry.* 262:17497–17503.

Alloisio, N., E. Morle, E. Dorleac, O. Gentilhomme, D. Bachir, D. Guetarni, P. Colonna, M. Bost, Z. Zouaoui, L. Roda, D. Roussel, and J. Delaunay. 1985. The heterozygous form of 4.1(−) hereditary elliptocytosis. *Blood.* 65:46–51.

Anderson, R. A., and R. E. Lovrien. 1984 Glycophorin is linked by band 4.1 to the human erythrocyte membrane skeleton. *Nature.* 307:655–657.

Anderson, R. A., and V. T. Marchesi. 1985. Regulation of the association of membrane skeletal 4.1 with glycophorin by a polyphosphoinositide. *Nature.* 318:295–297.

Becker, P.S., and S. E. Lux. 1985. Hereditary spherocytosis and related disorders. *Clinics in Haematology.* 14:15–43.

Bennett, V. 1985. The membrane skeleton of human erythrocytes and its implications for more complex cells. *Annual Review of Biochemistry.* 54:273–304.

Bennett, V., J. Davis, K. Gardner, and J. P. Steiner. 1988. The spectrin-based membrane skeleton: extensions of the current paradigm. *In* Cell Physiology of Blood. R. B. Gunn and J. C. Parker, editors. The Rockefeller University Press. New York, NY. 101–109.

Bloy, C., D. Blanchard, P. Lambin, and J. P. Cartron. 1987. Human monoclonal antibody against Rh(D) antigen. *Blood.* 69:1491–1498.

Bodine, D. M., C. S. Birkenmeier, and J. E. Barker. 1984. Spectrin deficient hemolytic anemias in the mouse. *Cell.* 37:721–729.

Byers, T. J., and D. Branton. 1985. Visualization of the protein associations in the erythrocyte membrane skeleton. *Proceedings of the National Academy of Sciences.* 82:6153–6157.

Chasis, J. A., P. Agre, and N. Mohandas. 1988. Decreased membrane mechanical stability and in vivo loss of surface area reflect spectrin deficiencies in hereditary spherocytosis. *Journal of Clinical Investigation.* 82:617–623.

Cohen, C. 1983. Molecular organization of the red cell membrane skeleton. *Seminars in Hematology.* 20:141–158.

Conboy, J., N. Mohandas, G. Tchernia, and Y. W. Kan. 1986. Molecular basis of hereditary elliptocytosis due to protein 4.1 deficiency. *New England Journal of Medicine.* 315:680–685.

Gahmberg, C. G. 1982. Molecular characterization of the human red cell $Rh_0(D)$ antigen. *EMBO Journal.* 2:223–230.

Gahmberg, C. G., and K. K. Karhi. 1984. Association of the $Rh_0(D)$ polypeptides with the membrane skeleton in $Rh_0(D)$-positive human red cells. *Journal of Immunology.* 133:334-337.

Goodman, S. R., K. A. Shiffer, L. A. Casoria, and M. E. Eyster. 1982. Identification of the molecular defect in the erythrocyte membrane skeleton of some kindreds with hereditary spherocytosis. *Blood.* 60:772–784.

Greenquist, A. C., S. B. Shohet, and S. E. Bernstein. 1978. Marked reductions of spectrin in hereditary spherocytosis in the common house mouse. *Blood.* 51:1149–1155.

Letsinger, J. T., P. Agre, and S. L. Marchesi. 1986. High molecular weight 4.1 in the cytoskeletons of hereditary elliptocytosis. *Blood.* 68:38. (Abstr.)

Liu, S. C., L. H. Derick, and J. Palek. 1987. Visualization of the hexogonal lattice in the erythrocyte membrane skeleton. *Journal of Cell Biology.* 104:527–536.

Lux, S. E. 1983. Disorders of the red cell membrane skeleton. In The Metabolic Basis of Inherited Disease. J. B. Stanbury, J. B. Wyngaarden, D. S. Fredrickson, J. L. Goldstein, and M. S. Brown, editors. McGraw-Hill, New York, NY. 1573–1605.

Lux, S. E., B. Pease, M. B. Tomaselli, K. M. John, and S. E. Bernstein. 1979. Hereditary anemias associated with deficient or dysfunctional spectrin. *In* Normal and Abnormal Red Cell Membranes. S. E. Lux, V. T. Marchesi, and C. F. Fox, editors. Alan R. Liss, Inc., New York, NY.

Marchesi, V. T. 1985. Stabilizing infrastructure of cell membranes. *Annual Review of Cell Biology*. 1:531–561.

Marchesi, S. L., J. T. Letsinger, D. T. Speicher, V. T. Marchesi, P. Agre, B. Hyun, and G. Gulati. 1987. Mutant forms of spectrin alpha subunits in hereditary elliptocytosis. *Journal of Clinical Investigation*. 80:191–198.

McGuire, M., and P. Agre. 1987. Three distinct variants of protein 4.1 in Caucasian hereditary elliptocytosis. *Clinical Research*. 35:428. (Abstr.)

Moore, S., C. F. Woodrow, and D. B. L. McClelland. 1982. Isolation of membrane components associated with human red cell antigens Rh(D), (c), and (E) and Fy^a. *Nature*. 295:529–531.

Mueller, T. J., and M. Morrison. 1981. Glycoconnectin (PAS 2), a membrane attachment site for the human erythrocyte cytoskeleton. *In* Erythrocyte Membranes 2. W. Kruckeberg, J. Eaton, and G. Brewer, editors. Alan R. Liss, Inc., New York, NY. p. 95.

Palek, J., and S. E. Lux. 1983. Red cell membrane skeletal defects in hereditary and acquired hemolytic anemias. *Seminars in Hematology*. 20:189–224.

Pasternack, G. R., R. A. Anderson, and V. T. Marchesi. 1985. Interactions between protein 4.1 and band 3. *Journal of Biological Chemistry*. 260:3676–3683.

Ridgwell, K., M. J. A. Tanner, and D. J. Anstee. 1984. The rhesus (D) polypeptide is linked to the human erythrocyte cytoskeleton. *FEBS Letters*. 174:7–10.

Rosse, W. F. 1988. Basic and Clinical Immunohematology. Blackwell, Boston, MA. In press.

Saboori, A. M., B. L. Smith, and P. Agre. 1987. Probable genetic basis of the Rh antigens. *Blood*. 70:310. (Abstr.)

Shen, B. W., R. Josephs, and T. L. Steck. 1986. Ultrastructure of the intact skeleton of the human erythrocyte membrane. *Journal of Cell Biology*. 102:997–1006.

Steck, T. 1974. The organization of proteins in the human red blood cell membrane. *Journal of Cell Biology*. 62:1–19.

Tchernia, G., N. Mohandas, and S. B. Shohet. 1981. Deficiency of skeletal membrane protein band 4.1 in homozygous hereditary elliptocytosis. *Journal of Clinical Investigation*. 68:454–460.

Waugh, R. E., and P. Agre. 1988. Reductions of erythrocyte viscoelastic coefficients reflect spectrin deficiencies in hereditary spherocytosis. *Journal of Clinical Investigation*. 81:133–141.

Winkelmann, J. C., S. L. Marchesi, P. Watkins, A. J. Linnenbach, P. Agre, and B. G. Forget. 1986*a*. Recessive hereditary spherocytosis is associated with an abnormal alpha spectrin subunit. *Clinical Research*. 34:474. (Abstr.)

Winkelmann, J. C., S. L. Marchesi, F. Gillespie, P. Agre, and B. Forget. 1986*b*. Dominant hereditary spherocytosis is not closely linked to the gene for alpha spectrin three kindreds. *Blood*. 68:42. (Abstr.)

Wolfe, L. C., K. M. John, J. C. Falcone, A. M. Byrne, and S. E. Lux. 1982. A genetic defect in the binding of protein 4.1 to spectrin in a kindred with hereditary spherocytosis. *New England Journal of Medicine*. 307:1367–1374.

Chapter 10

The Spectrin-based Membrane Skeleton: Extensions of the Current Paradigm

Vann Bennett, Jonathan Davis, Kevin Gardner, and Joseph P. Steiner

The Howard Hughes Medical Institute and Department of Biochemistry, Duke University Medical Center, Durham, North Carolina

Cell Physiology of Blood © 1988 by The Rockefeller University Press

Introduction

The spectrin-based membrane skeleton is a meshwork of proteins attached to the cytoplasmic surface of the erythrocyte membrane that stabilizes these cells and permits normal survival in the circulation. Spectrin, the most abundant component of this structure, is an elongated, flexible molecule that cross-links short actin filaments to form a continuous lattice beneath the plasma membrane. A high-affinity linkage between the lipid bilayer and the membrane skeleton is provided by ankyrin. Ankyrin contains binding sites for spectrin and for the cytoplasmic domain of the anion-exchange protein, band 3. An additional linkage between the membrane skeleton and the bilayer may be provided by protein 4.1, which stabilizes association between spectrin and actin and interacts with band 3 and at least one of the sialoglycoproteins. This assembly of membrane proteins was first discovered and characterized in the relatively simple human erythrocyte, where it has provided the basis for a molecular understanding of a number of hereditary hemolytic anemias. However, at least some aspects of the erythrocyte system are likely to represent general features of other cell membranes, since proteins closely related to spectrin are components of the plasma membrane of most eukaryotic cells (reviewed by Marchesi, 1985; Bennett, 1985; Goodman and Zagon, 1986). No function has yet been established for the spectrin skeleton, although possible roles include organization of integral membrane proteins into specialized domains on cell surfaces (Drenckhahn et al., 1985; Nelson and Veshnock, 1986, 1987a, b; Srinivasan et al., 1988), regulation of access of secretory vesicles to the inner surface of the plasma membrane (Perrin et al., 1987), and directed movement of membrane proteins (Levine and Willard, 1983; Nelson et al., 1983; Bourguignon et al., 1985).

A first step in elucidating cellular functions of the spectrin skeleton is to define the constituent proteins and their organization. This chapter will review recent progress in determining how spectrin is linked to the membrane in nonerythroid cells. A new membrane skeletal protein also will be discussed that has been purified from erythrocytes and brain. This protein, referred to as adducin, binds to calmodulin, is a major substrate for protein kinase C, and associates preferentially with spectrin-actin complexes rather than with spectrin or actin alone.

Ankyrin as a Possible Membrane Linkage for Spectrin

One mechanism for association of spectrin with membranes is likely, by analogy with the erythrocyte system, to involve ankyrin. An isoform of ankyrin has been purified from brain and demonstrated to bind to spectrin with high affinity at the midregion of spectrin tetramers, which is the same site recognized by erythrocyte ankyrin (Davis and Bennett, 1984a, b). Brain ankyrin also contains a binding site for the cytoplasmic domain of the erythrocyte anion-exchange protein. Brain ankyrin, like erythrocyte ankyrin, has separate domains that associate with spectrin and with the anion exchanger. Thus, brain ankyrin is likely to be able to bind both spectrin and an anion-exchange protein at the same time. Immunoreactive forms of ankyrin of similar M_r to erythrocyte and brain have been detected by immunoblot analysis in all tissues examined so far, including liver, kidney, intestinal epithelial cells, lens, prostate, and skeletal muscle (Davis and Bennett, 1984a; Nelson and Lazarides, 1984a, b; Drenckhahn and Bennett, 1987). The amount of ankyrin in brain is ~100 pmol/mg of membrane protein, which is easily enough to accommodate the amount of spectrin tetramers. Ankyrin in other tissues is present approximately in proportion to the amount of spectrin.

An important question in evaluating ankyrin as a possible linkage protein for spectrin is whether these proteins are present in the same regions of cells. Ankyrin has been localized by immunofluorescence to the basolateral domain of a variety of epithelial cells (kidney, prostate, pancreas, small intestine) and was missing from the apical domains of these cells (Drenckhahn et al., 1985; Drenckhahn and Bennett, 1987). Neuronal cells also exhibit a polarized distribution of ankyrin where staining was restricted to axons and cell bodies, but staining was absent from dendrites and photoreceptor inner and outer segments. Spectrin, in contrast, is present along all membrane surfaces in the epithelial tissues as well as in brain. These results indicate that ankyrin is a good candidate to provide one class of membrane linkage in basolateral domains of epithelial cells and in axons and cell bodies of neurons. Ankyrin-independent membrane linkages are presumably available for spectrin in the regions of these cells lacking ankyrin and may also be present in the domains where ankyrin is present (see below).

Ankyrin-binding Proteins in Membranes of Brain and Kidney

A simple extension of the erythrocyte model has not been possible in the identification of ankyrin-binding proteins in nonerythroid cells, except for a small population of cells in kidney. Proteins closely related to the erythrocyte anion exchanger have not been found in other cells or tissues using cDNA probes or antibodies, with the exception of intercalated cells of mammalian kidney collecting ducts (Cox et al., 1985; Drenckhahn et al., 1985). The protein in intercalated cells is a likely candidate to mediate chloride-bicarbonate exchange and to participate in the function of intercalated cells in acid-base balance. The kidney protein cross-reacts with the erythrocyte anion exchanger and is localized with ankyrin in the basolateral domains of these cells (Drenckhahn et al., 1985; Drenckhahn and Bennett, 1987). Thus, it is reasonable to propose that this specialized cell population recapitulates the erythrocyte with an anion exchanger associated with ankyrin, which in turn is linked to spectrin. The linkage of the anion exchanger to ankyrin may be important in targeting or maintaining this protein in a basolateral location in intercalated cells.

Measurements of binding of radiolabeled brain ankyrin with ankyrin- and spectrin-depleted brain membranes indicate that ankyrin has high-affinity protein-binding sites and that a major fraction of these sites may share conformational homology with the ankyrin-binding site of the erythrocyte anion exchanger (Davis and Bennett, 1986). Brain ankyrin associated with a K_d of 20–60 nM with sites present at ~25 pmol/mg of membrane protein. Binding was optimal at physiological pH and ionic strength and involved a protein, since association of ankyrin was prevented by digestion of membranes with protease. The cytoplasmic domain of the erythrocyte anion exchanger displaced ~60–70% of ankyrin binding to brain membranes. Thus, the brain membrane proteins and erythrocyte anion exchanger associate with the same region of ankyrin. One possibility is that these proteins share conformational homology with at least the ankyrin-binding region of the anion exchanger.

A major goal of current research in this laboratory is to purify and characterize these ankyrin-binding proteins from brain membranes. One approach has been to digest brain membranes under controlled conditions and isolate active proteolytic fragments of binding proteins by analogy with earlier studies with the erythrocyte membrane (Bennett, 1978; Bennett and Stenbuck, 1979). Proteolytic fragments are released from brain membranes that compete for binding to ankyrin to membrane sites and bind to ankyrin affinity columns (Davis and Bennett, 1986). A technical difficulty has been the diverse nature of these fragments, which indicates that multiple proteins may have ankyrin-binding domains. A solution to this problem of diverse binding proteins may be to prepare monoclonal antibodies and use them as probes to isolate individual proteins.

The tentative conclusions of this systematic effort to identify the ankyrin-binding site(s) in brain are that the brain does not contain proteins closely related to the erythrocyte anion exchanger, although ankyrin-binding proteins may share some homology with the ankyrin-binding site of the erythrocyte protein, and that multiple proteins are likely to interact with ankyrin.

Two membrane proteins have recently been identified that associate with ankyrin in vitro and support the idea that the ability to interact with ankyrin will turn out to be a property of a variety of membrane proteins. The sodium-potassium ATPase of kidney (Nelson and Veshnock, 1987b) and the voltage-dependent sodium channel (Srinivasan et al., 1988) interact with ankyrin in solution with apparent affinities of 1–20 nM. The association of the sodium channel with ankyrin is displaced by the cytoplasmic domain of the erythrocyte anion-exchange protein, which indicates that the sodium channel and anion exchanger may have some common sequence or at least conformational homology. The ATPase apparently interacts with ankyrin at a site distinct from the sodium channel, since the anion-exchange fragment does not inhibit the ATPase-ankyrin interaction. A feature the sodium channel and the sodium-potassium ATPase have in common is that both proteins require placement in precise domains of the plasma membrane for proper function: the sodium channel at the nodes of Ranvier of myelinated nerve, and the ATPase at the basolateral region of transporting epithelial cells. It is of interest that in brain an isoform of the ATPase is also localized at the nodes of Ranvier. These findings suggest that ankyrin provides a mechanism either to initially place these proteins at the appropriate location in cells or to stabilize their position after they have been targeted.

Ankyrin is present along the entire axon, which suggests that the sodium channel may bind preferentially to an isoform or modified form of ankyrin that is localized at the nodes of Ranvier. Brain tissue contains at least two isoforms of ankyrin with different localizations and different developmental expression (Nelson and Lazarides, 1984a). Ankyrin also is subject to post-translational modification by phosphorylation (Lu et al., 1985), fatty acid acylation (Staufenbiel and Lazarides, 1986), and proteolytic processing (Hall and Bennett, 1987). The details of when ankyrin associates with these proteins, the possible regulation of these interactions, and subsequent processing of ankyrin are important questions for future investigation.

Ankyrin-independent Association of Spectrin with Integral Membrane Proteins

Ankyrin is absent from the apical regions of epithelial cells and from dendrites and afferent processes of neuronal cells (Drenckhahn and Bennett, 1987). Moreover, as described above, ankyrin in nonerythroid cells may play a role that uses the spectrin skeleton rather than functioning as an integral structural unit as in erythrocytes. These considerations suggest the possibility that spectrin may be able to interact with the membrane by ankyrin-independent interactions. Experimental evidence has recently been obtained for a direct association of brain spectrin with membranes depleted of ankyrin and spectrin (Steiner, J. P., and V. Bennett, 1988). Association of radiolabeled spectrin with membranes is of high affinity and involves a protein, since the binding activity is destroyed by proteolysis of the membranes. The protein is operationally an integral protein, since binding is not extracted from membranes with sodium hydroxide but can be solubilized with detergent. The capacity of brain membranes for spectrin is ~50 pmol/mg of protein and thus is in the appropriate range to account for association of brain spectrin with membranes. Erythrocyte membranes bind very little erythrocyte spectrin after removal of ankyrin, which suggests that in this system the integral protein site for spectrin has been lost, and/or erythrocyte spectrin has lost the ability to bind to the integral protein. Membrane binding of spectrin exhibits isoform specificity in brain, where brain spectrin rather than erythrocyte spectrin is the preferred ligand.

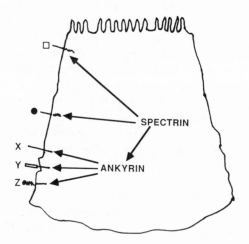

Figure 1. Highly schematic model for the possible association of spectrin to membrane proteins through ankyrin-dependent and -independent linkages in a polarized epithelial cell.

A Generalized View of Spectrin-Membrane Interactions

The observations described above suggest that the erythrocyte membrane represents a highly specialized variation of a general model for the organization of the spectrin skeleton (see Fig. 1). Spectrin in other cells can interact directly with integral proteins that remain to be purified. It is not yet known whether the integral spectrin-binding proteins are confined to the regions of cells lacking ankyrin (apical regions of epithelial cells, the dendritic portion of neurons) or whether both types of linkages can coexist within the same membrane domain. Ankyrin does not appear to be linked to a single class of sites as in erythrocytes, but rather seems likely to have the potential to bind to a variety of proteins. The more general function of ankyrin thus may not be as a primary membrane-attachment site for spectrin but as a mechanism for coupling a diverse group of integral proteins to the spectrin skeleton. The principal structural components involved in the linkage of spectrin to the membrane may thus be provided by direct attachment to integral proteins. According to this view, the skeleton, once it is assembled, is used by ankyrin for the localization of ion channels and other proteins whose function requires placement in a specific cellular domain. Evaluation of these ideas will require studies of localization and biosynthesis of spectrin-binding proteins, further characterization of ankyrin-binding proteins, and elucidation of steps in assembly of ankyrin and ankyrin-binding proteins in the membrane.

Adducin

The major components of the erythrocyte membrane skeleton (i.e., those present in 100,000 copies or more per cell) have been defined, but additional proteins associated with this structure probably remain to be discovered. One impetus for a search for additional proteins is to be able to explain mechanisms for regulation and assembly of the skeleton. For example, calcium ion has for many years been known to modulate erythrocyte deformability (Weed et al., 1969). None of the known proteins bind calcium or calmodulin with an affinity appropriate to explain these effects. Another puzzling question is related to how spectrin and actin assemble to form the remarkably regular lattice that has recently been visualized in negatively stained images of erythrocyte membrane skeletons (Byers and Branton, 1985; Shen et al., 1986; Liu et al., 1987). These images reveal an extended two-dimensional network of spectrin molecules interconnected by short actin filaments with five to seven spectrin molecules clustered about each actin filament. Spectrin associates with actin and cross-links actin filaments in vitro, but the isolated proteins do not form a network. Presumably, additional proteins are required to bring together multiple spectrin molecules at localized regions along actin filaments. A new membrane skeletal protein has been purified from erythrocytes (Gardner and Bennett, 1986) that binds calmodulin, is a substrate for protein kinase C (Ling et al., 1986), and may play a role in the assembly of spectrin and actin (Gardner and Bennett, 1987a, b). This protein has been named adducin (from *adducere,* to bring together) based on its ability to promote spectrin binding to actin. A protein closely related to erythrocyte adducin has been purified from brain that has similar activities and is a substrate for protein kinase C (Bennett, V., K. Gardner, and J. P. Steiner, manuscript submitted for publication).

Erythrocyte adducin exists in solution as a 200,000-dalton heterodimer with two subunits of M_r 103,000 (alpha) and 97,000 (beta). Brain adducin is slightly larger, with a molecular weight in solution of 207,000 and subunits of M_r 104,000 and a doublet of 107,000/109,000. In both proteins, the subunits can be distinguished by peptide maps, but exhibit some homology. Thus, it is likely that the subunits are products of different genes, but the genes are related. Erythrocyte and brain adducin contain a protease-resistant domain of M_r 48,000 and 48,500, respectively. Erythrocyte and brain adducin have nearly identical physical properties, with an asymmetric shape approximated by an oblate ellipsoid. Erythrocyte and brain adducin bind calmodulin, although the affinity of erythrocyte adducin from calmodulin ($K_d = 0.2$ μM) is about fivefold higher than the affinity of brain adducin. Sequence information will be required to establish the precise relationship between the brain and erythrocyte proteins, but available information supports the view that these proteins are products of distinct but closely related genes.

Adducin from erythrocytes and brain has the unusual property of binding with high affinity to spectrin-actin complexes but relatively weakly to spectrin or actin alone (Gardner and Bennett, 1987a, b; Bennett et al., 1988). There are two general possibilities for a basis for recognition of spectrin-actin complexes. (*a*) Adducin has two low-affinity binding sites, one for spectrin and one for actin. A spectrin-actin complex with both potential binding proteins present at a high local concentration would be greatly favored over the individual proteins. (*b*) Adducin recognizes a site on either spectrin or actin that is induced only when spectrin and actin form a complex. In either case, adducin provides a potential mechanism to selectively target proteins to spectrin-actin complexes.

A working hypothesis for cellular functions for brain and erythrocyte adducin is that these proteins have a role in mediating site-directed assembly of additional proteins at the spectrin-actin junction. One protein that may be recruited by adducin is spectrin itself. Erythrocyte and brain adducin both promote binding of spectrin to actin. Experiments with erythrocyte adducin suggest that the adducin-promoted binding of spectrin to actin is due to the creation of new sites for spectrin that are distinguished by their inhibition by calmodulin (Gardner and Bennett, 1987*a*, *b*). Adducin-mediated recruitment of additional spectrin molecules to spectrin-actin complexes could explain how multiple spectrin molecules become clustered about the same region on actin filaments, as occurs in the membrane skeleton. The spectrin-actin junction in erythrocytes and possibly other cells is likely to contain additional proteins such as tropomyosin and protein 4.9. It will be important to determine whether adducin associates with these and possibly other unidentified proteins and mediates their interaction with spectrin-actin complexes.

Erythrocyte and brain adducin are new additions to the list of substrates for protein kinase C, although functional consequences of phosphorylation remain to be elucidated. Adducin subunits closely resemble in electrophoretic mobility polypeptides that bind protein kinase C after transfer to nitrocellulose paper (Wolf and Sahyoun, 1986). Adducin may thus bind to protein kinase C and localize this enzyme to spectrin-actin complexes on the plasma membrane.

These initial studies suggest that adducin has the potential to play an important role in the assembly and regulation of the spectrin skeleton, although many questions remain to be answered. Information about activity of adducin is limited to simple in vitro assays that have only begun to explore the implications of ternary and higher-order protein interactions. The ultrastructural localization of adducin in erythrocyte membrane skeletons can be inferred to be at the spectrin-actin junction, although it will be important to document this prediction. The localization of adducin in brain is currently unknown, and may provide some clues as to its cellular function. Finally, it will be important to devise experiments to evaluate the function and regulation of adducin in living cells.

Acknowledgments

The technical assistance of James Erwin is gratefully acknowledged. This work was supported in part by grants from the National Institutes of Health (RO1 GM-33996, RO1 AM-198808) and by a hematology training grant (HL-07535).

References

Bennett, V. 1978. Purification of an active proteolytic fragment of the membrane attachment site for human erythrocyte spectrin. *Journal of Biological Chemistry.* 253:2292–2299.

Bennett, V. 1985. The membrane skeleton of human erythrocytes and its implications for more complex cells. *Annual Review of Biochemistry.* 54:273–304.

Bennett, V., and P. Stenbuck. 1979. Identification and partial purification of ankyrin, the high affinity membrane attachment site for human erythrocyte spectrin. *Journal of Biological Chemistry.* 254:2533–2541.

Bennett, V., K. Gardner, and J. P. Steiner. 1988. Brain adducin. A protein kinase C substrate that may mediate site-directed assembly at the spectrin-actin junction. *Journal of Biological Chemistry.* 263:5860–5869.

Bourguignon, L. Y., S. J. Suchard, M. L. Nagpal, and J. R. Glenney. 1985. A t-lymphoma transmembrane glycoprotein (gp 180) is linked to the cytoskeletal protein, fodrin. *Journal of Cell Biology.* 101:477–487.

Byers, T. J., and D. Branton. 1985. Visualization of the protein associations in the erythrocyte membrane skeleton. *Proceedings of the National Academy of Sciences.* 82:6153–6157.

Cox, J. V., R. T. Moon, and E. Lazarides. 1985. Anion transporter: highly cell-type-specific expression of distinct polypeptides and transcripts in erythroid and nonerythroid cells. *Journal of Cell Biology.* 100:1548–1557.

Davis, J., and V. Bennett. 1984*a*. Brain ankyrin-purification of a 72,000 M_r spectrin-binding domain. *Journal of Biological Chemistry.* 259:1874–1881.

Davis, J., and V. Bennett. 1984*b*. Brain ankyrin: a membrane associated protein with binding sites for spectrin, tubulin and the cytoplasmic domain of the erythrocyte anion channel. *Journal of Biological Chemistry.* 259:13550–13559.

Davis, J., and V. Bennett. 1986. Association of brain ankyrin with brain membranes and isolation of active proteolytic fragments of membrane-associated ankyrin-binding proteins. *Journal of Biological Chemistry.* 261:16198–16206.

Drenckhahn, D., and V. Bennett. 1987. Polarized distribution of M_r 210,000 and 190,000 analogs of erythrocyte ankyrin along the plasma membrane of transporting epithelia, neurons and photoreceptors. *European Journal of Cell Biology.* 43:479–486.

Drenckhahn, D., K. Schluter, D. Allen, and V. Bennett. 1985. Colocalization of band 3 with ankyrin and spectrin at the basal membrane of intercalated cells in the rat kidney. *Science.* 230:1287–1289.

Gardner, K., and V. Bennett. 1986. A new erythrocyte membrane-associated protein with calmodulin binding activity: identification and purification. *Journal of Biological Chemistry.* 261:1339–1348.

Gardner, K., and V. Bennett. 1987*a*. Modulation of spectrin-actin assembly by erythrocyte adducin. *Nature.* 328:359–362.

Gardner, K., and V. Bennett. 1987*b*. Erythrocyte adducin: a new calmodulin regulated membrane-skeletal protein that modulates spectrin-actin assembly. *Journal of Cellular Biochemistry.* In press.

Goodman, S. R., and I. S. Zagon. 1986. The neural cell spectrin skeleton: a review. *American Journal of Physiology.* 250:C347–C360.

Hall, T. G., and V. Bennett. 1987. Regulatory domains of erythrocyte ankyrin. *Journal of Biological Chemistry.* 262:10537–10545.

Levine, J., and M. Willard. 1983. Redistribution of fodrin (a component of the cortical cytoplasm) accompanying capping of cell surface molecules. *Proceedings of the National Academy of Sciences.* 80:191–195.

Ling, E., K. Gardner, and V. Bennett. 1986. Protein kinase C phosphorylates a recently identified membrane skeleton-associated calmodulin-binding protein in human erythrocytes. *Journal of Biological Chemistry.* 261:13875–13878.

Liu, S.-C., L. H. Derick, and J. Palek, 1987. Visualization of the hexagonal lattice in the erythrocyte membrane skeleton. *Journal of Cell Biology.* 104:527–536.

Lu, P. W., C. J. Soong, and M. Tao. 1985. Phosphorylation of ankyrin decreases its affinity for spectrin tetramer. *Journal of Biological Chemistry.* 260:14958–14964.

Marchesi, V. T. 1985. Stabilizing infrastructure of cell membranes. *Annual Review of Cell Biology.* 1:531–561.

Nelson, W. J., C. Colaco, and E. Lazarides. 1983. Involvement of spectrin in cell-surface receptor capping in lymphocytes. *Proceedings of the National Academy of Sciences.* 80:1626–1630.

Nelson, W. J., and E. Lazarides. 1984*a*. The patterns of expression of two ankyrin isoforms demonstrate distinct steps in the assembly of the membrane skeleton in neuronal morphogenesis. *Cell.* 39:309–320.

Nelson, W. J., and E. Lazarides. 1984*b*. Goblin (ankyrin) in striated muscle: identification of the potential membrane receptor for erythroid spectrin in muscle cells. *Proceedings of the National Academy of Sciences.* 81:3292–3296.

Nelson, W. J., and P. J. Veshnock. 1986. Dynamics of membrane-skeleton (fodrin) organization during development of polarity in Madrin-Darby canine kidney epithelial cells. *Journal of Cell Biology.* 103:1751–1766.

Nelson, W. J., and P. J. Veshnock. 1987*a*. Modulation of fodrin (membrane skeleton) stability by cell-cell contact in Madrin-Darby canine kidney epithelial cells. *Journal of Cell Biology.* 104:1527–1537.

Nelson, W. J., and P. J. Veshnock. 1987*b*. Ankyrin binding to the Na^+ and K^+) ATPase and implications for the organization of membrane domains in polarized cells. *Nature.* 328:533–535.

Perrin, D., O. K. Langely, and D. Aunis. 1987. Anti-alpha fodrin inhibits secretion from permeabilized chromaffin cells. *Nature.* 326:497–501.

Shen, B. W., R. Josephs, and T. L. Steck. 1986.Ultrastructure of the intact skeleton of the human erythrocyte membrane. *Journal of Cell Biology.* 102:997–1006.

Srinivasan, Y., L. Elmer, J. Davis, V. Bennett, and K. Angelides. 1988. Ankyrin and spectrin associate with voltage-dependent sodium channels in brain. *Nature* 333:177–180.

Staufenbiel, M., and E. Lazarides. 1986. Ankyrin is fatty acid acylated in erythrocytes. *Proceedings of the National Academy of Sciences.* 83:318–322.

Steiner, J. P., and V. Bennett. 1988. Ankyrin-independent membrane protein binding sites for brain and erythrocyte spectrin. *Journal of Biological Chemistry.* In press.

Weed, R., P. LaCelle, and E. Merril. 1969. Metabolic dependence of red cell deformability. *Journal of Clinical Investigation.* 48:795–809.

Wolf, M., and N. Sahyoun. 1986. Protein kinase C and phosphatidylserine bind to M_r 110,000/115,000 polypeptides enriched in cytoskeletal and postsynaptic density preparations. *Journal of Biological Chemistry.* 261:13327–13332.

Chapter 11

Structure and Function of the Platelet Membrane Skeleton

Joan E. B. Fox and Janet K. Boyles

*Gladstone Foundation Laboratories for Cardiovascular Disease,
Cardiovascular Research Institute, Department of Pathology,
University of California, San Francisco, California*

Cell Physiology of Blood © 1988 by The Rockefeller University Press

Introduction

Actin filaments exist throughout the cytoplasm of most eukaryotic cells. These filaments are in a dynamic state of flux, rapidly polymerizing and becoming cross-linked into a variety of three-dimensional structures in regions of cells where they are needed for a particular function (Korn, 1978). The functions of actin filaments are numerous. For example, by interacting with myosin, they generate the tension involved in contractile activities such as motility, cell division, and secretion. When the filaments are cross-linked into networks, they control the consistency of the cytoplasm. The rapid polymerization of new actin filaments can induce the formation of cellular protrusions.

One cell that does not contain actin filaments throughout its cytoplasm is the red blood cell. In this cell, actin is polymerized into short filaments ($\sim$50 nm long, compared with $>$1,000 nm for the cytoplasmic actin filaments of other cells) that are a component of a membrane skeleton (Cohen, 1983; Shen et al., 1986). In contrast to cytoplasmic actin filaments, the filaments of the red blood cell membrane skeleton are stable structures. They are cross-linked by spectrin into a network that adheres to the plasma membrane. The membrane skeleton of the red blood cell is a flexible but strong structure and has been shown to regulate several properties of the plasma membrane. For example, it stabilizes the plasma membrane, preventing the membrane from fragmenting; it gives the plasma membrane its contours; and it regulates the lateral mobility of the membrane glycoproteins to which it is attached. Since all cells presumably have mechanisms for regulating these properties of the plasma membrane, the question arises whether other cells contain a membrane skeleton composed of interlinked short actin filaments in addition to their previously described cytoplasmic actin filaments. Reports that various cultured cells contain a peripheral layer of Triton X-100–insoluble material that is composed in part of plasma membrane components (Ben-Ze'ev et al., 1979; Apgar et al., 1985) suggest that this is a possibility. Recently, the blood platelet has been used as a model system to test the idea that a membrane skeleton is present in other cell types. The evidence reviewed in this chapter strongly suggests that platelets do indeed contain a membrane skeleton, that its major components differ from those of the red blood cell membrane skeleton, and that it has some of the same functions as its counterpart in the red blood cell.

Identification of a Membrane Skeleton in Platelets

Unstimulated platelets have a characteristic discoid shape. When they are activated by agonists such as thrombin, they rapidly lose their discoid shape, extend filopodia, secrete their granule contents, and aggregate. It has often been suggested that the microtubule coil within platelets is responsible for maintaining the discoid shape of the unstimulated cell (White, 1971). However, the observation that platelets can lose their discoid shape even when the microtubule coil remains intact (White, 1982) suggests that additional mechanisms are operative. One potential mechanism might involve a membrane skeleton (Fox, 1985a). The first indication that platelets contain such a structure came from the observation (Fox, 1985a) that a small pool of the actin filaments in Triton X-100–lysed platelets are associated with specific membrane glycoproteins (GP). The membrane-bound filaments had properties different from the majority of the actin filaments in the Triton X-100 lysates. Whereas most of the filaments could be sedimented at low g forces, the membrane-bound ones required high g forces to be sedimented (Fox, 1985a). The membrane-bound filaments also differed from the rest of the actin filaments in that they were resistant to depolymerization by gelsolin under conditions in which the rest of the actin filaments were depolymerized (Fox, 1985a).

Morphological experiments support the idea that platelets contain a membrane skeleton. The platelet cytoskeleton has been visualized in cells that were solubilized with Triton X-100 and simultaneously fixed. As shown in Fig. 1 A, filaments are present throughout the cytoplasm and as a continuous layer external to the microtubule coil. However, an additional, more amorphous layer is present at the periphery of the cytoskeletons. When Ca^{2+} is present in the lysis buffer, essentially all of the visible actin filaments are lost, presumably as a result of depolymerization by gelsolin (Markey et al., 1981), but the amorphous layer remains (Fig. 1 B). This result suggests that the amorphous layer at the periphery of the cytoskeleton contains the membrane-bound actin filaments shown previously to be resistant to the action of gelsolin (Fox, 1985a). Immunocytochemical studies (Fox et al., 1988), showing that the amorphous material contains the GP Ib-IX complex, support this conclusion. Since the amorphous material is in close proximity to the lipid bilayer (Escolar et al., 1986) and follows the outline of the platelet (Fox et al., 1988), it appears to exist as a lining to the plasma membrane. A variety of lysis conditions have been defined under which the amorphous material retains the outline of the platelet for many hours, even though the lipid bilayer, microtubule coil, and the majority of other platelet proteins have been solubilized (Fox et al., 1988). Thus, the peripheral layer of amorphous material is a self-supporting structure. Because it lines the plasma membrane and is attached to membrane glycoproteins, this peripheral layer could represent a membrane skeleton that functions like that of the red blood cell.

Characterization of the Components of the Membrane Skeleton

Platelet membrane skeletons (i.e., the structures shown in Fib. 1 B) have been isolated by centrifugation and their components have been analyzed on sodium dodecyl sulfate–polyacrylamide gels. Actin and actin-binding protein are two of the major components (Fox, 1985a). Isolation of the membrane skeletons from platelets whose surface glycoproteins have been radiolabeled has identified the membrane glycoproteins to which the submembranous skeleton is attached (Fox, 1985a). As shown in Fig. 2, most of the radiolabeled glycoproteins are solubilized by Triton X-100. However, the three components of the GP Ib-IX complex (GP Ib$_\alpha$, GP Ib$_\beta$, and GP IX) remain associated with the Trition X-100–insoluble membrane skeleton. Other minor glycoproteins, including GP Ia, GP IIa, and an unidentified membrane glycoprotein of M_r 250,000, have also been identified as components of the membrane skeleton (Fox, 1985a).

Additional evidence that actin, actin-binding protein, and the GP Ib-IX complex are the major components of the membrane skeleton comes from the observation that these proteins co-isolate on sucrose density gradients of platelet lysates (Fox, 1985a) and bind to affinity columns containing a monoclonal antibody against GP Ib (Okita et al., 1985). Further, when plasma membrane vesicles are isolated from disrupted

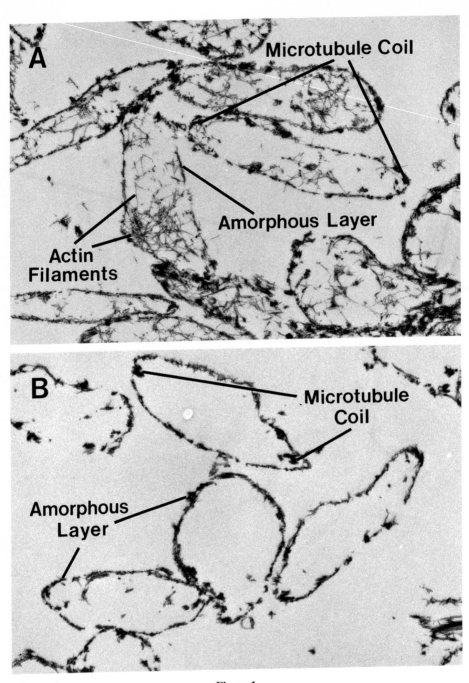

Figure 1.

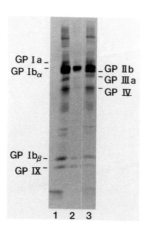

GP Ia —
GP Ibα —
— GP IIb
— GP IIIa
— GP IV

GP Ibβ —
GP IX —

1 2 3

Figure 2. Association of membrane glycoproteins with the membrane skeleton. Platelets were surface labeled by the sodium metaperiodate/sodium [³H]borohydride method and lysed with Triton X-100 in the presence of Ca²⁺ (Fox, 1985a). Intact platelets (lane 1), the Triton X-100–insoluble membrane skeleton (lane 2), and the Triton X-100–soluble fraction (lane 3) were solubilized in sodium dodecyl sulfate and analyzed on 5–20% polyacrylamide gels in the presence of reducing agent. ³H-labeled membrane glycoproteins were detected by fluorography.

platelets, small amounts of actin co-isolate with them (Fox, 1985a); the major proteins that remain associated with the actin when the membrane vesicles are solubilized in Triton X-100 are actin-binding protein and the GP Ib-IX complex (Fox, 1985a).

The membrane skeletons have a relatively amorphous appearance when visualized by thin-section microscopy (Fig. 1 B). Despite this morphological appearance, other criteria indicate that the actin in the membrane skeleton is in the form of filaments. For example, the membrane skeleton sediments at high *g* forces and is depolymerized by DNase I (Fox, 1985a). Further, the membrane skeletons bind phalloidin (Fig. 3), a molecule that binds only to the filamentous form of actin. The membrane skeleton of the red blood cell has a similar amorphous appearance when viewed in thin section, but has also been judged to contain actin filaments (Shen et al., 1986; Liu et al., 1987). The existence of actin filaments in the red blood cell membrane skeleton was convincingly demonstrated by the use of negative-stain techniques that allowed a direct visualization of the filaments and showed that these filaments are only ~50 nm long. This length is comparable to that of a single fibrinogen molecule or a single GP Ib-IX complex, a length that would preclude unambiguous visualization of the filaments by thin-section electron microscopy. It seems likely that the actin filaments in the platelet membrane skeleton are of similar dimensions.

A major component of the red blood cell membrane skeleton is spectrin, a

---

Figure 1. (*Opposite*) Electron micrographs showing the components of the platelet cytoskeleton. Platelets were lysed with Triton X-100 in the presence of alcian blue, either in the presence (*A*) or absence (*B*) of a Ca²⁺ chelator. Lysates were incubated for 4 min at 4°C and then fixed for electron microscopy by the method of Boyles et al. (1985). This method of preparation of cytoskeletons allows the solubilization of the lipid bilayer and most of the platelet protein, but maintains the organization of the components of the cytoskeleton. The cytoskeleton comprises a single microtubule coil at the periphery of the cell, networks of long actin filaments that are concentrated external to the microtubule coil but are also found throughout the cytoplasm, and a more amorphous material that exists as a continuous layer at the periphery of the cytoskeleton. When Ca²⁺ is present, endogenous gelsolin is active (Markey et al., 1981). Gelsolin depolymerizes most of the cytoplasmic actin filaments, but the peripheral layer of more amorphous material is resistant to the action of this protein. Magnifications: *A*, ×15,520; *B*, ×15,520. (Reproduced with permission from Fox and Boyles, 1988.)

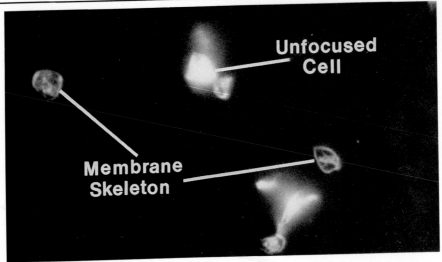

Figure 3. Detection of filamentous actin with rhodamine-labeled phalloidin. Platelets were lysed with Triton X-100 in the presence of Ca^{2+}. The resulting Triton-insoluble residues were incubated with rhodamine-labeled phalloidin and examined by fluorescence microscopy. Phalloidin stains the peripheral membrane skeleton, although actin filaments cannot be visualized in these structures by thin-section electron microscopy (see Fig. 1). This suggests that, as in the red blood cell, the platelet membrane skeleton contains actin filaments too short to visualize in thin section. ×3,169.

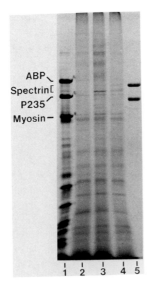

Figure 4. Sodium dodecyl sulfate–polyacrylamide gel showing the presence of spectrin in platelets. Platelets were solubilized directly into a sodium dodecyl sulfate–containing buffer (lane *1*) or were lysed with Triton X-100 and immunoprecipitated with control IgG (lane *2*), anti–red blood cell spectrin (lane *3*), or anti–brain spectrin (lane *4*). Lane *5* contains purified red blood cell spectrin. Spectrin antibodies specifically immunoprecipitated two polypeptides that were present in platelets in small amounts. The larger of the two polypeptides comigrated with the M_r 240,000 subunit of red blood cell spectrin. The smaller polypeptide had M_r 235,000, as judged by its co-migration with talin (P235) (platelet polypeptide of M_r 235,000). ABP, actin-binding protein. (Reproduced with permission from Fox, 1987.)

molecule that cross-links the actin filaments and also serves as a linkage protein between the filaments and glycoproteins on the plasma membrane (Cohen, 1983). As shown in Fig. 4, two polypeptides are specifically immunoprecipitated from platelet lysates by antibodies against red blood cell spectrin or brain spectrin (Fox et al., 1987*b*). These polypeptides have molecular weights similar to those of brain spectrin (M_r 240,000 and 235,000). They are associated primarily with the actin filaments of the membrane skeleton and presumably represent platelet spectrin. However, only very small amounts of spectrin are present in platelets (Fig. 4, lane *1*); the major component of the platelet membrane skeleton, actin-binding protein, is present in much greater amounts. Although spectrin and actin-binding protein have different amino acid sequences (Zweig and Singer, 1979), they do have similar morphologies and functions (Hartwig and Stossel, 1981; Cohen, 1983; Bennett, 1985). They are both elongated molecules that associate head to head. The free tails of both molecules bind to actin filaments, thus cross-linking them into networks. It appears likely, therefore, that actin-binding protein in platelets might function similarly to spectrin in red blood cells in that it may cross-link the actin filaments of the membrane skeleton.

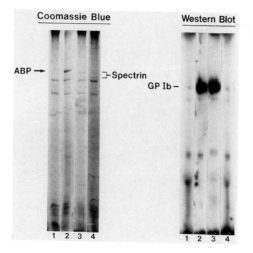

Figure 5. Sodium dodecyl sulfate–polyacrylamide gels showing the co-precipitation of actin-binding protein (ABP) and glycoprotein Ib. Platelets were lysed with Triton X-100 (lanes *2–4*). Actin filaments in the lysates were depolymerized with DNase I and removed by high-speed centrifugation. The resulting supernatant was immunoprecipitated with control rabbit IgG (lane *1*) or with antibodies against actin-binding protein (lane *2*), antibodies against GP Ib (lane *3*), or antibodies against spectrin (lane *4*).

Actin-binding protein also functions similarly to spectrin in that it links actin filaments to plasma membrane glycoproteins. This function was first suggested by the observation that the GP Ib-IX complex was released from Triton X-100–insoluble actin filaments under conditions in which the Ca^{2+}-dependent protease was active (Solum and Olsen, 1984). Both spectrin and actin-binding protein are substrates for the Ca^{2+}-dependent protease (Fox et al., 1985, 1987*b*), and were therefore considered as potential linkage proteins between actin and the GP Ib-IX complex (Fox, 1985*b*, 1987*b*). As shown in Fig. 5, spectrin antibodies did not immunoprecipitate GP Ib; thus, spectrin does not appear to be the linkage protein. In contrast, actin-binding protein antibodies precipitated both actin-binding protein and GP Ib. In the reciprocal experiment, GP Ib antibodies precipitated GP Ib and small amounts of actin-binding protein. Since these experiments were performed on platelet lysates in which actin filaments had been depolymerized and any remaining filaments had been removed by high-speed centrifugation (Fox, 1985*b*), the coprecipitation of actin-binding protein and GP Ib shows that actin-binding protein links the actin filaments of the membrane skeleton to the GP Ib-IX complex on the plasma membrane. Similar experiments have shown that actin-binding protein also serves to attach the submembranous actin to GP Ia and to a membrane glycoprotein of M_r 250,000 (Fox, 1985*b*), the additional membrane glycoproteins that have been identified as sites of attachment of the membrane skeleton to the plasma membrane (Fox, 1985*a*).

The studies reviewed above suggest the model shown in Fig. 6. In this model, the platelet membrane skeleton is distinct from the previously described cytoplasmic actin filaments. It is composed of a submembranous network of short actin filaments that are cross-linked by actin-binding protein. Actin-binding protein also links the membrane skeleton to the cytoplasmic domain of defined glycoproteins in the plasma membrane. The proposed model for the platelet membrane skeleton differs in some respects from that of the red blood cell membrane skeleton. For example, the membrane glycopro-

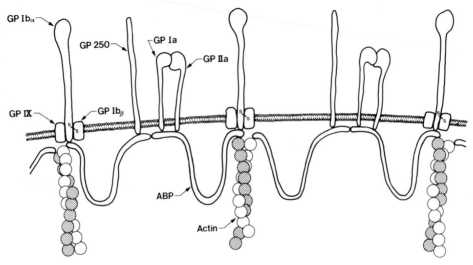

Figure 6. Schematic representation of the platelet membrane skeleton. ABP, actin-binding protein; GP 250, a plasma membrane glycoprotein of M_r 250,000; s-s, a disulfide bond. (Reproduced with permission from Fox, 1987.)

teins to which the skeleton is attached in platelets are not the same as those in the red blood cell. Another difference is that in the platelet, actin-binding protein substitutes for spectrin as the protein that cross-links the actin filaments of the membrane skeleton. Spectrin is present in the platelet membrane skeleton (Fox et al., 1987*b*), but only in small amounts, and its function is not known. In the red blood cell, protein 4.1 and ankyrin serve as linkage proteins between spectrin and the red blood cell membrane glycoproteins. Both protein 4.1 and ankyrin have been described in platelets (Bennett, 1979; Spiegel et al., 1984; Davies and Cohen, 1985); however, it is not known whether these proteins are components of the membrane skeleton. It remains to be determined whether these or other minor components of the membrane skeleton have a role in linking actin-binding protein to plasma membrane glycoproteins. Despite these differences, the platelet membrane skeleton appears to have a number of functions similar to those of the red blood cell membrane skeleton (see below).

Potential Functions of the Platelet Membrane Skeleton

In unstimulated platelets, the membrane skeleton lines the plasma membrane and is attached to it. This skeleton could regulate many of the properties of the plasma membrane in the unstimulated cell. During the early, reversible stages of platelet activation, the membrane skeleton remains intact (Fox, 1985a). At the later stages of platelet activation, after platelets have aggregated, the Ca^{2+}-dependent protease within platelets is activated (Fox et al., 1983). Because one of the major substrates for the Ca^{2+}-dependent protease is actin-binding protein (Fox et al., 1985), one consequence of activation of the Ca^{2+}-dependent protease within platelets is disruption of the membrane skeleton (Fox, 1985b). Disruption of this skeleton could lead to some of the events that occur in irreversibly activated platelets.

Table I lists possible functions of the platelet membrane skeleton. By analogy to the red blood cell, it is likely that the platelet membrane skeleton regulates the shape of the plasma membrane. Evidence for this function of the platelet membrane skeleton comes from the observation that the skeletons are self-supporting structures (Fox et al., 1988). It appears likely that they play a role in determining the discoid shape of the unstimulated platelet, and activation-induced modifications of components of the membrane skeleton could direct the reversible shape-change response.

TABLE I
Potential Functions of a Membrane Skeleton

Regulation of cell shape
Stabilization of the plasma membrane
Regulation of distribution of glycoproteins in the plasma membrane
Regulation of receptor function
Transmembrane signaling

Like the red blood cell membrane skeleton, the platelet membrane skeleton lines the plasma membrane and is attached to plasma membrane glycoproteins; therefore, the membrane skeleton probably stabilizes the plasma membrane, preventing it from fragmenting. In support of this idea, disruption of the membrane skeleton (for example, by exposure of platelets to agonists or other agents, such as dibucaine, that activate the Ca^{2+}-dependent protease within platelets) results in the formation of small membrane vesicles that bud and break away from the platelet (Boyles et al., 1987). These experiments show that the membrane skeleton stabilizes the plasma membrane in unstimulated platelets. Evidence that microvesiculation occurs physiologically when the platelets are activated in the circulation comes from the observation that platelet-free GP Ib is present in plasma (Coller et al., 1984) and that platelet microvesicles are present in the circulation of patients in whom platelet activation has occurred (George et al., 1986).

Again, by analogy to the red blood cell, it appears likely that the platelet membrane skeleton could regulate the lateral distribution of the plasma membrane glycoproteins to which it is attached. In support of this idea, the GP Ib-IX complex is present over the entire surface of the plasma membrane in unstimulated platelets, but clusters when the membrane skeleton is disrupted (Boyles et al., 1987). These experiments indicate that linkage of GP Ib-IX to the membrane skeleton maintains the glycoprotein in a uniform distribution.

Linkage of the membrane skeleton to the GP Ib-IX complex also appears to regulate the ability of GP Ib to bind von Willebrand factor. There is considerable evidence that the binding of von Willebrand factor is regulated by agents that affect the organization of the platelet cytoskeleton (Coller, 1981; Solum and Olsen, 1984). When GP Ib is released from actin filaments, as it is when actin-binding protein is hydrolyzed at the later stages of platelet activation, its ability to bind von Willebrand factor is lost (Fox and Boyles, 1985; George and Torres, 1986).

Finally, the platelet membrane skeleton may have a role in transmembrane signaling. GP Ib is the receptor for von Willebrand factor and also appears to bind the agonist thrombin (Harmon and Jamieson, 1986), while GP Ia appears to be the receptor for the agonist collagen (Nieuwenhuis et al., 1985; Santoro, 1986). Linkage of

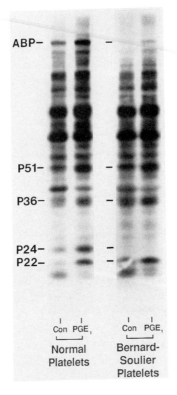

Figure 7. Autoradiograms of sodium dodecyl sulfate–polyacrylamide gels showing the absence of P24 from Bernard-Soulier platelets. Control platelets or those from a patient with Bernard-Soulier syndrome were labeled with [^{32}P]phosphate and incubated in the absence and/or presence (PGE$_1$) of 5 μM prostaglandin E$_1$ for 5 min. Platelets were solubilized in a sodium dodecyl sulfate–containing buffer in the presence of reducing agent and electrophoresed on sodium dodecyl sulfate–polyacrylamide gels. Phosphopolypeptides were detected by autoradiography. ABP, actin-binding protein; P51, P36, P24, P22, phosphopolypeptides of M_r 51,000, 36,000, 24,000, and 22,000, respectively.

these receptors to the membrane skeleton might provide a mechanism for transmembrane signaling. It is of interest in this regard that GP Ib$_\beta$ has been identified as one of the major proteins phosphorylated within platelets by agents that inhibit platelet function by elevating the intracellular concentrations of cyclic AMP (Fox et al., 1987a). One such agent is prostaglandin E$_1$. As shown in Fig. 7, exposure of platelets to this agent results in increased phosphorylation of actin-binding protein and polypeptides of M_r 51,000 (P51), 36,000 (P36), 24,000 (P24), and 22,000 (P22). Although the identities of P51, P36, and P22 are not known, P24 has recently been identified as GP Ib$_\beta$. This identification was based on the co-migration of GP Ib$_\beta$ and P24 on reduced and nonreduced polyacrylamide gels, the precipitation of P24 with anti-GP Ib, the association of P24 with actin filaments of the membrane skeleton, and the absence of

P24 in platelets from patients with Bernard-Soulier syndrome (these platelets lack the GP Ib-IX complex) (Fig. 7). Further studies on the functional consequences of phosphorylation of GP Ib$_\beta$ should provide insight into the possible role of GP Ib-IX and the platelet membrane skeleton to which it is attached in regulating agonist-induced events in platelets.

Conclusions

It has long been known that the red blood cell contains a membrane skeleton that stabilizes the plasma membrane, determines its shape, and regulates the lateral distribution of the membrane glycoproteins to which it is attached. The way in which these functions are regulated in other cells, however, has not been well understood. We have now shown that platelets also contain a membrane skeleton. In contrast to the membrane skeleton of the red blood cell, in which spectrin cross-links the actin filaments and links them to plasma membrane glycoproteins, actin-binding protein serves this function in the platelet. Actin-binding protein is present in many cell types and often has a submembranous location (Boxer and Richardson, 1976; Weihing, 1983). Furthermore, GP Ib–like proteins have been identified in cells other than platelets (Asch et al., 1986; Kieffer et al., 1986). Thus, it is conceivable that other cells may contain a membrane skeleton similar to that found in the platelet. The platelet membrane skeleton appears to regulate the same cellular functions as the red blood cell membrane skeleton. It is becoming increasingly apparent that the membrane skeleton plays an important role in maintaining the unstimulated platelet in a functional form and that disruption of the membrane skeleton in activated platelets is responsible for directing many of the cellular responses to activation.

Acknowledgments

We thank Vicki Vasquez for preparing the manuscript, James X. Warger for graphics, and Al Averbach and Sally Gullatt Seehafer for editorial assistance. This work was supported by research grant 30657-04 from the National Institutes of Health (J. E. B. Fox) and by Established Investigator awards from the American Heart Association (J. E. B. Fox and J. K. Boyles).

References

Apgar, J. R., S. H. Herrmann, J. M. Robinson, and M. F. Mescher. 1985. Triton X-100 extraction of P815 tumor cells: evidence for a plasma membrane skeleton structure. *Journal of Cell Biology.* 100:1369–1378.

Asch, A. S., M. Fujimoto, B. Adelman, and R. L. Nachman. 1986. An endothelial cell GPIb like molecule mediates von Willebrand factor binding. *Circulation.* 74:II–232. (Abstr.)

Bennett, V. 1979. Immunoreactive forms of human erythrocyte ankyrin are present in diverse cells and tissues. *Nature.* 281:597–599.

Bennett, V. 1985. The membrane skeleton of human erythrocytes and its implications for more complex cells. *Annual Review of Biochemistry.* 54:273–304.

Ben-Ze'ev, P., A. Duerr, F. Solomon, and S. Penman. 1979. The outer boundary of the cytoskeleton: a lamina derived from plasma membrane proteins. *Cell.* 17:856–859.

Boxer, L. A., and S. Richardson. 1976. Identification of actin-binding protein in membrane of polymorphonuclear leukocytes. *Nature.* 263:249–251.

Boyles, J. K., J. E. B. Fox, and M. C. Berndt. 1987. The distribution of GP Ib and the stability of the plasma membrane are dependent upon an intact membrane skeleton. *Thrombosis and Haemostasis.* 58:225a. (Abstr.)

Boyles, J., J. E. B. Fox, D. R. Phillips, and P. E. Stenberg. 1985. Organization of the cytoskeleton in resting, discoid platelets: preservation of actin filaments by a modified fixation that prevents osmium damage. *Journal of Cell Biology.* 101:1463–1472.

Cohen, C. M. 1983. The molecular organization of the red cell membrane skeleton. *Seminars in Hematology.* 20:141–158.

Coller, B. S. 1981. Inhibition of von Willebrand factor-dependent platelet function by increased platelet cyclic AMP and its prevention by cytoskeleton-disrupting agents. *Blood.* 57:846–855.

Coller, B. S., E. Kalomiris, M. Steinberg, and L. E. Scudder. 1984. Evidence that glycocalicin circulates in normal plasma. *Journal of Clinical Investigation.* 73:794–799.

Davies, G. E., and C. M. Cohen. 1985. Platelets contain proteins immunologically related to red cell spectrin and protein 4.1. *Blood.* 65:52–59.

Escolar, G., M. Krumwiede, and J. G. White. 1986. Organization of the actin cytoskeleton of resting and activated platelets in suspension. *American Journal of Pathology.* 123:86–94.

Fox, J. E. B. 1985a. Linkage of a membrane skeleton to integral membrane glycoproteins in human platelets. Identification of one of the glycoproteins as glycoprotein Ib. *Journal of Clinical Investigation.* 76:1673–1683.

Fox, J. E. B. 1985b. Identification of actin-binding protein as the protein linking the membrane skeleton to glycoproteins on platelet plasma membranes. *Journal of Biological Chemistry.* 260:11970–11977.

Fox, J. E. B. 1987. The platelet cytoskeleton. *In* Thrombosis and Hemostasis. M. Verstraete, J. Vermylen, R. Lijnen, and J. Arnout, editors. Leuven University Press, Leuven, Belgium. 175–225.

Fox, J. E. B., and J. K. Boyles. 1988. The membrane skeleton: a distinct structure that regulates the function of cells. *BioEssays.* 8:14–18.

Fox, J. E. B., J. K. Boyles, M. C. Berndt, P. K. Steffen, and L. Anderson. 1988. Identification of a membrane skeleton in platelets. *Journal of Cell Biology.* 106:1525–1538.

Fox, J. E. B., D. E. Goll, C. C. Reynolds, and D. R. Phillips. 1985. Identification of two proteins (actin-binding protein and P235) that are hydrolyzed by endogenous Ca^{2+}-dependent protease during platelet aggregation. *Journal of Biological Chemistry.* 260:1060–1066.

Fox, J. E. B., C. Reynolds, and M. M. Johnson. 1987a. Identification of glycoprotein Ib_β as one of the major proteins phosphorylated during exposure of intact platelets to agents that activate cyclic AMP-dependent protein kinase. *Journal of Biological Chemistry.* 262:12627–12631.

Fox, J. E. B., C. C. Reynolds, J. S. Morrow, and D. R. Phillips. 1987b. Spectrin is associated with membrane-bound actin filaments in platelets and is hydrolyzed by the Ca^{2+}-dependent protease during platelet activation. *Blood.* 69:537–545.

Fox, J. E. B., C. Reynolds, and D. R. Phillips. 1983. Calcium-dependent proteolysis occurs during platelet aggregation. *Journal of Biological Chemistry.* 258:9973–9981.

George, J. N., E. B. Pickett, S. Saucerman, R. P. McEver, T. J. Kunicki, N. Kieffer, and P. J. Newman. 1986. Platelet surface glycoproteins: studies on resting and activated platelets and platelet membrane microparticles in normal subjects, and observations in patients during adult respiratory distress syndrome and cardiac surgery. *Journal of Clinical Investigation.* 78:340–348.

George, J. N., and M. M. Torres. 1986. Thrombin alters platelet surface glycoprotein Ib, decreasing its von Willebrand factor receptor function. *Circulation.* 74:II–237. (Abstr.)

Harmon, J. T., and G. A. Jamieson. 1986. The glycocalicin portion of platelet glycoprotein Ib expresses both high and moderate affinity receptor sites for thrombin a soluble radioreceptor assay for the interaction of thrombin with platelets. *Journal of Biological Chemistry.* 261:13224–13229.

Hartwig, J. H., and T. P. Stossel. 1981. Structure of macrophage actin-binding protein molecules in solution and interacting with actin filaments. *Journal of Molecular Biology.* 145:563–581.

Kieffer, N., N. Debili, A. Wicki, M. Titeux, A. Henri, Z. Mishal, J. Breton-Gorius, W. Vainchenker, and K. J. Clemetson. 1986. Expression of platelet glycoprotein Ibα in HEL cells. *Journal of Biological Chemistry.* 261:15854–15862.

Korn, E. D. 1978. Biochemistry of actomyosin-dependent cell motility (a review). *Proceedings of the National Academy of Sciences.* 75:588–599.

Liu, S.-C., L. H. Derick, and J. Palek. 1987. Visualization of the hexagonal lattice in the erythrocyte membrane skeleton. *Journal of Cell Biology.* 104:527–536.

Markey, F., T. Persson, and U. Lindberg. 1981. Characterization of platelet extracts before and after stimulation with respect to the possible role of profilactin as microfilament precursor. *Cell.* 23:145–153.

Nieuwenhuis, H. K., J. W. N. Akkerman, W. P. M. Houdijk, and J. J. Sixma. 1985. Human blood platelets showing no response to collagen fail to express surface glycoprotein Ia. *Nature.* 318:470–472.

Okita, J. R., D. Pidard, P. J. Newman, R. R. Montgomery, and T. J. Kunicki. 1985. On the association of glycoprotein Ib and actin-binding protein in human platelets. *Journal of Cell Biology.* 100:317–321.

Santoro, S. A. 1986. Identification of a 160,000 dalton platelet membrane protein that mediates the initial divalent cation-dependent adhesion of platelets to collagen. *Cell.* 46:913–920.

Shen, B. W., R. Josephs, and T. L. Steck. 1986. Ultrastructure of the intact skeleton of the human erythrocyte membrane. *Journal of Cell Biology.* 102:997–1006.

Solum, N. O., and T. M. Olsen. 1984. Glycoprotein Ib in the Triton-insoluble (cytoskeletal) fraction of blood platelets. *Biochimica et Biophysica Acta.* 799:209–220.

Spiegel, J. E., D. S. Beardsley, F. S. Southwick, and S. E. Lux. 1984. An analogue of the erythroid membrane skeletal protein 4.1 in nonerythroid cells. *Journal of Cell Biology.* 99:886–893.

Weihing, R. R. 1983. Purification of a HeLa cell high molecular weight actin binding protein and its identification in HeLa cell plasma membrane ghosts and intact HeLa cells. *Biochemistry.* 22:1839–1847.

White, J. G. 1971. Platelet microtubules and microfilaments: effects of cytochalasin B on structure and function. *In* Platelet Aggregation. J. Caen, editor. Mason et Cie, Paris. 15–52.

White, J. G. 1982. Influence of taxol on the response of platelets to chilling. *American Journal of Pathology.* 108:184–195.

Zweig, S. E., and S. J. Singer. 1979. The two components of spectrin, filamin, and the heavy chain of smooth muscle myosin show no detectable homologies to one another by two-dimensional mapping of iodinated tryptic peptides. *Biochemical and Biophysical Research Communications.* 88:1147–1152.

Chapter 12

The Cortical Actin Gel of Macrophages

John H. Hartwig, Ken S. Zaner, and Paul A. Janmey

Hematology-Oncology Unit, Massachusetts General Hospital, Boston, Massachusetts

Cell Physiology of Blood © 1988 by The Rockefeller University Press

Introduction

Leukocytes extend portions of their cortical cytoplasm as pseudopods during locomotion. For over two centuries, microscopists have recognized that the cortical zone of cytoplasm of amoeboid cells is unique, describing it as hyaline in appearance and plastic in behavior (Von Rosenhof, 1792; Lewis, 1939; DeBruyn, 1946). These properties have been ascribed to the reversible gelation of peripheral cytoplasm: cytoplasm can be gel-like when excluding cellular organelles, or fluid (sol-like) when, for example, particulates approach the plasma membrane during secretion. It is now evident that actin filaments are principal elements in the cell periphery, and changes in the structure of cortical actin filaments have been suggested to be responsible for the dynamic mechanical behavior of this region of the cell. Alterations in actin structure are effected by a number of associated actin-binding proteins in vitro. Macrophage proteins that have been isolated and affect the assembly, stability, and length of actin filaments or cross-link them in vitro are listed in Table I. The functions of these components have been reviewed in detail elsewhere (Stossel et al., 1985). The purpose of this chapter is to review data on the mechanical properties of leukocyte cytoplasm.

Does Leukocyte Cytoplasm Have Gel-like Mechanical Properties?

Viscoelastic measurements can provide descriptive information about cytoplasm's material properties. The mechanical behavior of the cortex of leukocytes has been probed by application of small stresses to the edges of the cell and measurements of the resulting strains. Values for the shear modulus (ratio of stress to strain) between 300 and 1,500 dyn/cm^2, or greater under some conditions, have been obtained (Shapiro and Harvey, 1936; Chien et al., 1984; Skalak et al., 1984). These moduli are sufficient for the cells to withstand the shear stresses likely to be encountered in blood vessels in vivo. It is likely that high moduli are also required for these cells to move out of blood and into the connective tissue. Microaspiration experiments have also described a critical force at which the mechanical resistance of the cell decreased abruptly (Evans and Kukan, 1984). Many polymeric gels also display a rupture point corresponding to breakage of the polymer chains or interconnections between them and lose their elastic properties.

Cytoplasmic extracts from cells can also form rigid gels. In the absence of calcium (≤0.1 μM calcium), extracts from certain marine eggs, amoebae, and leukocytes form rigid gels in test tubes that do not flow in response to inversion of the tubes (Kane, 1975; Pollard, 1976; Stossel and Hartwig, 1976). A minimal estimate of the strength of these structures, based on the dimension and heights of the tubes, is $\geq1,000$ dyn/cm^2. Since the cytoplasmic volume is diluted by a factor of ~3 by buffers in preparing these extracts, the actual mechanical strength of cytoplasm in situ may be considerably greater than this estimate. Studies have further demonstrated that gelation of the extract requires the assembly of actin in it into filaments: the addition of cytochalasin B to the extracts inhibits solidification (Hartwig and Stossel, 1976; Weihing, 1976). Therefore, actin filaments can generate a structure responsible for rigid mechanical properties. The high rigidity of the extract gels is remarkable, in light of their relatively low (12–24 μM) actin concentration (0.05–0.1% volume fraction). The organization of filaments in such a gelled extract is shown in Fig. 1.

TABLE I
Purified Macrophage Actin-binding Proteins

Protein	M_r $(\times 10^{-3})$	Percent of cell	Regulation	Function
Filament cross-linking				
Actin-binding protein	2×270	0.6	None	Cytoskeletal architecture: orthogonal, filament branching in cortical cytoplasm; membrane-actin connection (?)
Alpha-actinin	2×100	1	Inhibited by micromolar calcium	Lateral filament association (?)
Filament assembly-disassembly				
Gelsolin	84	1	Activated by >0.1 μM calcium	Severs actin filaments, binds and blocks fast-growing filament ends, net filament disassembly, remains bound after severing in a calcium-insensitive fashion
			Inactivated by PIP$_2$ even in the presence of calcium	Net filament assembly
41K	41	?	Activated by micromolar calcium	Binds to fast-growing filament end, no filament-severing activity
Acumentin	65	1–5	None	End capping: new evidence suggests mechanism is not simply blockage of slow-growing filament end
Monomer binding				
Profilin	15		PIP$_2$ lowers the affinity of profilin-actin complexes	Binds actin monomer, preventing self-nucleation of cytoplasmic actin monomers
Contraction				
Myosin	2×200 2×20 2×15		Activated by micromolar calcium via calmodulin, myosin light-chain kinase	Pulls actin filaments
Tropomyosin	$2 \times 30*$			Stabilizes actin filaments to gelsolin, amplifies myosin's ATPase

*Subunits differ slightly in M_r.

The Organization of Actin Filaments in the Cortex of Leukocytes and Its Attachment to Membrane

As a first step to understanding how the mechanical properties and motile activity of cytoplasm derive from actin structure, we have analyzed the three-dimensional organization of actin filaments in the cortex of macrophages. Macrophages in the process of spreading on a glass surface were permeabilized with detergent. The detergent-insoluble residue, containing the polymerized fraction of cellular actin (the actin skeleton), was prepared for the electron microscope by rapid freezing and freeze

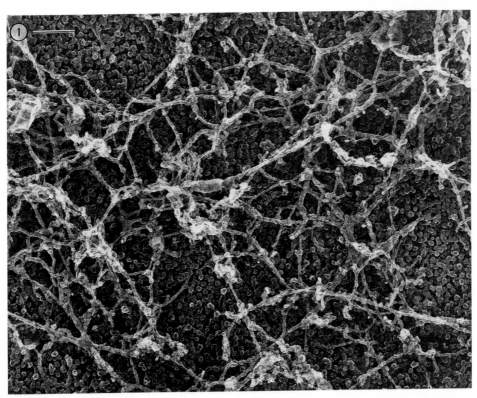

Figure 1. The organization of filaments in crude macrophage extract gels. 10 μl of a macrophage cytoplasmic extract in 0.34 M sucrose buffers (Stossel and Hartwig, 1976) containing 10 mM EGTA was placed on the surface of a coverslip and incubated for 60 min at 37°C. It was then fixed for 10 min with 1% glutaraldehyde, washed into water, and rapidly frozen and freeze-dried. Note that the filamentous structures found in these extracts resemble those in the cell cortex. Bar: 0.2 μm.

drying (Heuser and Kirschner, 1980; Hartwig and Shevlin, 1986; Fig. 2). Table II summarizes our analysis of the actin organization in these residues.

As shown in Fig. 2, the actin skeleton at the periphery of a macrophage is replete with short filaments, 0.6 μm in average length, that interconnect to form a continuous network. Greater than 95% of these filaments are actin, as demonstrated with myosin subfragment-1 labeling (Fig. 3). The structure of interfilament junctions is uniform throughout the actin skeleton: the ends of the actin filaments are almost always attached along the side of other actin filaments, forming T- and Y-shaped branches.

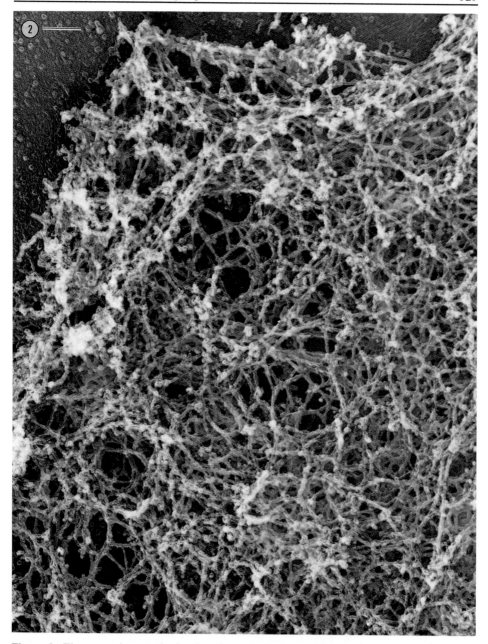

Figure 2. Electron micrograph showing the organization of filaments at the periphery of a macrophage cytoskeleton prepared by detergent treatment of an adherent cell, 1% glutaraldehyde fixation, and rapid freezing and freeze-drying. Bar: 0.2 μm.

Because all the ends of the filaments interconnect in this fashion, a nearly orthogonal network is established, and free filament ends are a rarity. The distance filaments transverse before intersecting other filaments to form X-shaped junctions is ~0.1 μm, which is considerably shorter than the average total filament length (Fig. 2). Therefore, the network pore size, rather than the total filament length, may be the

TABLE II
The Macrophage Cortical Actin Network: Ultrastructural Facts

Filament concentration	240–300 μM (monomers in filaments)
Average filament length*	0.6 μm
Intersection length	0.1 μm
Location and spacing of actin-binding protein	1 actin-binding protein: 0.35 μm of actin filament, located selectively at filament junctions
Free filament ends	Rare (see below)
Major type of filament intersection	Orthogonal branch: T- and Y-shaped filament junctions; these structures are formed when the end of one filament contacts the side of another filament; both the "barbed" and "pointed" ends of filaments form these end-on-side connections[‡]
Membrane-actin connections	"Barbed" end and lateral connections are the most common types found; many "pointed" end connections also found

*Filament length determined by counting ends with [H³]cytochalasin B and from morphometry.
[‡]With respect to the decoration of actin filaments with myosin fragments.

Figure 3. Identification of cortical 10-nm filaments as actin. A macrophage cytoskeleton was incubated with 1 mg/ml of subfragment-1 from skeletal muscle before fixation. The cytoskeleton was prepared for the electron microscope as in Fig. 1. Actin filaments decorated with myosin subfragment-1 have a twisted, cable-like appearance. The polarity of the filaments is most apparent near their ends, where they connect to the sides of other filaments. Bar: 0.1 μm.

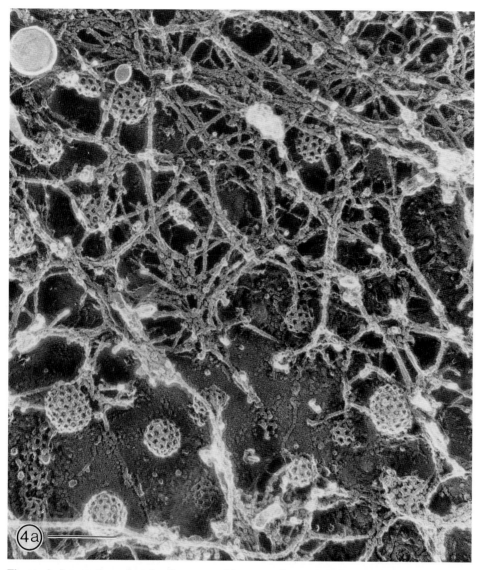

Figure 4. Interaction of actin filaments with macrophage plasma membrane. The plasma membrane–cytoplasm interface of cell spread on a glass coverslip was revealed by tearing it in half by adhering a polylysine-coated coverslip to its top surface and removing the coverslip. (*a*) The bulk of 10-nm fibers attaching to the membrane can be seen to derive from the three-dimensional filamentous network of the cell. Note that there are many regions of membrane that contain little filamentous material. Many clathrin-coated regions of the membrane are also apparent (Aggeler et al., 1983). (*b*) On next page. Identification of 10-nm filaments as actin using myosin subfragment-1. Bar: 0.2 μm.

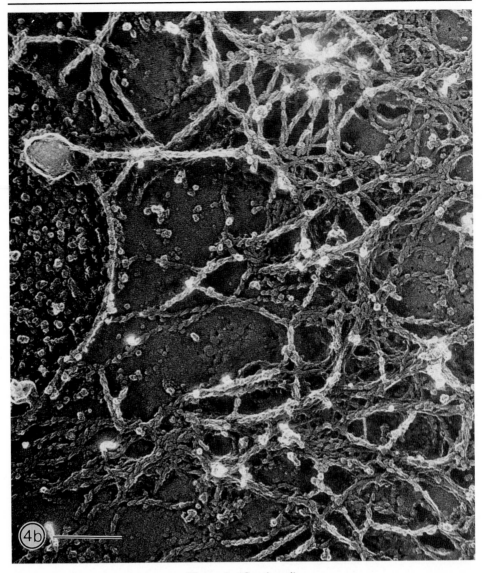

Figure 4. (Continued)

major determinant for understanding the mechanics of the cortical actin network. The mechanical properties of the network will therefore derive primarily from the number of X-shaped intersections, especially those stabilized by interfilament cross-links. As discussed below, the filament distance between cross-linked (rather than simply sterically overlapping) intersections in cytoplasm remains to be determined. It is clear, however, that many of these X-shaped junctions, as well many of the T- and Y-shaped ones, are points where actin-binding protein is located (see below). The actin filament concentration in the peripheral cytoplasm of a macrophage is 240 μM (1% volume fraction), considerably more concentrated than actin in the cellular extracts discussed above.

Do Leukocytes Contain a Two-Dimensional Actin Lamina That Supports the Plasma Membrane Analogous to the Spectrin-Actin Network of Erythrocytes That Participates in the Mechanics of Peripheral Cytoplasm?

We have studied the interaction of the orthogonal actin skeleton with the plasma membrane. The cytoplasmic side of the plasma membrane has been revealed by tearing adherent macrophages in half by attachment of polylysine-coated coverslips to their free surface, followed by their removal. As shown in Fig. 4, this approach provides an excellent view of this plasma membrane–cytoplasm interface. Fig. 4 *b* shows the orientation of actin filament–membrane associations, as revealed by myosin subfragment-1 binding. These micrographs fail to reveal a filamentous coat comparable to the specialized two-dimensional submembranous actin-spectrin networks of erythrocytes. In marked contrast, many large areas of clean membrane are apparent, dispersed between points where individual filaments, deriving from the three-dimensional actin network, contact the membrane.

The Mechanical Behavior of Purified Actins

Can Actin Filaments, at Concentrations and Lengths Found in the Cell Periphery, Account for the Mechanical Rigidity of Cells?

Solutions of macromolecules may be viscoelastic. For example, solutions of high-molecular-weight polymers (Ferry, 1980) show viscoelastic behavior even at very low concentrations, and at higher concentrations, where the domains of the coils overlap, much more prominent elastic effects are observed. The longer the molecule, the lower the concentration at which topological constraints between coils will become effective in restraining chain movement. Solutions containing polymers above this theoretical critical concentration are termed "semidilute." However, even with topological restraints, bulk flow (compliance) will occur in response to the application of a small constant shear stress (creep experiment) to a semidilute solution. Creep can occur because individual chains are not restricted by permanent cross-links and are free to disentangle. In this case, disentanglement is by Brownian motion of a flexible chain that can move in a contorted path defined by the positions of adjacent chains in what has been termed reptational movement (de Gennes and Leger, 1982). If the rate of reptation is restrained by increasing the concentration or chain length, a long time may elapse after the application of stress before the elastic response reaches a constant value and flow becomes steady. The flexibility of the macromolecular solute is also an important determinant of its viscoelasticity, and rigid macromolecules form semidilute solutions at much lower concentrations than do flexible coils of equivalent mass.

Purified actin filaments are extremely long (>5 μm) and rigid in vitro. The hydrodynamic behavior of actin filaments is more closely analogous to that of rods in solution than to random coil polymers such as polyacrylamide or polystyrene (Kasai et al., 1960). Unless filament length is decreased by sonication or addition of a severing protein, nearly all concentrations of f-actin above the critical monomer concentration in polymerizing buffers are semidilute (Janmey et al., 1985) and are predicted to display some degree of elastic behavior. A rod, in a semidilute solution, can move only by sliding along its axis or by waiting for its neighboring filaments to move out of the way by the same sliding motions, a process that is a very strong function of filament length (Doi and Edwards, 1986) (Fig. 5: note the interdigitation of individual actin polymers in an unperturbed actin filament solution). Therefore, the limited mobility of actin polymers in solution could be expected to decrease markedly, but not prevent totally, bulk flow in response to stress. Experimental verification for this type of actin behavior exists. Actin filament solutions can have high elasticity but always creep in response to an applied stress (Zaner and Stossel, 1982, 1983; Janmey, P. A., S. Huidt, J. Peetermans, J. Lamb, J. D. Ferry, and T. P. Stossel, manuscript submitted for publication). Decreasing the average size of filaments in a given solution with gelsolin, a molecule that stoichiometrically restricts filament length (Yin and Stossel, 1980; Yin et al., 1980; Janmey et al., 1986) to lengths near those found in cells, markedly

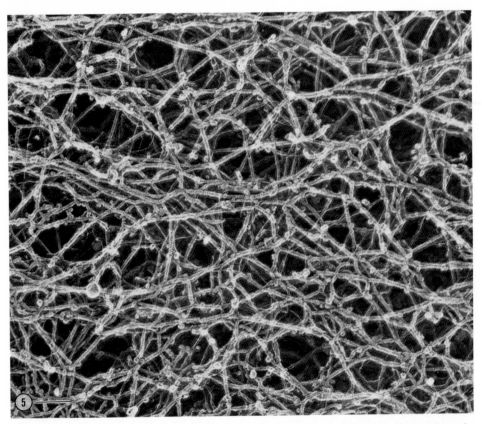

Figure 5. The three-dimensional organization of unperturbed, purified skeletal muscle actin filaments. Actin (48 μM in monomeric buffer) was polymerized on the surface of a coverslip by the addition of 2 mM $MgCl_2$, incubated at 37°C for 1 h, and then rapidly frozen. The frozen sample was mounted on the stage of a Cressington CFE-40 freeze-fracture machine and brought to −120°C, cleaved through the middle with a liquid nitrogen-cooled knife, freeze-dried for 20 min at −98°C, and rotary-shadowed with platinum at 20° and carbon at 90°. Bar: 0.1 μm.

decreases actin's elasticity and increases compliance (Zanev and Hartwig, 1988; Janmey, P.A., et al., manuscript submitted for publication). These experimental results are consistent with theoretical predictions that steric hindrance of the rotational motion of a macromolecule by its neighbors is a requisite for a high degree of viscoelasticity in relatively dilute solutions. Recent reports that monomeric actin is a viscoelastic solid at concentrations below those at which the protein is semidilute (Sato et al., 1985) are inconsistent with our experimental observations. It is therefore the degree of polymerization, and not an intrinsic chemical property, of actin from which its elasticity derives (Zaner et al., 1988).

Actin Solutions Containing Short Filaments Cross-Linked by Actin-binding Protein Have Mechanical Rigidity

The slow creep of actin filament solutions should be inhibited, as is the case for flexible polymers, by the addition of stable cross-links between the polymers. Three-dimensional networks of actin filaments, in association with other regulatory proteins

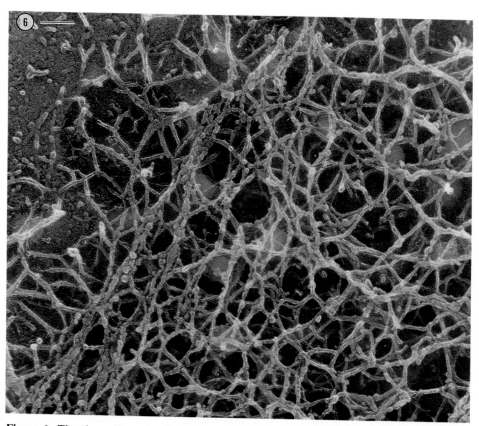

Figure 6. The three-dimensional structure of actin in the presence of actin-binding protein. Actin was assembled in the presence of actin-binding protein at a molar ratio of 1 actin-binding protein to 50 actin monomers. The sample was prepared for the electron microscope without fixation as described in Fig. 5. Bar: 0.1 μm.

(Table I), could therefore determine the viscoelastic properties of the cortical cytoplasm of leukocytes.

We have proposed that a molecule called actin-binding protein in white cells (Stossel and Hartwig, 1976; Hartwig and Stossel, 1981) and called filamin in smooth muscle (Shizuta et al., 1976; Wang, 1977) has an important role in the gelation of peripheral cytoplasm. The proposal is based on the evidence that actin-binding protein gels purified actin filaments in vitro. Actin-binding protein is an elongated protein of M_r 540,000. It is composed of two identical subunits that join together at one end,

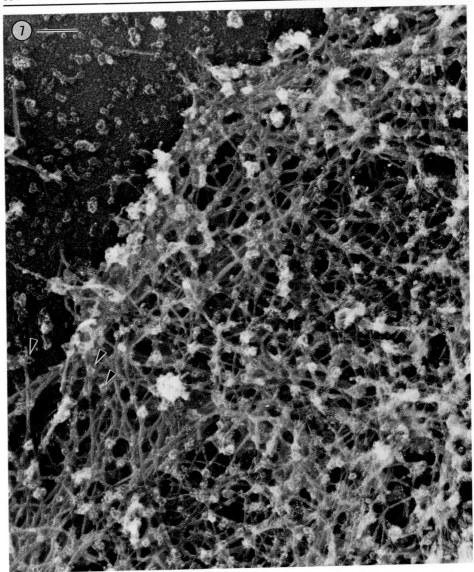

Figure 7. The location of actin-binding protein molecules in the periphery of a macrophage cytoskeleton. This electron micrograph demonstrates that actin-binding protein residues at points where actin filaments intersect and at the ends of actin filaments in the cortex of a macrophage cytoskeleton. Gold label is found in clusters and these clusters of gold particles decorate filament intersections and filaments that end at the cell margin. Some of the clusters are labeled with the arrowheads. Bar: 0.2 μm.

forming bipolar molecules. Since each subunit contains an actin filament–binding site at its end, the intact dimer can form stable cross-links between two actin filaments (Hartwig and Stossel, 1981). Actin-binding protein accounts for 1% of the total macrophage cellular protein and the bulk of actin filament cross-linking activity in these cells (Brotschi et al., 1978). Actin-binding protein–actin gels have high rigidities. Measurements at molar ratios of actin-binding protein to actin near those in the cell have yielded shear moduli 10 times greater than actin alone (Zaner, 1986; Stossel et al., 1987), and recently values as high as 2,000–8,000 dyn/cm^2 have been measured (Janmey, P. A., et al., manuscript submitted for publication). Further, these gel networks are stable structures, failing to creep in compliance measurements (Zaner, 1986).

Actin-binding protein is a very effective actin filament–cross-linking agent, with only 1.5 molecules required per actin filament for the incipient gelation of an actin solution. For a bivalent filament cross-linker, a minimum of one molecule (two cross-links) per filament is required before a macromolecular network can form (Flory, 1953). The effectiveness of this protein is due to its ability to promote orthogonal filament cross-linking (Hartwig et al., 1980; Niederman et al., 1983) rather than anisotropic bundles, as are formed by other cross-linkers such as alpha-actinin.

Actin-binding Protein Promotes Orthogonal Filament Branching In Vitro

Actin filaments assembled in the presence of actin-binding protein display a distinctive morphology in electron micrographs, being composed of perpendicular branching filaments (Hartwig et al., 1980; Niederman et al., 1983; Fig. 6). This configuration is similar to that of actin filaments in cortical cytoplasm and makes it reasonable to conclude that actin-binding protein might have a major architectural role in cytoplasmic organization.

Does Actin-binding Protein Locate at the Filament Junction In Situ?

The molar ratio of actin to actin-binding protein in the macrophage cytoskeleton is 150:1. From this ratio, if each actin-binding protein binds an actin filament at each end, the spacing between actin-binding protein–stabilized cross-links (the network pore size) is predicted to be 0.21 μm. We have located actin-binding protein molecules in the macrophage actin skeleton by immuno-electron microscopy (Hartwig and Shevlin, 1986). The reaction of detergent-prepared cytoskeletons with anti-rabbit actin-binding protein IgG and goat anti-rabbit IgG-coated 3–5-nm gold beads resulted in the deposition of clusters of gold selectively at points where filaments intersect and at the ends of filaments that may have been in contact with the membrane before its removal with detergent (Fig. 7). 40% of the the total number of intersections were labeled with gold, yielding a spacing of 0.3 μm between actin-binding protein–actin filament cross-links in the actin skeleton. This spacing between actin-binding protein connections is therefore greater than that found for intersections in the unlabeled actin skeletons. If the other 60% of the overlaps do not contain actin-binding protein molecules, then the actual lattice spacing between actin-binding protein–linked junctions is 0.26 μm, a value close to our determination of the actin-binding protein spacing in the cortical cytoskeleton.

Conclusion

The mechanical properties of peripheral macrophage cytoplasm derive from a three-dimensional network of short actin filaments that interconnect at 0.1-μm distances and occupy ~1% of the cytoplasmic volume. Solutions of short actin filaments, even at concentrations comparable to the cell, however, do not display gel-like mechanical properties in vitro. Therefore, the mechanical strength of cortical cytoplasm is conferred by cross-links that interconnect the actin filaments. We have demonstrated that actin-binding protein cross-links actin filaments into gel of high strength in vitro.

In addition, we have also localized the actin-binding protein molecules by immuno-electron microscopy and documented that the actin-binding protein label resides selectively at many points where filaments overlap in both macrophage cytoskeletons and in an actin gel made with actin-binding protein. The location of actin-binding protein molecules at points of filament intersection in the cell cortex strongly suggests that it may be responsible, in part, for the structure of cytoplasm and that cellular actin filaments are indeed cross-linked in situ into rigid ultrastructures. A two-dimensional membrane-actin lamina, analogous to that in erythrocytes, appears to play a minor role in leukocytes and is not apparent in electron micrographs of the cytoplasmic membrane face.

Acknowledgments

This work was supported by U.S. Public Health Service grants FM-36507 and AR-38910 and National Science Foundation grant DCB-8517973, and by grants from the Council for Tobacco Research, USA, the Edwin S. Webster Foundation, and the Whittaker Health Sciences Fund.

References

Aggeler, J. R., R. Takemura, and Z. Werb. 1983. High-resolution three dimensional views of membrane-associated clathrin and cytoskeleton in critical-point-dried macrophages. *Journal of Cell Biology*. 97:1542–1548.

Brotschi, E. A., J. H. Hartwig, and T. P. Stossel. 1978. The gelation of actin by actin-binding protein. *Journal of Biological Chemistry*. 253:8988–8993.

Chien, S., G. W. Schmid-Schonbein, K.-L. P. Sung, E. A. Schmalzer, and R. Skalak. 1984. Viscoelastic properties of leukocytes. *In* White Cell Mechanics: Basic Science and Clinical Aspects. P. LaCelle and M. A. Lichtman, editors. Alan R. Liss, Inc., New York, NY. 19–51.

DeBruyn, P. P. H. 1946. The amoeboid movement of the mammalian leucocyte in tissue culture. *Anatomical Record*. 95:117–191.

de Gennes, P. G., and L. Leger. 1982. Dynamics of entangled polymer chains. *Annual Review of Physical Chemistry*. 33:49–61.

Doi, M., and S. F. Edwards. 1986. The Theory of Polymer Dynamics. Clarendon Press, Oxford. 391 pp.

Evans, E., and B. Kukan. 1984. Passive material behavior of granulocytes based on large deformation and recovery after deformation tests. *Blood*. 64:1028–1035.

Ferry, J. D. 1980. Viscoelastic Properties of Polymers. 3rd edition. John Wiley & Sons, New York, NY. 641 pp.

Flory, P. 1953. Principles of Polymer Chemistry. Cornell University Press, Ithaca, NY. 672 pp.

Hartwig, J. H., and P. Shevlin. 1986. The architecture of actin filaments and the ultrastructural location of actin-binding protein in the periphery of lung macrophages. *Journal of Cell Biology*. 103:1007–1020.

Hartwig, J. H., and T. P. Stossel. 1976. Interactions of actin, myosin, and actin-binding protein of rabbit alveolar macrophages. III. Effects of cytochalasin b. *Journal of Cell Biology*. 71:295–303.

Hartwig, J. H., and T. P. Stossel. 1981. The structure of actin-binding protein molecules in solution and interacting with actin filaments. *Journal of Molecular Biology.* 145:563–581.

Hartwig, J. H., J. Tyler, and T. P. Stossel. 1980. Actin-binding protein promotes the bipolar and perpendicular branching of actin filaments. *Journal of Cell Biology.* 87:841–848.

Heuser, J. E., and M. W. Kirschner. 1980. Filament organization revealed in platinum replicas of freeze-dried cytoskeletons. *Journal of Cell Biology.* 86:212–234.

Janmey, P. A., S. E. Lind, H. L. Yin, and T. P. Stossel. 1985. Effects of semi-dilute actin solutions on the mobility of fibrin protofibrils during clot formation. *Biochimica et Biophysica Acta.* 841:151–158.

Janmey, P. A., J. Peetermans, K. S. Zaner, T. P. Stossel, and T. Tanaka. 1986. Structure and mobility of actin filaments as measure by quasielastic light scattering, viscometry, and electron(microscopy. *Journal of Biology Chemistry.* 261:8357–8362.

Kane, R. E. 1975. Preparation and purification of polymerized actin from sea urchin egg extracts. *Journal of Cell Biology.* 66:305–315.

Kasai, M., H. Kawashima, and F. Oosawa. 1960. Structure of f-actin solutions. *Polymer Science.* 44:51–69.

Lewis, W. H. 1939. The role of a superficial plasmagel layer in changes of form, locomotion and division of cells in tissue cultures. *Archiv für Experimentelle Zellforschung.* 23:7–13.

Niederman, R., P. Amrein, and J. H. Hartwig. 1983. The three dimensional structure of actin filaments in solution and an actin gel made with actin-binding protein. *Journal of Cell Biology.* 96:1400–1413.

Pollard, T. D. 1976. The role of actin in the temperature-dependent gelation and contraction of extracts of acanthamoeba. *Journal of Cell Biology.* 68:579–601.

Sato, M., G. Leimbach, W. H. Schwarz, and T. D. Pollard. 1985. Mechanical properties of actin. *Journal of Biological Chemistry.* 260:8585–8592.

Shapiro, H., and E. N. Harvey. 1936. The tension at the surface of macrophages. *Journal of Cellular and Comparative Physiology.* 8:21–30.

Shizuta, Y., H. Shizuta, M. Gallo, P. Davies, and I. Pastan. 1976. Purification and properties of filamin, an actin-binding protein from chicken gizzard. *Journal of Biological Chemistry.* 251:6562–6567.

Skalak, R., S. Chien, and G. W. Schmid-Schonbein. 1984. Viscoelastic deformation of white cells: theory and analysis. *In* White Cell Mechanics: Basic Science and Clinical Aspects. P. LaCelle and M. A. Lichtman, editors. Alan R. Liss, Inc., New York, NY. 3–18.

Stossel, T. P., C. Chaponnier, R. M. Ezzell, J. H. Hartwig, P. A. Janmey, D. J. Kwiatkowski, S. E. Lind, D. Smith, F. S. Southwick, H. L. Yin, and K. S. Zaner. 1985. Nonmuscle actin-binding proteins. *Annual Review of Cell Biology.* 1:353–402.

Stossel, T. P., and J. H. Hartwig. 1976. Interaction of actin, myosin, and a new actin-binding protein of rabbit pulmonary macrophages. II. Role in cytoplasmic movement and phagocytosis. *Journal of Cell Biology.* 68:602–619.

Stossel, T. P., P. A. Janmey, and K. S. Zaner. 1987. Cytomechanics. Springer-Verlag, Berlin.

Von Rosenhof, A. J. Roesel. 1792. Insecten Belustigungen. Kleeman, Nuremberg. (Neue edition.) 151 pp.

Wang, K. 1977. Filamin, a new high-molecular-weight protein found in smooth muscle and nonmuscle cells. Purification and properties of chicken gizzard filamin. *Biochemistry.* 16:1857–1865.

Weihing, R. R. 1976. Cytochalasin b inhibits actin-related gelation of Hela cell extracts. *Journal of Cell Biology.* 71:303–307.

Yin, H. L., and T. P. Stossel. 1980. Purification and structural properties of gelsolin, a Ca^{2+}-activated regulatory protein of macrophages. *Journal of Biological Chemistry.* 255:9490–9493.

Yin, H. L., K. S. Zaner, and T. P. Stossel. 1980. Ca^{2+} control of actin gelation. *Journal of Biological Chemistry.* 255:9494–9500.

Zaner, K. S. 1986. The effect of the 540-kilodalton actin cross-linking protein, actin-binding protein, on the mechanical properties of F-actin. *Journal of Biological Chemistry.* 261:7615–7620.

Zaner, K. S., and T. P. Stossel. 1982. Some perspectives on the viscosity of actin filaments. *Journal of Cell Biology.* 93:987–991.

Zaner, K. S., and T. P. Stossel. 1983. Physical basis of the rheologic properties of F-actin. *Journal of Biological Chemistry.* 258:11004–11009.

Zaner, K. S., and J. H. Hartwig. 1988. The effect of filament shortening on the mechanical properties of gel-filtered actin. *Journal of Biological Chemistry.* 263:4532–4536.

Zaner, K. S., R. G. King, J. Newman, K. L. Schick, R. Furukawa, and B. R. Ware. 1988. Rheology of g-actin solutions. *Journal of Biological Chemistry.* In press.

Anion Transport

Chapter 13

The Red Cell Anion-Transport System: Kinetics and Physiological Implications

Jesper Brahm

Department of General Physiology and Biophysics, The Panum Institute, University of Copenhagen, Copenhagen, Denmark

Cell Physiology of Blood © 1988 by The Rockefeller University Press

Introduction

All biological membranes constitute a barrier to free diffusion of solutes and solvent between the two separated phases. The magnitude of the barrier to a solute depends strongly on the membrane considered, since the permeability of the membrane to the solute may vary greatly, depending on the specialized transport mechanisms of the different membranes. In transport physiology, it is therefore of central importance that permeability characteristics of biological membranes be referenced to permeability characteristics of artificial bilayer lipid membranes. Such membranes are devoid of any of the special transport systems found in many biological membranes and show solute permeability coefficients that can be considered basic or ground permeabilities. Biological membranes, however, are still a hindrance to free diffusion of solutes, even though transport systems are incorporated in the membrane. This is illustrated in Table I, which summarizes the permeability coefficients of different solutes seen for lipid bilayers and for human red cells, compared with the "permeability" of a layer of water of 10^{-6} cm, which is approximately the thickness of a biological membrane. For a chloride ion, it takes ~ 50 ns to diffuse through such a water layer, whereas the

TABLE I
The Passive Permeability of Some Solutes

	Lipid bilayer	Red cell membrane	10^{-6}-cm layer of water
		$cm \cdot s^{-1}$	
Hemoglobin	0	0	1×10^{-2}
K^+	10^{-9}–10^{-10}	10^{-10}	20
Cl^- net flux	10^{-9}–10^{-10}	10^{-8}–10^{-9}	20
Exchange	10^{-9}–10^{-10}	10^{-4}	20

The values of diffusion through a 10^{-6}-cm layer of water, which is similar to the thickness of a membrane, were calculated from the relation $P = D/L$, where D is the diffusion coefficient in water ($cm^2 \cdot s^{-1}$) and L (cm) is the thickness of the water layer. The values apply to room temperature.

diffusion time is $\sim 10^5$ times greater if the water layer is replaced by a red cell membrane, and $\sim 10^{11}$ times greater if a lipid bilayer membrane constitutes the diffusion barrier. The low permeability of the lipid bilayer membrane to chloride varies, depending on the lipid composition of the membrane, as is the case for nonionic solutes (Finkelstein, 1976).

The 10^6-fold-higher permeability of erythrocyte membranes as compared with the permeability of artificial lipid bilayer membranes results from the presence of a highly specialized anion-transport system associated with the integral membrane protein designated capnophorin.

The Morphological Asymmetry of Capnophorin

The 100-kD integral membrane capnophorin molecule (band 3) is asymmetric with respect to both its position in the membrane and its function. The protein's function is characterized by a tightly coupled 1:1 anion exchange (for recent reviews, see Lowe and Lambert, 1983; Macara and Cantley, 1983; Jennings, 1984, 1985; Knauf, 1985;

Wieth and Brahm, 1985; Brahm, 1986; Fröhlich and Gunn, 1986; Jay and Cantley, 1986). The asymmetrical orientation of the protein in the membrane is established by studies showing that certain groups of the peptide chain, including the NH_2-terminal amino acid, are located intracellularly, whereas other groups appear to be accessible only from the outside. The position of the COOH-terminal amino acid is still an open question, as pointed out by Kopito and Lodish (1985), who recently sequenced the murine capnophorin molecule. However, Jay and Cantley (1986) suggest that a cytosolic location of the COOH-terminal is more likely than an extracellular or intramembranous location. The asymmetry is probably fixed as the molecule traverses the membrane ~ 12 times, making a reorientation of the entire structure too costly. This does not exclude the possibility that part of the molecule, e.g., the 17-kD fragment essential for an intact anion transport, may undergo a structural reorientation during the transport of anions. However, so far no evidence suggests such a major change in the orientation of the transport molecule. Consequently, any presumed conformational change in the transport protein as it carries out anion exchange can only be detected functionally.

The Functional Asymmetry of Capnophorin

Under steady state conditions, there is no net movement of solvent or solute across the red cell membrane, so any transport of solutes is restricted to self-exchange. The monovalent anions, such as chloride and bicarbonate, which are transported across the membrane, are not distributed evenly between the cytosolic and the external phases, because the macroanion hemoglobin contributes to the cytosolic negative charge balance. Since the unidirectional flux of an anion in either direction (J^{uni}, mol $\cdot$ cm^{-2} $\cdot$ s^{-1}) is proportional to the anion concentration (C, mol $\cdot$ cm^{-3}) by $J^{uni} = P^{app} \cdot C$, the proportion factor, P^{app} (cm $\cdot$ s^{-1}), known as the apparent permeability coefficient, must vary inversely with C if J^{uni} is to be kept constant. In order to keep the unidirectional fluxes equal under steady state conditions, the ratio of the respective apparent permeabilities must vary inversely with the ratio of the anion concentration in phases "i" and "o," as indicated by the relation $P_i^{app}/P_o^{app} = C_o/C_i$.

The external factors creating the functional asymmetry described here can be eliminated under experimental conditions in which the external and internal anion concentrations are made equal, as in ghosts or highly anion-loaded intact red cells. The functional asymmetry that is detectable under such conditions must arise from intrinsic asymmetry.

The formal kinetic description and experimental evidence for the anion-exchange process being carrier-mediated transport exhibiting Michaelis-Menten kinetics have been given elsewhere (see e.g. Knauf, 1985; Fröhlich and Gunn, 1986). Again, it should be emphasized that the commonly used term "carrier kinetics" in no way implies that carriers in the classic sense are involved. The kinetics for a carrier-mediated transport, however, also apply to the transport in which a conformational change of the transport protein is involved (Fröhlich and Gunn, 1986; Knauf and Brahm, 1988). It should suffice here to indicate that the principal reaction scheme describing the transport of a monovalent anion, such as chloride, from the cytosolic phase (i) to the extracellular solution (o), or vice versa, by the anion transporter, E, according to the ping-pong model, is

$$Cl_i + E_i \rightleftharpoons ECl_i \rightleftharpoons ECl_o \rightleftharpoons E_o + Cl_o. \tag{1}$$

There are three restrictions. (*a*) Both mechanisms contributing to the chloride conductance of the membrane, namely the interconversion of the two unloaded forms of the transporter ("slippage"),

$$E_i \rightleftharpoons E_o, \tag{2}$$

or the net transmembrane movement of an anion with no configurational change in the transporter ("tunneling"; see Fröhlich, 1984), must be extremely slow with respect to the rate of the exchange process shown in Eq. 1. (*b*) The transporter, unloaded or loaded, must face either side (i or o) of the membrane at any time. (*c*) The transport of an anion in one direction cannot be linked to a concomitant, oppositely directed transport of an anion by another transporter (coupling), or by the simultaneous binding of an anion to the transporter at the other side of the membrane, as predicted by the sequential model. (The apparent confusion in terminology refers to the fact that in enzyme kinetics, the simultaneous mode of action is termed "sequential.") A recent nuclear magnetic resonance study of chloride binding and transport (Falke et al., 1984) strongly suggests that the translocation step $ECl_i \rightleftharpoons ECl_o$, and not the association or dissociation of chloride, is the rate-limiting step. This greatly simplifies the kinetics and the mathematical expression that describes the distribution of unloaded E_i and E_o forms. This term, known as the asymmetry factor, A, under the given conditions of $Cl_i = Cl_o$ can be defined as (see Knauf, 1985)

$$A = E_o/E_i = K_o \cdot k/(K_i \cdot k'), \tag{3}$$

where K_o and K_i are the dissociation constants for chloride at the outside and inside surfaces, respectively, and k and k' are, respectively, the rate coefficients for the i $\rightarrow$ o and o $\rightarrow$ i conformational changes.

In practice, the value of A can be obtained from

$$A = K_{1/2_{o,max}} / K_{1/2_{i,max}} \tag{4}$$

where $K_{1/2_{o,max}}$ and $K_{1/2_{i,max}}$ are the outside and inside chloride concentrations, which cause a half-maximal flux when the concentration on the opposite side of the membrane saturates the transport system. Alternatively, A can be determined as

$$A = (K_{1/2_o} + K_{1/2_o} \cdot K_{1/2}/Cl_i)/(K_{1/2} + K_{1/2_o}), \tag{5}$$

where Cl_i is the cell chloride concentration at which $K_{1/2_o}$ is measured, and $K_{1/2}$ is the concentration that half-saturates the transport system when the chloride concentration is varied similarly on both sides of the membrane. For any transport mediated by a ping-pong model process, the overall half-saturation constant $K_{1/2}$ is related to $K_{1/2_{o,max}}$ and $K_{1/2_{i,max}}$ by

$$K_{1/2} = K_{1/2_{o,max}} + K_{1/2_{i,max}}. \tag{6}$$

$K_{1/2_o}$ can be determined in intact red cells (or resealed ghosts) with a physiological chloride concentration of ~100 mM by replacing the extracellular chloride with a solute that keeps the osmolarity of the medium constant without interacting with or being transported by the anion-exchange system. In this respect, sucrose is the solute of choice from a list of solutes that includes buffers such as citrate and 2-[*N*-morpholino]ethanesulfonic acid (MES) (Gunn, 1987; Brahm J., and P. A. Knauf, unpublished results).

From experiments such as those outlined above and inhibition studies (see Knauf and Brahm, 1988), the intrinsic asymmetry of the anion transporter has been determined for chloride exchange mostly at 0°C, whereas the asymmetry has been determined in only one study at body temperature, where refined techniques, such as the continuous flow tube method, are required, because the transport process is completed within a fraction of a second. Results from such studies show that the asymmetry, measured by $1/A$, increases from ~16 at 0°C to ~29 at body temperature (Table II). Thus, it appears that ~29 times more transporters face the cytosolic side of the membrane (E_i) than face the external side of the membrane (E_o). Recent experiments on bicarbonate transport in human red cells show a similar magnitude and temperature dependence of the intrinsic asymmetry (Gasbjerg, P., and J. Brahm, manuscript in preparation). However, much higher affinities to the two physiologically important anions, which only qualitatively confirm the asymmetry, have been reported in a fluorescent probe study of chloride-bicarbonate exchange in red cell ghosts (London et al., 1987).

TABLE II
The Functional Asymmetry of Capnophorin Mediating Chloride Self-Exchange

Temperature	$K_{1/2}$	$K_{1/2_{0,max}}$	$K_{1/2_0}$	$K_{1/2_{1,max}}$	A	$1/A$
°C		mM				
0*	65	3.9	—	61.1	0.064[‡]	16
38[§]	340	—	2.8	—	0.035[‖]	29

*From Brazy and Gunn (1976) and Gunn and Fröhlich (1979).
[‡]Calculated from Eq. 4.
[§]From Knauf and Brahm (1986).
[‖]Calculated from Eq. 5.

The Physiology of Anion Transport

The anion-exchange process is one of several steps involved in the transport of CO_2 produced in the tissues by the blood. Table III summarizes the physiological conditions at rest, where a pressure gradient for CO_2 as small as 0.8 kPa (6 mmHg) drives ~2 mmol of CO_2 from the tissue, with a CO_2 partial pressure of 6.1 kPa (46 mmHg) into 1 liter of blood passing through the tissue capillaries each minute, to enter the different reaction steps listed in the table. The distribution of CO_2 either in the forms of CO_2 or bicarbonate in plasma and cell water is also shown.

The relatively lipid-soluble CO_2 molecule diffuses easily through the plasma and across the cell membrane and into the cell water (reaction step 1) because the cell membranes are very permeable to the solute and because the diffusion distances are short. Very little CO_2 is physically dissolved in the plasma and cytosol since its solubility in water is low. A small fraction of the cellular CO_2 reacts with hemoglobin and forms carbaminohemoglobin (reaction step 2), whereas the major part is hydrated to carbonic acid and subsequently undergoes a dissociation into hydrogen ions and bicarbonate (reaction steps 3 and 4). The presence of carbonic anhydrase within the cells but not in the plasma implies that this reaction rate is ~10^4-fold higher in the cells. About 40% of the hydrogen ions bind to oxyhemoglobin, which promotes the liberation of O_2 (the Bohr effect) and reduces the acidification of the cell (reaction step 5). By the chloride-bicarbonate exchange process (reaction step 6), a fraction of the HCO_3^- formed in the cells is transported to the plasma, thereby increasing the CO_2-transport capacity of the blood, because the spontaneous hydration in plasma (reaction step 3*b*) is too slow to form any significant amount of HCO_3^- during the residence time of the blood in the capillary. As the blood enters the lung capillary, the reverse sequence of reaction steps takes place, driven by a pressure gradient of ~0.8 kPa (6 mmHg) for CO_2 between the blood and the lung alveoli (here the reverse reaction step 5—binding of O_2 to and liberation of H^+ from hemoglobin—is termed the Haldane effect). The overall forward and backward reaction sequences are called the Jacobs-Stewart cycle (Jacobs and Stewart, 1942) (although they more properly represent a half-cycle reaction sequence).

The distribution of total CO_2 shown in Table III clearly demonstrates the importance of CO_2 conversion into HCO_3^-, (reaction steps 3 and 4) and the subsequent exchange of plasma chloride for intracellularly formed bicarbonate (reaction step 6),

TABLE III
The Time Course of the Partial Reaction Steps of the Uptake/Release, and the Fractional Distribution of the Different Forms of 2 mmol CO_2 by 1 Liter of Blood (45% Hematocrit)

Reaction Step	90% Completion	Plasma	Cell Water
	ms	%	%
1. Diffusion of CO_2	3–4	4.5	3.5
2. $Hb + Co_2 \rightleftharpoons HbCO_2$	250	0	12
3. $CO_2 + H_2O \rightleftharpoons H_2CO_3$			
a) Cell (+ CA)	<10	—	—
b) Plasma (− CA)	4×10^4	—	—
4. $H_2CO_3 \rightleftharpoons H^+ + HCO_3^-$	<1	—	—
5. $HbO_2 + 0.4\ H^+ \rightleftharpoons HbH + O_2$	250	—	—
6. Cl^-/HCO_3^- Exchange	400–500	50	30

The values are obtained from Hoffman and Henkens (1987), and from reviews of Wagner (1977), Klocke (1978), and Brahm (1986). + CA indicates that the intracellular hydration/dehydration of CO_2 is enzymatically governed by the presence of carbonic anhydrase, which is absent in plasma (− CA).

the Hamburger shift (Hamburger, 1891). The question is whether the distribution of CO_2 shown here can be obtained within the <1-s residence time of the blood in the tissue capillary. This capillary residence time is somewhat shorter than the residence time of the blood in the lung capillary. A crude estimate of whether the blood exploits the CO_2-transport capacity (Wieth and Brahm, 1980) showed that the anion exchange, known as the Hamburger shift, is the bottleneck with respect to the use of the CO_2-transport capacity, and is 90% complete after 350–380 ms, and 99% complete after 700–800 ms.

Consequently, it appears that there is sufficient time for CO_2 to equilibrate between the tissue and the blood at rest during the residence time in the capillary. Were this not the case, a greater driving force (that is, a greater difference between the CO_2 tension of the tissue and the blood and the blood and the alveoli) would be necessary to ensure the uptake of CO_2 in the tissue capillary and subsequent release in the alveolar capillary. The higher CO_2 gradient between the tissue and the blood as it enters the capillary is created predominantly by an increase of the tissue CO_2 tension, which is a stimulating factor for the cells in the respiratory center controlling ventilation. Hence, an increase of CO_2 tension in the center stimulates hyperventilation. This mechanism is the main factor for the hyperventilation that follows exercise, as up to ~6 mmol CO_2 is taken up by 1 liter of blood per minute (at heavy exercise, 180 mmol CO_2/min [4 liters/min] is removed from the tissues by the blood, whose cardiac output increases to ~30 liter/min).

Pathophysiological Considerations

It appears that the anion-exchange mechanism is the rate-limiting step in the transport of CO_2 from tissues to lungs. Had nature made it so that the plasma also contained carbonic anhydrase, the CO_2 uptake by the blood would not be dependent on anion exchange across the red cell membrane. Although ~20 ml of the old cell population is destroyed each day by the organism, the carbonic anhydrase concentration in the plasma is negligible. Hence, the CO_2 has to enter the Jacobs-Stewart cycle to exploit the CO_2-transport capacity of the plasma. Were the anion exchange absent, the organism would have to increase the CO_2 gradient between the tissue and the blood as described above and the individual would have to hyperventilate, even at rest (see Wieth and Brahm, 1985). In accordance with this concept, red cells from all warm-blooded animals (mammal and birds) so far investigated have an anion exchange of comparable capacity (Wieth et al., 1974; Brahm and Wieth, 1977), and the anion-exchange mechanism (capnophorin) appears to be present at an early stage of life in fetal red cells (Brahm, J., and P. D. Wimberley, manuscript in preparation) (cf. Table IV). Interestingly, it has recently been shown that the lamprey, which is located at the root of the phylogenetic tree of the vertebrates, has red cells with no anion transport (Ohnishi and Asai, 1985). This indicates that vertebrate life is not incompatible with the lack of a red cell anion transporter.

As discussed elsewhere (Wieth and Brahm, 1980, 1985; Wieth et al., 1982), a pathophysiological hyperventilation is connected to anemia, where the total cell mass of the blood is reduced, thus reducing the CO_2-transport capacity. We also suggested that the hyperventilation following salicylate poisoning is related to the inhibition of anion transport rather than being a direct stimulatory effect of salicylate on the respiratory center. We have tried to show a similar effect of salicylate on red cell anion transport in perfused dogs (Brahm et al., 1987). However, the effect appeared to be much smaller than that obtained in vitro with dog red cells, as did the effect of the anion-transport inhibitor 4,4'-dinitro-stilbene-2,2'-disulfonate (DNDS). This suggests that the clearance in the kidneys and/or binding to plasma proteins of the inhibitors in the dog is pronounced.

Among the drugs of clinical interest, the diuretics have been tested for their inhibitory effect on anion transport in red cells. They do inhibit anion transport, but at concentrations that exceed by far the clinically relevant concentrations of the free drug in plasma, which cause a diuretic effect in the kidney (Wieth and Brahm, 1985).

In diabetic patients, the increased glucose concentration induces an increase of the fraction of nonenzymatically glycated hemoglobin (HbA_{1c}), and the nonenzymatic glycation of membrane proteins parallels the levels of HbA_{1c}. Recently, it has been shown that the glycation site of capnophorin is ~280 residues from the COOH-terminal (see Jay and Cantley, 1986). We (Brahm and Mortensen, 1988) have tested whether glycation of capnophorin impairs anion transport in diabetic patients. As shown in Table IV, our results with red cells from healthy and diabetic children do not show significant differences. Hence, the acidosis observed in diabetics is not aggravated by an impaired anion-transport system.

The red cell as a test object for hereditary diseases is an attractive diagnostic approach. With respect to anion transport, an alteration of chloride self-exchange in red cells from patients with the neurological disorder Huntington's disease has been reported (Bialas, 1984). It was found that the rate of chloride self-exchange decreased more rapidly and was more pronounced in young red cells than in old red cells during in vitro aging. In cystic fibrosis, the suggested general defect of increased chloride permeability of cultured tracheal epithelial cells appears not to be related to the chloride conductance channels of the membrane, but rather seems to be associated with a metabolic pathway that regulates the open or closed state of the channels, which depends on calcium (Pedersen et al., 1987). However, neither chloride conductance (Boucher et al., 1984) nor chloride self-exchange in red cells from these patients, either

TABLE IV
Chloride and Bicarbonate Transport in Red Blood Cells at 38°C, pH 7.4.

	J_C	J_{bic}	Blood glucose	HbA$_{1c}$
	$mol \cdot cm^{-2} \cdot s^{-1} \cdot 10^9$		mM	$\%$
Fetal	30	8	—*	—*
Children				
Control	31	7	4.6	5.4
Diabetic	26	8	10.1‡	11.1‡
Adult	31	9	—*	—*

*Not determined.
‡The values are significantly different from the respective control values ($p < 0.001$). Modified from Brahm and Mortensen (1987).

at low and physiological temperatures (Berghout et al., 1984; Brahm, J., unpublished results), is different from the values of control red cells.

Thus, it appears that the red cell anion exchange is present in all species of warm-blooded animals, and that the transport system is probably little affected by heredity or environmental conditions that may have pronounced effects on other cells and tissues.

References

Berghout, A. G. R. V., and S. W. Bender. 1984. Anion transport and 2,3-diphosphoglycerate in cystic fibrosis red blood cells. *Pediatric Research.* 18:1017–1020.

Bialas, W. A. 1984. Alteration of Cl⁻ transport in erythrocytes from patients with Huntington's disease. *General Physiology and Biophysics.* 5:403–411.

Boucher, R. C., D. W. Ross, M. R. Knowles, J. T. Gatzy, and J. C. Parker. 1984. Cl⁻ permeabilities in red blood cells and peripheral blood lymphocytes from cystic fibrosis and control subjects. *Pediatric Research.* 18:1336–1339.

Brahm, J. 1986. The physiology of anion transport in red cells. *Progress in Hematology.* 14:1–21.

Brahm, J., J. Grønlund, M. P. Hlastala, J. Ohlsson, and E. Swenson. 1987. Effects of in vivo inhibition of red cell anion exchange and Haldane effect on carbon dioxide output (VCO_2) in the dog lung. *Journal of Physiology.* 390:245P. (Abstr.)

Brahm, J., and H. B. Mortensen. 1988. Anion transport as related to hemoglobin A_{1c} in red cells of diabetic children. *Clinical Chemistry.* 34: In press.

Brahm, J., and J. O. Wieth. 1977. Separate pathways for urea and water, and for chloride in chicken erythrocytes. *Journal of Physiology.* 266:727–749.

Falke, J. J., R. J. Pace, and S. I. Chan. 1984. Chloride binding to the anion transport binding sites of band 3: a ^{35}Cl NMR study. *Journal of Biological Chemistry.* 259:6472–6480.

Finkelstein, A. 1976. Water and nonelectrolyte permeability of lipid bilayer membranes. *Journal of General Physiology.* 68:127–135.

Fröhlich, O. 1984. Relative contributions of the slippage and tunneling mechanisms to anion net efflux from human erythrocytes. *Journal of General Physiology.* 84:877–893.

Fröhlich, O., and R. B. Gunn. 1986. Erythrocyte anion transport: the kinetics of a single-site obligatory exchange system. *Biochimica et Biophysica Acta.* 864:169–194.

Gunn, R. B. 1987. The mess with MES: a good buffer gone bad. *Journal of General Physiology.* 90:20a (Abstr.)

Gunn, R. B., and O. Fröhlich. 1979. Asymmetry in the mechanism for anion exchange in human red blood cell membranes. Evidence for reciprocating sites that react with one transported anion at a time. *Journal of General Physiology.* 74:351–374.

Hamburger, H. J. 1891. Über den Einfluss der Athmung auf die Permeabilitat der Blutkörperchen. *Zeitschriftung der Biologie.* 28:405–416.

Hoffman, D. W., and R. W. Henkens. 1987. The rates of fast reactions of carbon dioxide and bicarbonate in human erythrocytes measured by carbon-13 NMR. *Biochemical and Biophysical Research Communications.* 143:67–73.

Jacobs, M. H., and D. R. Stewart. 1942. The role of carbonic anhydrase in certain ionic exchanges involving the erythrocyte. *Journal of General Physiology.* 25:539–552.

Jay, D., and L. Cantley. 1986. Structural aspects of the red cell anion exchange protein. *Annual Review of Biochemistry.* 55:511–538.

Jennings, M. L. 1984. Oligomeric structure and the anion transport function of human erythrocyte band 3 protein. *Journal of Membrane Biology.* 80:105–119.

Jennings, M. L. 1985. Kinetics and mechanism of anion transport in red blood cells. *Annual Review of Physiology.* 47:519–523.

Klocke, R. A. 1978. Carbon dioxide transport. *In* Extrapulmonary Manifestations of Respiratory Disease. E. D. Robins, editor. Marcel Dekker, New York, NY. 315–343.

Knauf, P. A. 1985. Anion transport in erythrocytes. *In* Physiology of Membrane Disorders. T. Andreoli, S. Schultz, J. F. Hoffman, and D. Fanestil, editors. Plenum Publishing Corp., New York, NY. 191–220.

Knauf, P. A., and J. Brahm. 1986. Asymmetry of the human red blood cell anion transport system at 38°C. *Biophysical Journal.* 49:579a. (Abstr.)

Knauf, P. A., and J. Brahm. 1988. Functional asymmetry of the anion exchange protein, capnophorin: effects on substrate and inhibitor binding. *Methods in Enzymology.* In press.

Kopito, R. R., and H. F. Lodish. 1985. Primary structure and transmembrane orientation of the murin anion exchange protein. *Nature*. 316:234–238.

London, R. D., M. S. Lipkowitz, and R. G. Abramson. 1987. Cl^-/HCO_3^- antiporter in red cell ghosts: a kinetic assessment with fluorescent probes. *American Journal of Physiology*. 21:F844–F855.

Lowe, A. G., and A. Lambert. 1983. Chloride-bicarbonate exchange and related transport processes. *Biochimica et Biophysica Acta*. 694:353–374.

Macara, I. G., and L. C. Cantley. 1983. The structure and function of band 3. *In* Cell Membranes: Methods and Reviews. E. Elson, W. Fracier, and L. Glaser, editors. Plenum Publishing Corp., New York, NY. 1:41–87.

Ohnishi, S. T., and H. Asai. 1985. Lamprey erythrocytes lack glycoproteins and anion transport. *Comparative Biochemistry and Physiology*. 81B:405–407.

Pedersen, S. P., E. H. Larsen, and N. J. Brandt. 1987. Restitution of chloride permeability in cystic fibrosis. *Medical Science Research*. 15:151–152.

Wagner, P. D. 1977. Diffusion and chemical reaction in pulmonary gas exchange. *Physiological Review*. 57:257–312.

Wieth, J. O., O. S. Andersen, J. Brahm, and P. J. Bjerrum. 1982. Chloride-bicarbonate exchange in red blood cells: physiology of transport and chemical modification of binding sites. *Philosophical Transactions of the Royal Society of London, Series B*. 299:383–399.

Wieth, J. O., and J. Brahm. 1980. Kinetics of bicarbonate exchange in human red cells: physiological implications. *In* Membrane Transport in Erythrocytes. U. V. Lassen, H. H. Ussing, and J. O. Wieth, editors. Munksgaard, Copenhagen. 467–482.

Wieth, J. O., and J. Brahm. 1985. Cellular anion transport. *In* The Kidney: Physiology and Pathophysiology. D. W. Seldin and G. Giebisch, editors. Raven Press, New York, NY. 49–89.

Wieth, J. O., J. Funder, R. B. Gunn, and J. Brahm. 1974. Passive transport pathways for chloride and urea through the red cell membrane. *In* Comparative Biochemistry and Physiology of Transport. L. Bolis, K. Bloch, S. E. Luria, and F. Lynen, editors. Elsevier/North-Holland, Amsterdam. 317–337.

Chapter 14

Structure and Tissue-specific Expression of the Mouse Anion-Exchanger Gene in Erythroid and Renal Cells

Ron R. Kopito, Mona M. Andersson, Doris A. Herzlinger, Qais Al-Awqati, and Harvey F. Lodish

Whitehead Institute for Biomedical Research, Cambridge, Massachusetts; Department of Medicine, Columbia University College of Physicians and Surgeons, New York, New York; Department of Biology, Massachusetts Institute of Technology, Cambridge, Massachusetts

Introduction

In mammals, metabolically produced CO_2 is transported in the blood principally as the soluble bicarbonate anion, a process that facilitates its elimination and provides a pH buffer system linked to respiration. Two proteins in the erythrocyte critical to this process are intracellular carbonic anhydrase, which catalyzes the reversible hydration of CO_2, and the integral membrane anion exchanger, band 3, which mediates the movement of HCO_3^- in exchange for chloride across the plasma membrane.

Band 3 is the most abundant integral protein of the erythrocyte membrane, where it constitutes >25% of the total. This ~95-kD glycoprotein possesses two structurally and functionally distinct domains. The ~420 NH_2-terminal amino acids of band 3 (reviewed in Low, 1986) form a domain that is polar and highly charged and projects into the cytoplasm. This domain possesses binding sites for hemoglobin (Walder et al., 1984), glycolytic enzymes (Strapazon and Steck, 1976), and ankyrin (Bennett and Stenbuck, 1979, 1980). By virtue of this latter association, band 3 functions to "anchor" the mesh-like spectrin cytoskeleton. Anion-transport activity resides within the COOH-terminal half of band 3 (Grinstein et al., 1978) and occurs via an obligatory, electroneutral, one-for-one exchange driven by the concentration gradients of the transported species (reviewed in Cabantchik et al., 1978; Knauf, 1986). These functional domains of the band 3 polypeptide correlate well with structural features identified in the primary sequence, deduced from a full-length cDNA clone of the mouse protein (Kopito and Lodish, 1985a).

Biosynthetic labeling studies in the mouse (Chang et al., 1976) and in murine erythroleukemia (MEL) cells (Sabban et al., 1981; Patel and Lodish, 1984) demonstrate that the appearance of band 3 coincides temporally with the synthesis of hemoglobin during erythropoiesis. The parallel rise in steady state mRNA levels for band 3 in differentiating MEL cells suggests that the expression of the band 3 gene is regulated in concert with other erythroid-specific genes such as globin (Kopito and Lodish, 1985b). Several distinct *cis*-acting sequences, located within the 5' flanking region and 3' to the mRNA cap site, are both necessary and sufficient for tissue-specific globin gene regulation in differentiating murine (Charnay et al., 1985) and avian (Choi and Engel, 1986) erythroid precursors. In addition, *trans*-acting factors are capable of activating the expression of globin genes in a stage- and tissue-specific manner (Baron and Maniatis, 1986). It is therefore of interest to determine whether common structural elements within flanking sequences or transcriptional units are shared among genes such as band 3, whose expression is regulated in concert with globin during terminal erythroid differentiation. The recent identification in renal epithelia of an mRNA transcript that cross-hybridizes with band 3 cDNA at high stringency (Alper et al., 1987), and of proteins antigenically related to band 3 (Drenckhahn et al., 1985; Schuster et al., 1986), suggests the involvement of more than one tissue-specific activator element in the regulation of the band 3 gene.

Results and Discussion

Characterization of the Mouse Band 3 Gene

The exact correspondence between the restriction map of the overlapping genomic clones shown in Fig. 1 and the restriction pattern in genomic DNA (Fig. 2) argues for the presence in the mouse genome of a single copy of the band 3 gene. This gene spans ~17 kb of genomic DNA and is split by 19 intervening sequences. With the exception

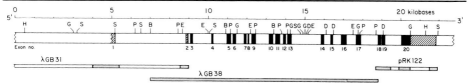

Figure 1. Structure of the mouse band 3 gene. (Top) Nucleotide scale in kilobases. (Middle) Structure of band 3 gene compiled from restriction mapping and sequence analysis of genomic clones. The black vertical bars denote the location and sizes of the exons within the protein coding (translated) region. The hatched bars denote untranslated regions of exons. The 5′ boundary of exon 1 is jagged to denote the heterogeneity in the locations of the mRNA cap sites in transcripts of the gene (see below). Restriction sites were determined by analysis of sequence and by restriction mapping of subclones of the genomic clones. Enzymes are Eco RI (E); HindIII (H); BglII (G); SstI (S); PstI (P); Bam HI (B); and SphI (D). (Bottom) Overlapping bacteriophage λ clones λGB31 and λGB38 and plasmid clone pRK122 are shown relative to the gene map and kilobase scale. The stippled areas have been completely sequenced.

of exon 20, all of the exons in the band 3 gene range between 73 and 253 bp, with a median size of 170 bp. Exon 1 consists entirely of nucleotides within the 5′ untranslated region (UT), while exon 2 contains 68 bp of 5′ UT and the first five codons of the mature band 3 protein, which is synthesized without a cleaved leader sequence (Braell and Lodish, 1982; Kopito and Lodish, 1985a). All of the other exons contain only protein coding sequences, except exon 20, which comprises the COOH-terminal 27 codons and the entire 3′ UT (1,463 bp).

The introns in the band 3 gene range in length from 79 bp to 3.9 kb, with a median size of 475 bp. Among the 18 introns that interrupt the protein coding sequence, there are 10 that occur in between codons, 6 that interrupt at frame 1, and 2 at frame 2. These split-codon introns place restrictions as to possible alternative splicing combinations.

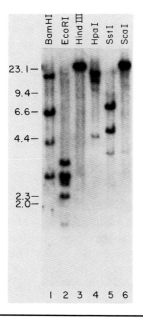

Figure 2. Genomic Southern blot of the mouse band 3 gene. Each lane contains 10 μg mouse genomic DNA, digested to completion with the enzymes indicated, and probed with the full-length cDNA probe pB3SP4 (Kopito and Lodish, 1985a).

Fig. 3 shows a striking correspondence between the location of the intron/exon splice junctions in the band 3 mRNA and structural features of the band 3 polypeptide. In the upper panel, the location of the band 3 introns is indicated on a hydrophobicity plot of the protein, showing that the nine introns that interrupt the membrane-associated domain (between exons 11 and 19) all occur at positions that encode the hydrophilic sequences, which are presumed to form the "loops" that lie between the long, hydrophobic, membrane-spanning segments (Kopito and Lodish, 1985a). The bottom panel of Fig. 3 reveals that the location of the introns in band 3 also correlates with the predicted structural features of the cytoplasmic, membrane skeleton–binding domain. Within this region are three sequences that have a high degree of β-turn

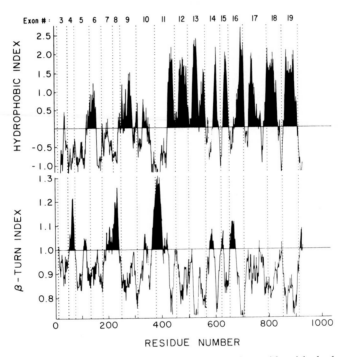

Figure 3. Correction of structural features of the band 3 polypeptide with the locations of the introns in the mouse band 3 gene. The vertical broken lines indicate the positions of the splice junctions of the introns, which interrupt the coding region exons (numbered on top) of the band 3 mRNA. The plots were generated using the algorithm and values of Kyte and Doolittle (1982) for hydrophobicity (upper blot) and Chou and Fasman (1978) for β-turn index (lower) using a window of 18 residues.

character (Chou and Fasman, 1978). The two NH_2-terminal sequences are bounded by distinct introns. The third is located predominantly within exon 11, which it shares with the first hydrophobic, membrane-spanning sequence. While the three-dimensional correlates of these β-rich regions are unknown, they correspond to sequencs rich in proline and other helix-breaking residues that may form flexible "hinges" that separate the functional domains of the band 3 cytoplasmic "tail." The coincidence of the locations of the introns that interrupt the sequence of the band 3 gene in this region, and the locations of the hydrophilic "loops" connecting these segments, is similar to the pattern that has been observed in the structures of the genes encoding rhodopsin

(Nathans and Hogness, 1983) and 3-hydroxy-3-methylglutaryl (HMG) coenzyme A reductase (Liscum et al., 1985). The products of these genes encode integral membrane proteins that are not apparently involved in any transport activity, but which, like band 3, span the bilayer more than once. While the precise three-dimensional structures of these proteins are not known, it is possible that this correlation may provide useful insight into the topology and structure of this class of proteins.

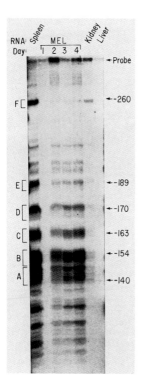

Figure 4. (*Top*) S1 nuclease protection analysis of the 5' end of mouse band 3 mRNA in multiple tissues. Poly(A$^+$) RNA from mouse spleen (2 µg), MEL cells 1–4 d post-DMSO induction (10 µg), kidney (22 µg), and liver (18 µg) was annealed with 5×10^4 cpm of a single-stranded, ^{32}P 5'-end–labeled DNA probe corresponding to nucleotide positions −95 to −320, digested with S1 nuclease, and resolved on a 6% denaturing polyacrylamide gel. The clusters of protected fragments are designated A–F. Autoradiographic exposure was 28 h. (*Bottom*) Schematic representation of nuclease protection assay using a 5'-end–labeled (asterisk) DNA probe. The bottom line shows the location of exon 1 (black), intron 1, and the 5' flanking region (stippled) in genomic clone λGB31 (Fig. 1). The 5' terminus of cDNA clone pB3SP4 (Kopito and Lodish, 1985a) is designated "C."

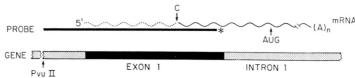

Multiple Sites of Transcriptional Initiation of Band 3 mRNA in Erythroid and Nonerythroid Tissues

Nuclease mapping with an end-labeled, single-stranded DNA probe was used to analyze the 5' end of the band 3 mRNA isolated from several mouse tissues. These results reveal the presence of multiple transcriptional start sites for band 3 message in anemic mouse spleen. Five clusters of protected fragments were generated with S1 nuclease, labeled A–E in Fig. 4 (between −146 and −189). A sixth band, F, mapping to −260, was also reproducibly observed.

Anemic mouse spleen is rich in hematopoietic cells and abundant in mRNA encoding band 3 (Braell and Lodish, 1981, 1982). An identical set of nuclease-protected bands was observed in RNA from the spleens of normal mice (not shown). Spleen, however, is a heterogeneous tissue; it contains several other cell types, including those of myeloid and lymphoid lineage. We therefore examined the initiation sites of band 3 mRNA in MEL cells, which can undergo terminal erythroid differentiation when induced with DMSO or other agents (reviewed in Marks and Rifkind, 1978), and in two mouse tissues. Fig. 4 (*bottom*) shows that in MEL cells, the S1-protected fragments increase in abundance after erythroid induction, and correspond exactly in size to the clusters A–E observed in the anemic spleen. Strikingly, the largest transcript, F (at -260), is virtually absent in RNA from even 4-d induced cells. In contrast, the predominant protected transcript in the kidney is the one that maps to position -260 (F), with only faintly detectable bands at the more 3' sites. No protected fragments were detectable in liver RNA.

These observations suggest that at least two distinct promoters are active, in a tissue-specific manner, in the band 3 gene (Fig. 5). The initiation of band 3

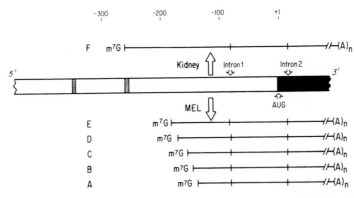

Figure 5. Model for tissue-specific transcription initiation sites in the mouse band 3 gene. The nucleotide scale refers to position in the 5' untranslated region of processed mRNA. The transcriptional initiation sites are designated A–F, as in the text. The black shaded region denotes the protein coding sequence, and the two hatched boxes represent the locations of hypothetical promoter elements.

transcription at both the erythroid and kidney sites in RNA from the spleens of anemic and normal mice may be due to the presence, in this heterogeneous tissue, of both erythroid and nonerythroid cells, the latter type sharing with the kidney usage of the same promoter. Alternatively, the regulation of the band 3 gene in erythroid precursor cells in the spleen may differ from its regulation in these cells in the bone marrow or in MEL cells. This pattern of erythroid- and kidney-specific transcription of the mouse band 3 gene implies the presence of at least two cis-acting elements that direct the expression of the band 3 gene in the appropriate cell types.

Most eukaryotic mRNAs, like globin, are expressed in specific tissues and have a single distinct cap site, which is found ~30 bases downstream of a highly conserved promoter region containing TATA and CAAT boxes, which serve to direct the faithful initiation of RNA polymerase II transcription (Breathnach and Chambon, 1981). Examination of the 1,700 nucleotides of mouse genomic sequence upstream of the multiple cap sites of band 3 mRNA (Fig. 6) reveals no suitable TATA boxes or homologies to any other consensus sequences typical of eukaryotic promoter regions. Similarly, there are no detectable sequence homologies to globin or any other genes present in available databases.

A common feature among genes that initiate mRNA transcription at multiple sites and lack TATA or CAAT sequences is the presence of a G+C-rich region

```
-1710  TCCCTTTAGCTTTGCTGGGTCTCAGCCATCACGAGCTAGTGAGGGAGACCACACAATATG  -1651
-1650  TTCCATTTGCCCATCTCTGAGTTTTTCTATTGTTGGAGCCTGAACTTAGTGGGCAAAGAG  -1591
-1590  TCAGGGTCACTTGAGTGCTCACTCTCATTCACACTCTCTCCCTCTCTCCTCATATATCTC  -1531
-1530  AGGCTGGCCTCGAACTCCCTAAGTATGTAATGGAGGATACCTTGAACTTCTGATCCCCCT  -1471
-1470  GTCTCCACCTTTCGAGCTTTGGGATACCCCTACCCCACCCCCAAGTGCTGGAATTACAAG  -1411
-1410  TGAGCAGCACTAGCCAGGCGGTGGTGGCGCACACCTTTAATCCCAGCACTCGGGAGGCAG  -1351
-1350  AGGCAGGTGGATTTCTGAGTTCGAGGCCAGCCTGGTCTACAGAGTGAGTTCCAGGACAGC  -1291
-1290  CAGGGCTACACAGAGAAACCCTGTCTCTAAAAAAAAAAAAGAAGACAAGTGAGCAGCACCA  -1231
-1230  CACCCGGTTTTACGTGTTGCTGGGGATTAAAGCAGGGCTTCATGCATGCTGGAGAAGGAC  -1171
-1170  TCTGCCATCCGAGCTACATCCTCAGCCCTGGGAAGTTTCTCTTAATAGAAGAGGAAACAG  -1111
-1110  AAGCTCCAAGAGACCAGGAGGCTGCCCTGGTCCCATGGTCAGTACAGGAGCCATGCAGAG  -1051
-1050  CCTTCTCATCCCTCCAGGCCACAGTACCCCACTTAGTCTGCTTTAAACCATTTCATTCCC  -991
 -990  ACCCAGAACTCATCTGTCAGTCAGAGATAGGTGCCCTCCCAGTGTTCACCCCCCAGCAAG  -931
 -930  AGGAGAGAACAGTCCTGTTGGAGGTGTCGGGGGAGGGGTATGCGGGGGGCCAAGAATGCAG  -871
 -870  GCACTCTCTGCCTAGGGGCCTGGGCTGGGCTGCCACAGAGGAGCCAGTGTGCCCAGACAT  -811
 -810  TGGTGCTTTGTGAGCCGAGGCTCGTGTGGGGACAGTTGTTTTCTTGACCTGACATTGTTT  -751
 -750  CAGGCTGACTGTCAGTGTCTTAGACCCTAGGGGCCAGCTCTCAGGAGCCAGTGGGGTCTT  -691
 -690  TGGGAGAGGGTAGCTAAAGTCCTCTAGTTAGGGGCAAGAAGAAGAGAGGACTAATGGGGT  -631
 -570  GACAATGGTAGGAGAGAGAAGATGTCTTGAGACTTGGGGCCAGAGAAGACCTGGCCATAG  -511
 -510  TGAAGAGTGGTGACAGAGGCCAGGCAGACATGTCCAGGGGTCAGTGCCCTCCTACCAGAG  -451
 -450  ACCTGGAAGTTCTGAGAACTGAGCAGTCAAGCCTTAGTTCACAAAGGCAGATCGGAGAAA  -391
 -390  TCCCCAGGGGCCTGAAGGGCCCAAAGACCTTAGGGACTGGACAGGGTGGGTGGGGTGGGC  -331
 -330  CTGATGCACAGCTGTCCAGATGTGGGTGAGAGATAACTCTGTTTATGGGGCACCCGGTCG  -271
 -270  GAGCAGGGCAAGGGGGTGGGGTGGGAAGGGGCTCCCGGCTGTGGGACGGGCAGATTAGGA  -211
 -210  GGCAGGGCATGAGTCAGGGGTTTGGGAGCCTGCCCCAGGGCACCTGCCTGTGCTGTTAGA  -151
 -150  GCAGTTGGGAGCTCAGCCAGTCACAGAGGGCACCAGTGGCGGCAGGCAGGACTTCCCGAA  -91
  -90  GACCCCCGAGGCTCCTGAGCAG(GTGGG..INTR #1 3.94KB..TCCAG)GGACCTGA  -61
  -60  GGTGCAGGTTATGCTGGCGGCCCAGAGAGCAGAGAACGACTGGACACTGAGGACGGCATC  -1
  +1  ATG GGG GAC ATG CGG(GTAAG..INTR #2...100BP..CACAG)GAC CAC GAG...3'
       MET GLY ASP MET ARG............................ASP HIS GLU...
```

Figure 6. Sequence of the 5' end of the mouse band 3 gene. The nucleotide sequence of the 5' end of the mouse band 3 gene is numbered by consecutive negative numbers relative to the A of the initiator AUG triplet of the mature band 3 protein (+1). The sequences of introns 1 and 2 are omitted from the numbering scheme. The clusters of S1-protected bands A–F are overlined and labeled accordingly.

immediately upstream of these sites. Most of these genes also have single or multiple copies of the sequence GGCGGG, which binds to the transcription factor Sp_1 (Dynan and Tijan, 1983). Although no perfect matches to the Sp_1 binding sequence are found within the 250-bp sequence upstream of the 3'-most initiation site of the band 3 gene, despite a G+C content of 65%, two possibly related G-rich sequence elements are located at positions −245 and −332, and contain imperfect tandem repeats of the sequence GGTGGG (Fig. 6). These elements are 70 bases upstream of the 5′-most transcriptional start sites F and D, respectively.

In MEL cells, which have been widely used as a model for erythroid gene regulation, the steady state level of band 3 mRNA increases concurrently with that of globin after induction (Kopito and Lodish, 1985b), and in parallel with the appearance of detectable band 3 polypeptide (Sabban et al., 1981; Patel and Lodish, 1984). The lack of overt similarity between the structure and usage of the promoters of band 3 and globin genes reinforces the observation that *cis*-acting elements 3′ to the globin gene mRNA cap site are responsible for erythroid-specific transcription (Charnay et al., 1984; Wright et al., 1984; Choi and Engel, 1986). Choi and Engel (1986) have recently identified a 400-bp Pvu II fragment, located ∼600 bp downstream of the chicken adult β-globin gene, which acts as an erythroid-specific enhancer. A 12-bp sequence within this fragment (TTTGCATCGAAC), in which the central 10 bp are palindromic, is also found within intron 14 of the mouse band 3 gene sequence (not shown).

Expression of the Band 3 Gene in Nonerythroid Tissues

The finding that renal and erythroid transcripts of the single-copy band 3 gene share a common 5′ upstream sequence, but initiate at different sites, establishes that this gene is expressed in the kidney. We used an antibody raised against a synthetic peptide corresponding to the COOH-terminus (anti-CT) of the mouse band 3 protein to localize the site of expression of the band 3 gene in thin sections of mouse kidney by indirect immunoelectron microscopy (Fig. 7). Rats were perfused with 2% paraformaldehyde, and their kidneys were removed, cut into 1-mm cubes, and infiltrated overnight in 30% sucrose. Thick sections (10 μm) were cut with a cryostat at −20°C, collected on gelatin-coated slides, and incubated with a 1:100 dilution of anti-CT for 2 h, followed by peroxidase-labeled goat anti-rabbit IgG. The reaction product was developed with diaminobenzidine; sections were then processed for electron microscopy by conventional techniques. Unstained sections were examined and no staining was observed on sections treated with preimmune serum (not shown).

Electron-dense staining is restricted to the basolateral surface of intercalated cells in the medullary collecting tubule (left). The intercalated cell is characterized by abundant mitochondria, subapical cytoplasmic vesicles, and limited basal plasma membrane infoldings. The adjacent principal cell (right), characterized by extensive basal plasma membrane infoldings, is negative for antibody binding.

These data are consistent with previous studies localizing proteins cross-reactive with polyclonal (Drenckhahn et al., 1985; Wagner et al., 1987) or monoclonal (Schuster et al., 1986) antibodies to rat (Drenckhahn et al., 1985) and human (Schuster et al., 1986; Wagner et al., 1987) erythroid band 3. Anion transporter was first immunolocalized in the avian kidney, albeit in proximal tubule (Cox et al., 1985).

The S1 nuclease protection analysis (Fig. 4 *top*) establishes that the authentic band 3 gene is expressed in mouse kidney mRNA. Northern blots of mouse kidney mRNA indicate that the high-stringency transcript detected with a band 3 cDNA probe migrates slightly faster than its erythroid counterpart (Alper et al., 1986), which suggests the possibility of alternative RNA processing. Wagner et al. (1987) have recently demonstrated reactivity of polyclonal but not monoclonal antibodies raised against the 43-kD NH$_2$-terminal fragment of human erythrocyte band 3 with a

110-kD polypeptide in human renal membranes. These results are also consistent with the suggestion that part of the band 3 cytoplasmic domain is expressed in kidney, and is possibly a consequence of alternative processing of this domain of the band 3 gene.

The identification in the mouse kidney of multiple band 3 cross-hybridizing mRNAs (Alper et al., 1986) supports the existence, in the mouse genome, of a family of genes homologous to band 3. The partial sequence of cDNAs encoding one of these homologues in human K562 cells (Demuth et al., 1986) reveals that the homology to

Figure 7. Immunoelectron micrographic localization of band 3 in kidney. A section was stained with anti-CT antiserum and developed with peroxidase-conjugated goat anti-rabbit antiserum at 1:100 as described in the text. × ~12,000.

the band 3 amino acid sequence varies from 30% at the NH_2 terminus to 80% in the membrane-spanning domain. This degree of similarity among the homologous genes indicates that antibody localization data must be interpreted with caution, since similar but distinct gene products may share some epitopes. In particular, of the 13 COOH-terminal amino acids to which the anti-CT antiserum was raised, 8 are identical to the corresponding domain of the human "nonerythroid" band 3 homologue (Demuth et al., 1986), leading to the prediction that this antibody is likely to recognize both proteins. Therefore, the band 3 antibody–cross-reactive proteins identified in renal collecting tubule cannot be unequivocally identified as the product of the erythroid gene. cDNA cloning and antipeptide antibodies raised against unique domains of the different band 3 family homologues will be necessary to fully characterize the expression of the multiple members of the anion-exchanger gene family in the kidney.

Acknowledgments

This work was supported by grants from the National Institutes of Health (RO1-GM-35011) and the Cystic Fibrosis Foundation (G122 6-01). R. Kopito is a Lucille P. Markey scholar in Biomedical Science.

References

Alper, S. L., R. R. Kopito, and H. F. Lodish. 1987. A molecular biological approach to the study of anion transport. *Kidney International*. 32(Suppl. 23):S117–S118.

Baron, M. H., and T. Maniatis. 1986. Rapid reprogramming of globin gene expression in transient heterokaryons. *Cell*. 46:591–602.

Bennett, V., and P. J. Stenbuck. 1979. The membrane attachment protein for spectrin is associated with band 3 in human erythrocyte membranes. *Nature*. 280:268–473.

Bennett, V., and P. J. Stenbuck. 1980. Association between ankyrin and the cytoplasmic domain of band 3 isolated from the human erythrocyte membrane. *Journal of Biological Chemistry*. 255:6424–6432.

Braell, A., and H. F. Lodish. 1981. Biosynthesis of the erythrocyte anion transport protein. *Journal of Biological Chemistry*. 256:11337–11344.

Braell, A., and H. F. Lodish. 1982. The erythrocyte anion transport protein is cotranslationally inserted into microsome. *Cell*. 28:23–31.

Breathnach, R., and P. Chambon. 1981. Organization and expression of eucaryotic split genes coding for proteins. *Annual Review of Biochemistry*. 50:349–383.

Cabantchik, Z. I., P. A. Knauf, and A. Rothstein. 1978. The anion transport system of the red blood cell: the role of membrane proteins evaluated by the use of "probes." *Biochimica et Biophysica Acta*. 515:239–302.

Chang, H., P. J. Langer, and H. F. Lodish. 1976. Asynchronous synthesis of erythrocyte membrane proteins. *Proceedings of the National Academy of Sciences*. 73:3206–3210.

Charnay, P., P. Mellon, and T. Maniatis. 1985. Linker scanning mutagenesis of the 5'-flanking region of the mouse β-major-globin gene: sequence requirements for transcription in erythroid and nonerythroid cells. *Molecular and Cellular Biology*. 5:1498–1511.

Choi, O.-R., and J. D. Engel. 1986. A 3' enhancer is required for temporal and tissue-specific transcriptional activation of the chicken adult β-globin gene. *Nature*. 323:731–734.

Chou, P. Y., and G. D. Fasman. 1978. Empirical predictions of protein conformation. *Annual Review of Biochemistry*. 47:251–276.

Cox, J. V., R. T. Moon, and E. Lazarides. 1985. Anion transporter: highly cell-type-specific expression of distinct polypeptides and transcripts in erythroid and nonerythroid cells. *Journal of Cell Biology*. 100:1548–1577.

Demuth, D. R., L. C. Showe, M. Ballantine, A. Palumbo, P. J. Fraser, L. Cioe, G. Rovera, and

P. J. Curtis. 1986. Cloning and structural characterization of a human non-erythoid band 3-like protein. *EMBO Journal.* 5:1205–1214.

Drenckhahn, D., K. Schluter, D. P. Allen, and V. Bennett. 1985. Colocalization of band 3 with ankyrin and spectrin at the basal membrane of intercalated cells in the rat kidney. *Science.* 230:1287–1289.

Dynan, W. S., and R. Tijan. 1983. The promoter-specific transcription factor SP1 binds to upstream sequences in the SV40 early promoter. *Cell.* 35:79–87.

Grinstein, S., S. Ship, and A. Rothstein. 1978. Anion transport in relation to proteolytic dissection of band 3 protein. *Biochimica et Biophysica Acta.* 507:294–304.

Knauf, P. A. 1986. Anion transport in erythrocytes. *In* Physiology of Membrane Disorders. Andreoli, T. E., J. F. Hoffman, D. D. Fanestil, and S. G. Schultz, editors. Plenum Medical Book Co., New York, NY. 191–220.

Kopito, R. R., and H. F. Lodish. 1985a. Primary structure and transmembrane orientation of the murine anion exchange protein. *Nature.* 316:234–238.

Kopito, R. R., and H. F. Lodish. 1985b. Structure of the murine anion exchange protein. *Journal of Cellular Biochemistry.* 29:1–17.

Kyte, J., and R. F. Doolittle. 1982. A simple method for displaying the hydropathic character of a protein. *Journal of Molecular Biology.* 157:105–132.

Liscum, L., J. Finer-Moore, R. M. Stroud, K. L. Luskey, M. S. Brown, and J. L. Goldstein. 1985. Domain structure of 3-hydroxy-3-methylglutaryl coenzyme A reductase, a glycoprotein of the endoplasmic reticulum. *Journal of Biological Chemistry.* 260:552–530.

Low, P. S. 1986. Structure and function of the cytoplasmic domain of band 3: center of erythrocyte membrane-peripheral protein interaction. *Biochimica et Biophysica Acta.* 864:145–168.

Marks, P. A., and R. A. Rifkind. 1978. Erythroleukemic differentiation. *Annual Review of Biochemistry.* 47:419–426.

Nathans, J., and D. S. Hogness. 1983. Isolation, sequence analysis, and intron-exon arrangement of the gene encoding bovine rhodopsin. *Cell.* 34:807–814.

Patel, V. P., and H. F. Lodish. 1984. Loss of adhesion of murine erythroleukemia cells to fibronectin during erythroid differentiation. *Science.* 224:996–998.

Sabban, E., V. Marchesi, M. Adesnik, and D. D. Sabatini. 1981. Erythrocyte membrane protein band 3: its biosynthesis and incorporation into membranes. *Journal of Cell Biology.* 91:637–646.

Schuster, V. L., S. M. Bonsib, and M. L. Jennings. 1986. Two types of collecting duct mitochondria-rich (intercalated) cells: lectin and band 3 cytochemistry. *American Journal of Physiology.* 251:C347–C355.

Strapazon, E., and T. L. Steck. 1976. Binding of rabbit muscle aldolase to band 3, the predominant polypeptide of the human erythrocyte membrane. *Biochemistry.* 16:2966–2971.

Wagner, S., R. Vogel, R. Lietzke, R. Koob, and D. Drenkhahn. 1987. Immunochemical localization of a band 3-like anion exchanger in collecting duct of human kidney. *American Journal of Physiology.* 253:F213–F221.

Walder, J. A., R. Chatterjee, T. L. Steck, P. S. Low, G. F. Musso, E. T. Kaiser, P. H. Rogers, and A. Arnone. 1984. The interaction of hemoglobin with the cytoplasmic domain of band 3 of the human erythrocyte membrane. *Journal of Biological Chemistry.* 259:10238–10246.

Chapter 15

Functional Roles of Carboxyl Groups in Human Red Blood Cell Anion Exchange

Michael L. Jennings, Matthew P. Anderson, and
Sally J. McCormick

*Department of Physiology and Biophysics, University of Texas
Medical Branch, Galveston, Texas; and the Department of
Physiology and Biophysics, The University of Iowa, Iowa City,
Iowa*

Introduction

The previous two chapters in this volume have described the physiology and kinetics of red blood cell Cl/HCO_3 exchange, and the structure and expression of the gene for the anion-exchange protein in erythroid and nonerythroid cells. This chapter concerns the transport pathway itself, and the relationships between the structure of the protein and its transport function. After a brief summary of some general issues related to band 3 structure/function relations, recent data from our laboratory on the role of carboxyl groups in anion exchange will be presented.

Oligomeric Structure

It is reasonably certain that no polypeptide other than the 95-kD band 3 is necessary for inorganic anion transport (Lukacovic et al., 1981). Given that band 3 is the only polypeptide required, the most basic question about structure/function relationships is that of how many band 3 subunits make up one functional unit of transport. Numerous studies using a variety of techniques have provided evidence that band 3 is a dimer or a tetramer (a double dimer, not a rotational tetramer) in the membrane (see Passow, 1986). The existence of the dimeric structure raises the possibility that the transport pathway consists of the interface between subunits, and that the catalytic cycle involves converted interactions between functioning subunits. However, each subunit has a single binding site for the stilbene disulfonate class of anion-transport inhibitors (Jennings and Passow, 1979). For covalently binding stilbene disulfonates such as DIDS and H_2DIDS, there is a linear relation between bound inhibitor and irreversible inhibition of transport (Wieth et al., 1982a; Passow, 1986). The reversible binding of stilbene disulfonates to the band 3 dimer does not show any evidence of positive cooperativity; in fact, for the bulkiest of the stilbene disulfonates, the binding displays negative cooperativity (Macara and Cantley, 1981). Therefore, the linear relation between covalent binding and irreversible inhibition implies that a subunit can transport anions normally, even when the other member of the dimer has been completely and irreversibly inactivated. Accordingly, the prevailing view of band 3 is that the oligomeric structure may be of importance in the assembly or stability of the molecule (see Boodhoo and Reithmeier, 1983), but the catalytic cycle for anion exchange does not require interactions between functioning subunits of a dimer.

Cytoplasmic and Membrane Domains

Each band 3 subunit consists of two distinct domains. A water-soluble cytoplasmic domain (43 kD) serves as an attachment site for the membrane skeleton (Bennett and Stenbuck, 1980; Low, 1986). The work of Lepke and Passow (1976) and Grinstein et al. (1978) some years ago showed that the cytoplasmic domain is not absolutely necessary for anion transport. However, it is an overstatement to say that the cytoplasmic domain has no role in the transport function of the protein. The detailed characteristics of anion exchange have not been determined in preparations that lack the cytoplasmic domain, and there is a preliminary report that the expression of the mRNA for the entire band 3 causes a higher anion flux in frog oocytes than does the mRNA for the membrane domain alone (Garcia et al., 1987). It is possible, then, that the cytoplasmic domain, though not essential for transport, does have some influence on transport. A physiological regulatory role is unlikely, because there is no evidence that red cell anion exchange is regulated in mature cells, and there is no theoretical reason to expect such regulation.

The COOH-terminal 52-kD membrane domain, which contains the catalytic machinery for anion exchange, is very hydrophobic and high in α-helix content (Oikawa et al., 1985). Most structural models of this part of the protein contain several membrane-spanning, hydrophobic segments connected by hydrophilic loops at one or the other membrane-water interface (Kopito and Lodish, 1985). The exact number and arrangement of the membrane crossings is unknown, but chemical modification and proteolysis experiments have provided direct evidence for at least eight portions of the sequence that cross the permeability barrier (Jennings et al., 1986). Because the COOH-terminus (Lieberman, D. M., and R. A. F. Reithmeier, manuscript submitted for publication) and the NH_2-terminus (Steck et al., 1978) of the protein are both cytoplasmic, there must be an even number of crossings.

Amino Acid Residues at the Stilbene Disulfonate Site

Much of what is known about structure/function relationships in band 3 is derived from chemical modification studies. The general approach that we and others have used is to perform chemical modification of the protein and examine the functional consequences of that modification in intact membranes (cells, resealed ghosts, or inside-out vesicles). The protein or fragments thereof can then be isolated in an attempt to determine which amino acid residue is associated with the functional change. This approach, using a variety of chemical modification reagents, has provided considerable information regarding the roles of particular amino acid residues in the catalysis of anion transport.

Many chemical modification studies have been devoted to modifying and labeling amino acid residues at the binding site for the stilbene disulfonate class of anion-transport inhibitors. The site can be reached by stilbene disulfonates only from the extracellular water (Kaplan et al., 1976; Barzilay and Cabantchik, 1979), and reversibly binding stilbene disulfonates are competitive inhibitors of anion exchange (Shami et al., 1978; Fröhlich, 1982). It is generally believed, therefore, that the exofacial anion-translocation site (i.e., the site occupied by a transported anion just before it is translocated inward across the permeability barrier) is contained in the stilbene disulfonate site (Passow, 1986; Fröhlich and Gunn, 1987). Although the site is accessible from the extracellular water, it appears to exist in a cleft that extends some distance from the actual outer surface of the membrane (Rao et al., 1979).

There are at least two lysine residues at the stilbene disulfonate site. One of these ("Lys a") is very likely to be either Lys 558 or Lys 561 (in the mouse sequence; Kopito and Lodish, 1985), and the other is located in the COOH-terminal 150 residues (Jennings et al., 1986). H_2DIDS can react covalently with both of these lysines to form an intramolecular cross-link (Jennings and Passow, 1979). Passow and co-workers (see Passow, 1986) have characterized the reactivity of Lys a and have shown that although the residue is allosterically influenced by a variety of ligands, the residue itself is not the binding site for substrate anions. When the two lysines are cross-linked by H_2DIDS, the transport is completely inhibited. However, when the same two lysines are cross-linked by the less bulky, uncharged residue of bis(sulfosuccinimidyl)suberate, the transport is not completely inhibited and can be reactivated by lowering the extracellular pH (Jennings et al., 1985). Therefore, neither of the lysine residues at the stilbene disulfonate site is absolutely necessary for anion transport. (Again, they may have a more subtle role in the process, without being essential.)

Arginine Residues

The extracellular pH dependence of monovalent anion exchange indicates that an arginine residue is of central importance in the catalysis of transport (Wieth and Bjerrum, 1982). Chemical modification studies with phenylglyoxal and cyclohexane-dione support this idea (Zaki, 1981; Bjerrum et al., 1983). When labeling with [^{14}C]phenylglyoxal is carried out under conditions in which only about one residue is modified per copy of the protein, the label is on the COOH-terminal third of the protein (Bjerrum et al., 1983). The exact location of the reactive arginine is not yet known, nor is the precise role of the arginine in transport. The simplest and most attractive hypothesis is that the positively charged side-chain forms a complex with transported anions and that the complex somehow crosses a permeability barrier. However, the rate of modification of the residue does not appear to be sensitive to the "inward-facing" vs. "outward-facing" conformation of the protein (Passow, 1986). These observations are hard to reconcile with a simple model of the protein in which the substrate anions bind to a site that contains a single arginine side-chain. Although the details are very far from being understood, it is likely that one or more arginine residues are very closely involved with the substrate anion-binding and translocation events.

Titratable Groups in the Acid pH Range

In addition to the evidence that positively charged residues are of significance in band 3–catalyzed anion exchange, there is also evidence that at least one carboxyl group is associated with the transport pathway. First, the extracellular pH dependence of monovalent anion exchange shows inhibition of transport resulting from protonation of a group of pK_a ~5 (Milanick and Gunn, 1982; Wieth et al., 1982a). Net influx of SO_4 into chloride-loaded cells is very strongly accelerated by extracellular protons (Milanick and Gunn, 1984). The pK_a of the proton activation depends on the SO_4 concentration in a manner that indicates binding of proton and sulfate in random order (Milanick and Gunn, 1984). The reciprocal pH dependences of monovalent and divalent anion exchange had earlier led Gunn (1973) to postulate that low pH converts the transport site from a state that transports Cl 40,000 times more rapidly than SO_4 to a state that transports Cl and SO_4 at much more similar rates. In keeping with the prediction of this titratable carrier model, the net influx of SO_4 into Cl-loaded cells is accompanied by a stoichiometric (at pH <7) influx of protons (Jennings, 1976). Therefore, the H ion that is bound at acid pH can be translocated by band 3. The pK_a of the proton binding that stimulates SO_4 influx and inhibits monovalent anion transport is consistent with the idea that a carboxylate is the titratable group. In addition to the carboxylate, there is evidence for involvement of a histidine residue in red cell anion exchange (Chiba et al., 1986; Matsuyama et al., 1986); the relationship between histidine(s) and the titratable carboxyl group is not yet clear.

Labeling of Carboxyl Groups

The first attempts to modify and label functionally important carboxyl groups in band 3 employed positively charged carbodiimides (Bjerrum, 1983; Craik and Reithmeier, 1984, 1985). These agents cause inhibition of anion transport, but the inhibition is partial unless very high concentrations of the carbodiimide are used. One limitation of the use of positively charged carbodiimides is that they may not have access to a carboxyl group that is located in a region of net positive charge. The carboxyl group that is involved in anion-proton cotransport may be in such an environment. As an alternative approach, we investigated Woodward's reagent K (N-ethyl-5-phenylisoxazolium-3'-sulfonate) as a modifier of carboxyl groups (Jennings and Anderson, 1987). We found that the reagent is a reasonably potent, irreversibly acting inhibitor of monovalent anion exchange. The chemistry of the interaction of Woodward's reagent K with carboxyl groups is summarized in Fig. 1 (Dunn et al., 1974; Jennings and Anderson, 1987). The alkali-catalyzed ring opening (not shown) is followed by reaction with a protein-bound carboxyl group to form the enol ester, which can then rearrange to the N-acyl derivative. We showed that it is possible to use [^{3}H]BH$_4$ as a secondary label of the modified carboxyl groups. If [^{3}H]BH$_4$ is added after the N-acyl shift has taken place, the reagent is labeled but still covalently bound to the protein. If [^{3}H]BH$_4$ is added before the N-acyl shift, the enol ester is reductively cleaved, leaving the aldehyde, which is then reduced to the primary alcohol (Jennings and Anderson,

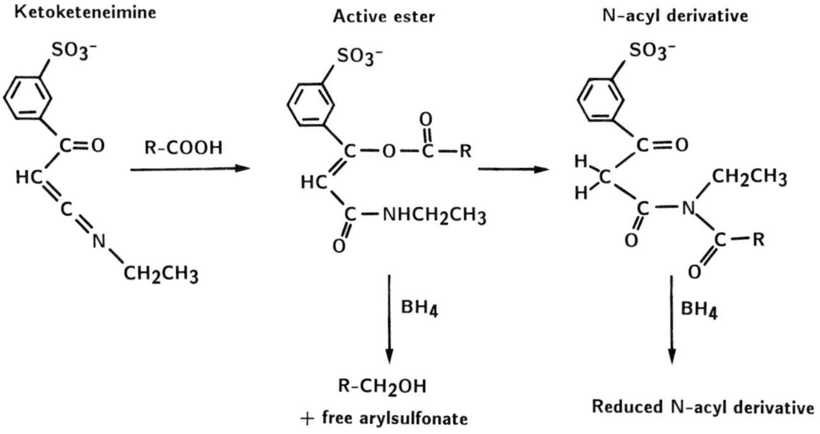

Figure 1. Partial scheme of the chemistry of the reaction of Woodward's reagent K with carboxyl groups. See text.

1987). That is, the original carboxyl group on the protein is converted to a hydroxyl group.

Polyacrylamide gel electrophoresis of the labeled membranes demonstrates that the method labels only extracellular residues (as indicated by the absence of spectrin labeling). Both chymotryptic fragments of band 3 are labeled, and the labeling of each is decreased by the presence of stilbene disulfonate during the exposure to Woodward's reagent K. Amino acid analysis of band 3 isolated from cells labeled under conditions in which the carboxyls are converted to alcohol demonstrates that glutamate, but not aspartate, residues are labeled (Jennings and Anderson, 1987).

From these studies, we concluded that there are at least two extracellular glutamate residues that are sufficiently near the stilbene disulfonate site that labeling is protected by bound DNDS or H$_2$DIDS. The experiments described below were designed to investigate further the functional role, if any, of these glutamate residues, and to compare our results using Woodward's reagent K with those of Bjerrum (1983) using carbodiimides.

Materials and Methods

Materials

Human blood was obtained either from the Lipid Research Clinic of the University of Iowa or from Dr. Claude Benedict, University of Texas Medical Branch. Woodward's reagent K, MOPS, and $NaBH_4$ were purchased from Sigma Chemical Co., St. Louis, MO. $[^3H]NaBH_4$ (319 mCi/mmol) and $[^{35}S]SO_4$ were purchased from New England Nuclear, Boston, MA. H_2DIDS was prepared as described previously (Jennings et al., 1984). Chymotrypsin and papain were from Boehringer-Mannheim, Indianapolis, IN. All other buffers and salts were from Fisher Scientific Co., Pittsburgh, PA.

Modification of Cells

Cells were washed three times in 150 mM KCl, 10 mM MOPS, pH 7.0, and resuspended at a 10% hematocrit in the same medium. The suspension was chilled on ice, and once the temperature was <2°C, solid Woodward's reagent K was added to a final concentration of 2 mM. After 10 min, $NaBH_4$ was added to a concentration of 2 mM from a 100-mM stock that had been freshly prepared. 5 min later, a second $NaBH_4$ addition was made. To prevent the pH from rising above ~7.2, 10 mM MOPS acid was added between the BH_4 additions. 5 min after the second BH_4 addition, the cells were washed twice in 150 mM KCl, 10 mM MOPS, pH 7. The above procedure was used with slight variations for all the chemical modifications (see figure and table legends). For the labeling studies, $[^3H]BH_4$ was used instead of BH_4, and the enzyme treatments, membrane isolation, and gel electrophoresis were as described previously (Jennings et al., 1984), except that no endo-β-galactosidase was used, and the membranes were not alkali-stripped.

Flux Measurements

The initial influx of $[^{35}S]SO_4$ was measured by resuspending the Cl-containing cells in media (20°C) consisting of 100 mM K_2SO_4, buffered with 10–20 mM MOPS and/or bis-Tris to pH values between 5.5 and 7.0. At timed intervals, 1 ml of the suspension was removed and mixed with 10 ml ice-cold KCl/MOPS. The combination of the Cl, temperature, and dilution effectively stops the influx, and the extracellular radioactivity was removed by washing, without loss of cells. The influx was calculated from the number of milliliter cells per milliliter of suspension, the intracellular radioactivity, and the initial extracellular specific activity. The proton influx associated with the sulfate influx was measured with the glass electrode, pH meter, and strip-chart recorder. The pH recordings were performed in media that had been purged of CO_2 by N_2 bubbling. The tracer influx measurements were done in an air atmosphere. (CO_2, of course, accelerates the Cl efflux, but, under our initial influx conditions, traces of CO_2 do not alter the SO_4 influx.)

Estimates of Conductance

Cells were treated for 10 min at 0°C either with 2 mM Woodward's reagent K alone or with the reagent followed by 2 mM BH_4. The cells were washed in HEPES-buffered physiological saline (5 mM K, 10 mM glucose) and loaded with ^{86}Rb by incubation for 90 min at 37°C with 5 μCi ^{86}Rb/ml suspension. The cells were then washed three times in 150 mM NaCl, 10 mM MOPS, pH 7, and resuspended at 20°C, 4% hematocrit, in either 150 mM NaCl, 10 mM MOPS/NaOH, pH 7, or 150 mM KCl, 10 mM

MOPS/NaOH, pH 7. At $t = 0$, valinomycin was added from a 1-mM ethanolic stock solution to a final concentration of 2 μM, and the initial efflux of ^{86}Rb was measured. H$_2$DIDS, when used, was present in the flux medium at a total concentration of 12.5 μM.

The ^{86}Rb efflux was assumed to follow the constant field equation. In the 150 mM KCl medium, the membrane potential is near 0 and the rate of ^{86}Rb efflux is a measure of P_{Rb} in the valinomycin-treated cells. In the 150 mM NaCl medium, the ^{86}Rb efflux is given by the following equation:

$$J_{Rb} = P_{Rb}[Rb_i][\ln B]/[B - 1],$$

where

$$B = [P_K K_i + P_{Cl}Cl_o]/P_{Cl}Cl_i.$$

The relative permeabilities P_K and P_{Cl} were calculated without assuming that P_{Rb} and P_K are equal. The above treatment is essentially identical to that used by Knauf et al. (1977) and Hunter (1977) for the special case of zero extracellular K. There is, of course, finite extracellular K in these experiments (coming out of the cells), but the progressive increase in extracellular K during the experiment did not appear to affect the membrane potential, as indicated by a very good exponential loss of ^{86}Rb over the times measured. Although the valinomycin method gives only a relative and semiquantitative estimate of P_{Cl} (and may be subject to systematic errors; see Bennekou and Christophersen, 1986), we believe the method is a reliable indicator of large effects of the chemical modification on P_{Cl}.

Results

Location of the Modified Residue Relative to the Papain Cleavage Site

Bjerrum (1983) has presented evidence that papain treatment of intact cells removes the carboxyl group that is modified by carbodiimides. Extracellular papain is known to cleave band 3 at two sites (Gln 569 and Gln 582) very near the chymotrypsin cleavage site (Tyr 572) and also at another extracellular site (Thr 647) nearer the COOH-terminus (Jennings et al., 1984). Papain may also cleave at sites farther toward the COOH-terminus, but not beyond the site of glycosylation, which in mouse band 3 is Asn 660 and in human band 3 is in a similar location (Jay, 1986). To determine whether papain removes residues that are labeled by Woodward's reagent K/BH$_4$, cells were labeled and then treated with papain. Fig. 2 shows that much of the radioactivity that had been in the 35-kD COOH-terminal chymotryptic fragment moves to the 28-kD papain fragment, which suggests that the main labeled residue is not removed from the membrane by papain. It is also noteworthy that very little label is in the 7-kD papain fragment, which contains an extracellular aspartate residue but no glutamate (Brock et al., 1983). The lack of label in this fragment is consistent with our finding that the main residues labeled by this method are glutamates, not aspartates (Jennings and Anderson, 1987).

We had shown some years ago that papain, although it inhibits monovalent anion exchange, actually accelerates stilbene disulfonate–sensitive SO$_4$ influx into Cl-loaded cells (Jennings and Adams, 1981). We interpreted this result as evidence that papain stabilizes the inward-facing conformation (and destabilizes the outward-facing conformation). The catalytic cycle for Cl/SO$_4$ exchange is limited by the inward SO$_4$

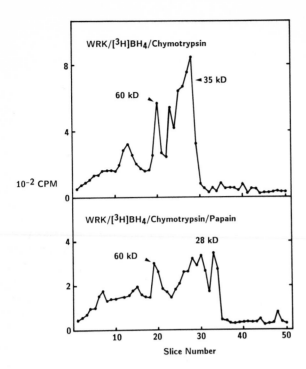

10⁻² CPM

Figure 2. Labeling of band 3 fragments by treatment of intact cells with 2 mM Woodward's reagent K (WRK) followed by 2 mM [³H]BH₄, pH 6.5–7.0, 0°C. (*Top*) Cells were treated with chymotrypsin before membrane isolation, solubilization, and sodium dodecyl sulfate–polyacrylamide gel electrophoresis. The positions of the NH₂ terminal 60-kD fragment and the leading edge of the glycosylated 35-kD fragment are indicated. (*Bottom*) Cells were treated with chymotrypsin followed by papain, and then processed further as in the upper portion of the figure. The position of the leading edge of the COOH-terminal papain fragment (28 kD) is indicated. The 7-kD fragment runs near the tracking dye, and it is difficult to determine whether the small amount of radioactivity in slice 48 is associated with the papain fragment of band 3.

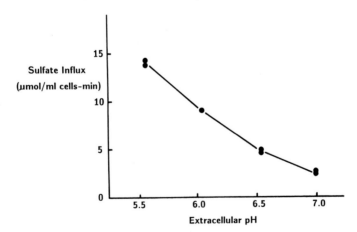

Figure 3. Extracellular pH dependence of SO_4 influx into Cl-loaded cells that had been pretreated with papain at pH 7. After papain treatment, cells were washed in 150 mM KCl, 10 mM MOPS, pH 7, and the initial influx of [³⁵S]SO_4 was determined at 20°C in a medium containing 100 mM K_2SO_4 buffered with 20 mM bis-Tris/MOPS at the indicated pH. The extracellular pH changed by at most 0.15 units during the 1-min flux measurement.

translocation step. In the papain-treated cells, the outward Cl translocation step is inhibited by a factor of ~10, but the Cl/SO_4 exchange rate is still limited by the much slower SO_4 influx. Therefore, papain does not inhibit the SO_4 influx and in fact accelerates it slightly (by a factor of 1.5–2). Because net Cl/SO_4 exchange is not inhibited in papain-treated cells, it is possible to determine whether papain alters the requirement for extracellular protons.

Fig. 3 shows that, even after papain treatment, lowering the extracellular pH strongly accelerates the net SO_4 influx into Cl-loaded cells. In addition, there is still a net proton influx associated with Cl/SO_4 exchange in the papain-treated cells (Fig. 4). The papain treatment was with 1 mg/ml for 1 h, 37°C, pH 7, which we previously showed is sufficient to affect well over 75% of the copies of band 3 (Jennings and Adams, 1981). We conclude that papain does not remove the carboxyl group that participates in proton-SO_4 cotransport.

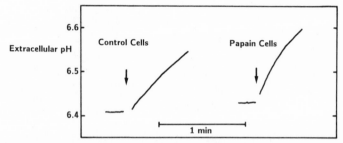

Figure 4. Proton influx associated with Cl/SO_4 exchange in control and papain-treated cells. Cells were washed and incubated with or without papain (1 mg/ml) in 150 mM KCl, 10 mM MOPS, for 1 h at 37°C, and then washed in the same medium. Cells were then resuspended in 100 mM K_2SO_4 buffered with 2 mM MOPS at pH 6.5, and the extracellular pH was monitored before and after adding the cells (final hematocrit, 1%). The medium had been bubbled vigorously with nitrogen for 30 min before addition of cells, and the CO_2 had been depleted from the cells by blowing N_2 gently over the final cell pellet before adding them to the SO_4 medium. Because of the higher temperature of the medium (27°C), the initial influx of protons is higher than the initial SO_4 influx at pH 6.5 in Fig. 2.

Possible Role of Carboxyl Groups in a "Molecular Zipper" Mechanism

The known facts about band 3 structure and function are consistent with a variety of physical pictures of the transport pathway itself. One such picture, termed the "molecular zipper," was put forth independently by two laboratories five years ago (Weith et al., 1982b; Brock et al., 1983). This class of models postulates that the transport pathway consists of an array of paired charges, and that anion translocation takes place as the transported anion trades places with a protein-bound negative charge. Fig. 5 represents one of many possible versions of the molecular zipper. In this version, for simplicity, the positive charges are fixed and the transported anion exchanges with a succession of protein-bound carboxyl groups. In other (kinetically equivalent) versions, the transported anion, complexed with a positive charge, moves to a position in which the transported anion is replaced by a protein-bound negative charge. The important features are that the model formally gives rise to the observed "ping-pong" kinetics (Gunn and Fröhlich, 1979), and the protein-bound negative

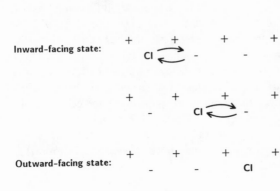

Figure 5. One version of the molecular zipper model of anion exchange. A Cl ion bound to the inward-facing state (top) changes places with a carboxyl group to form an intermediate conformation (middle), and then changes places again with another carboxyl group to produce the outward-facing state (bottom), from which the Cl can diffuse into the extracellular water. In this representation, the carboxyl groups appear to move rather large distances, but, in reality, the positive charges could move part of the way with the Cl ion, which could exchange for a carboxyl group without the carboxyl group itself moving more than a few angstroms.

charges are carboxyl groups. Because anion translocation involves the exchange of a transported anion for a protein-bound carboxyl group, conversion of the protein from the outward-facing to the inward-facing state should involve a shift of at least one carboxyl group outward relative to the protein-bound positive charges. That is, the carboxyl group and the transported anion should appear to move in opposite directions.

If either of the Woodward's reagent K–reactive carboxyl groups participate in a zipper-like mechanism such as that in Fig. 5, the accessibility of the carboxyl group might be expected to be greater when the protein is in the inward-facing state than when it is in the outward-facing state. To determine the effect of the band 3 conformation on the accessibility of the reactive carboxyl groups, cells were labeled with Woodward's reagent K/BH_4 in the presence of either an inward or an outward Cl gradient (Table I). In both cases, the extracellular medium is the same: 15 mM KCl, 10 mM MOPS, pH 7, osmotically balanced with sucrose. A 10-fold outward Cl gradient is produced simply by washing cells in a 150 mM KCl medium (the Cl distribution ratio is about unity at this pH) and suspending them in the 15 mM Cl medium. An inward Cl gradient is produced by first washing the cells in a Cl-free 100 mM SO_4 medium, with 10-min incubations between washes to allow the SO_4 to replace the intracellular Cl. Because of the intrinsic asymmetry of the transporter (see Knauf

TABLE I
Initial Intracellular and Extracellular Anion Concentrations during Labeling/"Recruitment" Experiments.

	Outward Cl gradient	
$Cl_i = 150$ mM		$Cl_0 = 15$ mM
	Inward Cl gradient	
$Cl_i < 0.1$ mM		$Cl_0 = 15$ mM
$SO_{4i} = 100$ mM		

et al., 1984; Knauf and Mann, 1984; Fröhlich and Gunn, 1986; Brahm, 1988), it is difficult to calculate the exact expected number of transporters in each state. It is clear, however, that there should be many more outward-facing states in the Cl-loaded than in the SO_4-loaded cells.

Table II shows that neither of the labeled carboxyl groups is protected when the protein is recruited into the outward-facing conformation. In fact, both residues are more exposed when the protein is in the outward-facing state. The enhanced labeling in the presence of the outward Cl gradient could have explanations other than the actual movement of carboxyl groups (see Discussion), but the experiment certainly provides no evidence that either of the reactive carboxyl groups takes part in a zipper-type exchange with a transported anion.

TABLE II
Effect of the Cl Gradient on the Labeling of the 60-kD and 35-kD Chymotryptic Fragments by Woodward's Reagent K/[^{3}H]BH$_4$.

Experiment	cpm (outward Cl gradient)/cpm (inward Cl gradient)	
	60-kD fragment	35-kD fragment
10-23	1.40	1.34
10-30	1.46	1.63
11-6	1.63	1.42
11-10	1.69	1.35

Cells were loaded with either Cl or SO_4 and resuspended at 5% hematocrit, 0°C, in 15 mM KCl, 30 mM MOPS, pH 7, 200 mM sucrose. Woodward's reagent K was immediately added to a final concentration of 0.25 mM from a 25-mM stock solution that had been prepared <60 s previously. The cells were incubated with reagent for 10 min, washed, and then incubated for 30 min on ice with 1 mM [^{3}H]BH$_4$. The cells were then washed and treated with chymotrypsin (and with endo-β-galactosidase in two of the experiments), and membranes were isolated, solubilized, and run on gels. The extent of the effect of the Cl gradient is represented as the ratio of counts per minute (cpm) per microgram membrane protein from cells labeled in the presence of an outward Cl gradient to that from cells labeled in the presence of an inward Cl gradient.

Effect of Woodward's Reagent K/BH$_4$ on Cl Conductance

Work in three laboratories has provided convincing evidence that the conductive Cl flux that is catalyzed by band 3 is largely the result of the "tunneling" of the anion through the permeability barrier in the absence of the usual conformational change associated with outward translocation (Fröhlich et al., 1983; Knauf et al., 1983; Kaplan and Passow, 1984). Fig. 6 summarizes the effects of Woodward's reagent K on the Cl conductance of the membrane. The reagent alone (10 min, 0°C) has little effect. However, reductive cleavage with BH$_4$ causes a 10-fold increase in the calculated P_{Cl}, all of which is inhibitable by 12 μM H$_2$DIDS in the flux medium. The reductive cleavage was performed under conditions in which the main modified residue is in the 35-kD COOH-terminal chymotryptic fragment (Fig. 2). The large effect of the modification on the Cl conductance is evidence that the modified carboxyl group normally constitutes part of the permeability barrier in the band 3 protein.

Discussion

The experiments presented here were designed to examine three aspects of the role of carboxyl groups in red cell anion exchange: the location of reactive carboxyl groups relative to papain cleavage sites; the effect of the Cl gradient on the availability of reactive groups; and the role of carboxyl groups in the permeability barrier to conductive anion movements.

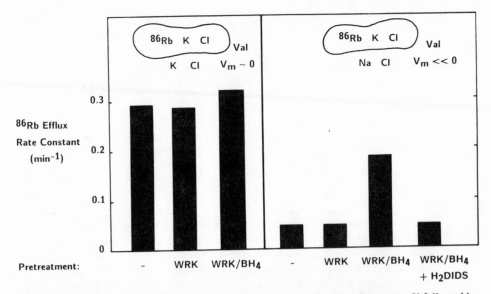

Figure 6. Effect of Woodward's reagent K alone (WRK), or Woodward's reagent K followed by BH$_4$ (WRK/BH$_4$), on the apparent Cl conductance of the red cell membrane. Cells were treated, as indicated, for 10 min at 0°C, with either 2 mM Woodward's reagent K alone or the reagent followed by 2 mM BH$_4$. The cells were washed in HEPES-buffered physiological saline (10 mM glucose) and loaded with ^{86}Rb. The cells were than washed in NaCl/MOPS, pH 7, and resuspended at 20°C, 4% hematocrit, in either 150 mM NaCl, 10 mM MOPS, pH 7, or 150 mM KCl, 10 mM MOPS, pH 7. At $t = 0$, valinomycin was added from a 1-mM ethanolic stock solution to a final concentration of 2 μM, and the initial efflux of ^{86}Rb was measured. H$_2$DIDS, when used, was present in the flux medium at a total concentration of 12.5 μM. The data depict the results of one experiment (two fluxes under each condition). The increase in P_{Cl} caused by Woodward's reagent K/BH$_4$ was confirmed in at least six separate experiments.

Effect of Papain on the Titratable Carboxyl Group

The experiments in Figs. 2–4 provide three independent pieces of evidence that the titratable group involved in proton-SO$_4$ cotransport is not removed from the membrane by extracellular papain under conditions in which the COOH-terminal 35-kD fragment is cleaved into 7-kD and 28-kD peptides. This conclusion differs from that of Bjerrum (1983), who presented evidence that the main carboxyl group labeled by carbodiimide/nucleophile is released from the membrane by papain. We believe that both experimental results are correct, but that the positively charged carbodiimide does not react with the carboxyl group that is involved in anion-proton cotransport. Our own labeling experiments cannot be interpreted quantitatively, because of the breadth of the 28-kD papain fragment (Fig. 2), but the flux studies (which do not involve any chemical modifications) demonstrate that the titratable carboxyl is preserved after papain treatment. Of course, we do not know exactly which carboxyl group is the titratable group, but it appears to be a glutamate residue in the 28-kD COOH-terminal fragment (Fig. 2).

Outward Cl Gradient Exposes Reactive Carboxyl Groups

The recruitment (by an outward Cl gradient) of transport systems into the outward-facing configuration does not protect either of the carboxyl groups that can be labeled by extracellular Woodward's reagent $K/[^3H]BH_4$. Therefore, neither of the reactive groups appears to move inward as the Cl moves outward. We thus find no evidence for zipper-type models in which the transported anion exchanges places with a protein-bound carboxyl group. Of course, there could be carboxyls that do exchange with the transported anion, but no such carboxyl group is labeled by our methods.

The kind of labeling/recruitment experiment presented in Table II has some intrinsic limitations that should be pointed out. First, we do not know the affinity of reversible binding of the reagent to band 3. If reversible binding precedes covalent reaction, and if reversible binding to the outward-facing state is of higher affinity than that to the inward-facing state, then the Cl gradient could affect the labeling, even if the immediate environment of the labeled residue is not affected by the translocation event. Second, the presence of the labeling reagent could distort the proportions of inward- and outward-facing states of the protein, or the agent could induce a conformation of the protein that is not identical to either the inward- or the outward-facing state (see Passow, 1986).

From the data in Table II it appears that the labeling of the carboxyl group in the 60-kD fragment may be slightly more sensitive to the Cl gradient than is the labeling of the 35-kD fragment. The random errors in this kind of experiment are considerable (the uncertainty in the listed ratios is probably ~20%), and there may also be systematic errors in the estimate of background counts in the two portions of the gel. We therefore hesitate to interpret the experiments as evidence that one or the other fragment is affected more by the Cl gradient. Separate experiments indicate that a glutamate residue on the 35-kD fragment is more likely to be the one associated with proton-SO_4 cotransport (Jennings et al., manuscript submitted for publication). The main point of the recruitment experiments is that we find no evidence for a carboxyl group that appears to participate in a "molecular zipper" type of exchange mechanism.

Relation to Conductance Pathway

The conversion by Woodward's reagent K/BH_4 of the carboxyl group to the alcohol accelerates the stilbene disulfonate–sensitive Cl conductance by a factor of ~10. This large effect on the conductance suggests that the reactive carboxyl group may actually constitute part of the permeability barrier. The acceleration of the conductive anion flux by the removal of a negative charge is further evidence in favor of the "tunneling" concept of the conductive anion flux through band 3 (Knauf et al., 1983; Fröhlich et al., 1983; Kaplan and Passow, 1984). A very simple mechanism for the accelerated net

Cl flux would be the removal of coulombic repulsion at a rate-determining step in the tunneling process.

If removal of the charge on the reactive carboxyl group accelerates conductance, then lowering the pH should have the same effect. The experimental data on this point are incomplete. Knauf et al. (1977) reported that lowering the extracellular pH from 7.1 to 6.8 nearly doubles P_{Cl}. This finding is consistent with the idea that a carboxyl group represents a coulombic barrier to net anion flow, but a thorough study of the effects of intracellular and extracellular pH on red cell Cl conductance has not yet been published. It should be emphasized that the reactive carboxyl group does not constitute the only barrier to net anion movement. Removal of the charge on this group does increase conductance by a large factor, but the Cl conductance of the modified band 3 is still far below that of most ion channels. Therefore, although we believe that the reactive carboxyl group does represent an important part of the barrier to net anion flow, the barrier is still substantial, even when the charge is removed.

How Many Carboxyl Groups Are There at the Stilbene Disulfonate Site?

At least two glutamate residues can be labeled by Woodward's reagent K/BH_4 in intact cells; both these residues are protected by stilbene disulfonate (Jennings and Anderson, 1987). The labeling is performed under conditions in which the reagent cannot penetrate the membrane (as indicated by the lack of spectrin labeling). Both the residues are therefore sufficiently close to the stilbene disulfonate site for their labeling to be affected by bound DNDS or H_2DIDS. We believe that one of these residues (in the COOH-terminal portion of the molecule) is responsible for proton-SO_4 cotransport and is part of the permeability barrier to conductive anion flux. We have no information regarding the functional role, if any, of the other glutamate residue. The extracellular pH dependences of both Cl and SO_4 transport in normal cells suggest that there is only one titration in the pH range 4–7 that has a major effect on fluxes (Milanick and Gunn, 1982, 1984). The only indication of a second extracellular titration in the acid range comes from experiments with cells modified with bis-sulfosuccinimidyl suberate (Jennings et al., 1985). This agent cross-links two amino groups at the stilbene disulfonate site but does not completely inhibit transport because it does not introduce as much bulk and negative charge as H_2DIDS. Cl-Br exchange in modified cells has a biphasic extracellular pH dependence in the range below 8; one protonation stimulates and one inhibits. We do not know the relationship between either of these titrations and the reactive carboxyls, but all our data are consistent with the presence of two carboxyls at the stilbene disulfonate site in band 3. The "second" carboxyl may be in such a positively charged environment in normal band 3 that it remains unprotonated at extracellular pH above 3.

Milanick and Gunn (1986) have made the novel suggestion that states of the protein are possible in which the anion-binding site and the titratable carboxyl group are on opposite sides of the permeability barrier. These states could come about from the asynchronous movement of the anion without the carboxyl. Our present data suggest that the carboxyl groups are normally recruited in the same direction as the transport site, but a small number of copies of band 3 with anion- and proton-binding sites facing opposite directions would not be detectable by our methods. Our experiments therefore do not argue either for or against such states of the protein, but asynchronous movement of anion- and proton-binding sites is a possibility that should be considered in future work on this system.

Summary

Chemical modification and proteolytic digestion have been used to examine three aspects of the role of carboxyl groups in the anion transport catalyzed by the red cell inorganic anion-exchange protein (known as band 3 or capnophorin). The modification employs the negatively charged Woodward's reagent K and BH_4, which together can convert protein carboxyl groups to alcohols (Jennings and Anderson, 1987). The reagent, with radioactive BH_4, can be used to label reactive carboxyl groups in intact cells under conditions in which only extracellular residues are labeled. Carboxyl groups in both major chymotryptic fragments are labeled, with most of the label in the 35-kD COOH-terminal fragment. There is significant labeling in the 28-kD papain fragment, but not in the 7-kD papain fragment. Papain itself does not inhibit the proton-SO_4 cotransport that takes place during Cl/SO_4 exchange. These results all indicate that the carboxyl group associated with proton-SO_4 cotransport is not removed by papain and is in the COOH-terminal 28-kD fragment. A set of experiments was performed to determine the effect of the Cl gradient on the labeling of carboxyl groups in intact cells. An outward Cl gradient increased the labeling of both the 60-kD and 35-kD chymotrytic fragments. This result provides evidence against the idea that either of the reactive carboxyl groups participates in a "zipper" transport mechanism, in which anions and protein-bound negative charge move in opposite directions. Finally, conversion of band 3 carboxyl groups to alcohols (again under conditions of minimal permeation of the reagent) causes a nearly 10-fold increase in stilbene disulfonate–sensitive Cl conductance. This indicates that a reactive carboxyl group constitutes part of the permeability barrier to net conductive anion transport through band 3. This carboxyl group may be the same as the one associated with proton-anion cotransport.

Acknowledgments

This work was supported by National Institutes of Health grant RO1 GM-26861 and National Institutes of Health Research Career Development Award DK-01137.

References

Barzilay, M., and Z. I. Cabantchik. 1979. Anion transport in red blood cells. III. Sites and sidedness of inhibition by high affinity reversibly binding probes. *Membrane Biochemistry.* 2:297–322.

Bennekou, P., and P. Christophersen. 1986. Flux ratio of valinomycin-mediated K fluxes across the human red cell membrane in the presence of the protonophore CCCP. *Journal of Membrane Biology.* 93:221–227.

Bennett, V., and P. J. Stenbuck. 1980. Association between ankyrin and the cytoplasmic domain of band 3 isolated from the human erythrocyte membrane. *Journal of Biological Chemistry.* 255:6424–6432.

Bjerrum, P. 1983. Identification and location of amino acid residues essential for anion transport in red cell membranes. *In* Structure and Function of Membrane Proteins. F. Quagliariello, and F. Palmieri, editors. Elsevier/North Holland, Amsterdam. 105–117.

Bjerrum, P. J., J. O. Wieth, and C. L. Borders, Jr. 1983. Selective phenylgly-oxalation of functionally essential arginine residues in the erythrocyte anion transport protein. *Journal of General Physiology.* 81:453–484.

Boodhoo, A., and R. A. F. Reithmeier. 1983. Characterization of matrix-bound band 3, the anion transport protein from human erythrocyte membranes. *Journal of Biological Chemistry.* 259:785–790.

Brahm, J. 1988. The red cell anion-transport system: kinetics and physiological implications. *In* Cell Physiology of Blood. R. B. Gunn and J. C. Parker, editors. The Rockefeller University Press, New York, NY. *Society of General Physiologists Series.* 43:000–000.

Brock, C. J., M. J. A. Tanner, and C. Kempf. 1983. The human erythrocyte anion-transport protein. *Biochemical Journal.* 213:577–586.

Chiba, T., Y. Sato, and Y. Suzuki. 1986. Amino acid residues complexed with eosin 5-isothiocyanate in band 3 protein of the human erythrocyte. *Biochimica et Biophysica Acta.* 858:107–117.

Craik, J. D., and R. A. F. Reithmeier. 1984. Inhibition of phosphate transport in human erythrocytes by water-soluble carbodiimides. *Biochimica et Biophysica Acta.* 778:429–434.

Craik, J. D., and R. A. F. Reithmeier. 1985. Reversible and irreversible inhibition of phosphate transport in human erythrocytes by a membrane impermeant carbodiimide. *Journal of Biological Chemistry.* 260:2404–2408.

Dunn, B. M., C. B. Anfinsen, and R. I. Shrager. 1974. Kinetics of Woodward's reagent K hydrolysis and reaction with staphlococcal nuclease. *Journal of Biological Chemistry.* 249:3717–3723.

Fröhlich, O. 1982. The external anion binding site of the human erythrocyte anion transport protein: DNDS binding and competition with chloride. *Journal of Membrane Biology.* 65:111–123.

Fröhlich, O., and R. B. Gunn. 1986. Erythrocyte anion transport: the kinetics of a single-site obligatory exchange system. *Biochimica et Biophysica Acta.* 864:169–194.

Fröhlich, O., and R. B. Gunn. 1987. Interaction of inhibitors on anion transporter of human erythrocyte. *American Journal of Physiology.* 252:C153–C162.

Fröhlich, O., C. Leibson, and R. B. Gunn. 1983. Chloride net efflux from intact erythrocytes under slippage conditions. Evidence for a positive charge on the anion binding/transport site. *Journal of General Physiology.* 81:127–152.

Garcia, A. M., R. Kopito, and H. F. Lodish. 1987. Expression of band 3 in *Xenopus laevis* oocytes microinjected with in vitro prepared mRNA. *Biophysical Journal.* 51:567a. (Abstr.)

Grinstein, S., S. Ship, and A. Rothstein. 1978. Anion transport in relation to proteolytic dissection of band 3 protein. *Biochimica et Biophysica Acta.* 507:294–304.

Gunn, R. B. 1973. A titratable carrier model for monovalent and divalent inorganic anions in red blood cells. *In* Erythrocytes, Thrombocytes, Leukocytes. E. Gerlach, K. Moser, E. Deutsch, and W. Wilmanns, editors. Georg Thieme, Stuttgart. 77–79.

Gunn, R. B., and O. Fröhlich. 1979. Asymmetry in the mechanism of anion exchange in human red blood cell membranes. Evidence for reciprocating sites that react with one transported anion at a time. *Journal of General Physiology.* 74:351–374.

Hunter, M. J. 1977. Human erythrocyte anion permeabilities measured under conditions of net charge transfer. *Journal of Physiology.* 268:35–49.

Jay, D. G. 1986. Glycosylation site of band 3, the human erythrocyte anion-exchange protein. *Biochemistry.* 25:554–556.

Jennings, M. L. 1976. Proton fluxes associated with erythrocyte membrane anion exchange. *Journal of Membrane Biology.* 28:187–205.

Jennings, M. L., and M. F. Adams. 1981. Modification by papain of the structure and function of band 3, the erythrocyte anion transport protein. *Biochemistry*. 20:7118–7122.

Jennings, M. L., M. Adams-Lackey, and G. H. Denny. 1984. Peptides of human erythroycte band 3 protein produced by extracellular papain cleavage. *Journal of Biological Chemistry*. 259:4652–4660.

Jennings, M. L., and M. P. Anderson. 1987. Chemical modification and labeling of glutamate residues at the stilbenedisulfonate site of human red blood cell band 3 protein. *Journal of Biological Chemistry*. 262:1691–1697.

Jennings, M. L., M. P. Anderson, and R. Monaghan. 1986. Monoclonal antibodies against human erythrocyte band 3 protein. *Journal of Biological Chemistry*. 261:9002–9010.

Jennings, M. L., R. Monaghan, S. M. Douglas, and J. S. Nicknish. 1985. Functions of extracellular lysine residues in the human erythrocyte anion transport protein. *Journal of General Physiology*. 86:653–669.

Jennings, M. L., and H. Passow. 1979. Anion transport across the erythrocyte membrane, *in situ* proteolysis of band 3 protein, and cross-linking of the proteolytic fragments by 4,4′-diisothiocya-no-dihydrostilbene-2,2′-disulfonate. *Biochimica et Biophysica Acta*. 554:498–519.

Kaplan, J. H., M. Pring, and H. Passow. 1983. Band 3-mediated anion conductance of the red blood cell membrane. *FEBS Letters*. 156:175–179.

Kaplan, J. H., K. Skorah, H. Fasold, and H. Passow. 1976. Sidedness of the inhibitory action of disulfonic acids on chloride equilibrium exchange and net transport across the human erythro-cyte membrane. *FEBS Letters*. 62:182–185.

Knauf, P. A., G. F. Fuhrmann, S. Rothstein, and A. Rothstein. 1977. The relationship between anion exchange and net anion flow across the human red blood cell membrane. *Journal of General Physiology*. 69:363–368.

Knauf, P. A., F.-Y. Law, and P. J. Marchant. 1983. Relationship of net chloride flow across the human erythrocyte membrane to the anion exchange mechanism. *Journal of General Physiology*. 81:95–126.

Knauf, P. A., F.-Y. Law, T. Tarshis, and W. Furuya. 1984. Effects of the transport site conformation on the binding of external NAP-taurine to the human erythrocyte anion exchange system. Evidence for intrinsic asymmetry. *Journal of General Physiology*. 83:683–701.

Knauf, P. A., and N. Mann. 1984. Use of niflumic acid to determine the nature of the asymmetry of the human erythrocyte anion exchange system. *Journal of General Physiology*. 83:703–725.

Kopito, R. R., and H. F. Lodish. 1985. Primary structure and transmembrane orientation of the murine anion exchange protein. *Nature*. 316:234–238.

Lepke, S., and H. Passow. 1976. Effect of incorporated trypsin on anion exchange and membrane proteins in human red blood cell ghosts. *Biochimica et Biophysica Acta*. 455:353–370.

Low, P. S. 1986. Structure and functino of the cytoplasmic domain of band 3: center of erythrocyte membrane-peripheral protein interactions. *Biochimica et Biophysica Acta*. 864:145–167.

Lukacovic, M. F., M. B. Feinstein, R. I. Sha'afi, and S. Perrie. 1981. Purification of stabilized band 3 protein at the human erythrocyte membrane and its reconstitution into liposomes. *Biochemistry*. 20:3145–3151.

Macara, I. G., and L. C. Cantley. 1981. Interactions between transport inhibitors at the anion binding sites of the band 3 dimer. *Biochemistry.* 20:5095–5105.

Matsuyama, H., Y. Kawano, and N. Hamasaki. 1986. Involvement of a histidine residue in inorganic phosphate and phosphoenolpyruvate transport across the human erythrocyte membrane. *Journal of Biochemistry.* 99:495–501.

Milanick, M. A., and R. B. Gunn. 1982. Proton-sulfate cotransport: mechanism of H^+ and sulfate addition to the chloride transporter of human red blood cells. *Journal of General Physiology.* 79:87–113.

Milanick, M. A., and R. B. Gunn. 1984. Proton-sulfate cotransport: external proton activation of sulfate influx into human red blood cells. *American Journal of Physiology.* 247:C247–C259.

Milanick, M. A., and R. B. Gunn. 1986. Proton inhibition of chloride exchange: asynchrony of band 3 proton and inhibitory sites? *American Journal of Physiology.* 250:C955–C969.

Oikawa, K., D. M. Lieberman, and R. A. F. Reithmeier. 1985. Conformation and stability of the anion transport protein of human erythrocyte membranes. *Biochemistry.* 24:2843–2848.

Passow, H. 1986. Molecular aspects of band 3 protein-mediated anion transport across the red blood cell membrane. *Reviews of Physiology, Pharmacology, and Biochemistry.* 103:61–203.

Rao, A., P. Martin, R. A. F. Reithmeier, and L. C. Cantley. 1979. Location of the stilbene disulonate binding site of the human erythrocyte membrane by resonance energy transfer. *Biochemistry.* 18:4505–4516.

Shami, Y., A. Rothstein, and P. A. Knauf. 1978. Identification of the Cl-transport site of human red blood cells by a kinetic analysis of the inhibitory effects of a chemical probe. *Biochimica et Biophysica Acta.* 508:357–363.

Steck, T. L., J. J. Koziarz, M. K. Singh, G. Reddy, and H. Köhler. 1978. Preparation and analysis of seven major, topographically defined fragments of band 3, the predominant transmembrane polypeptide of human erythrocyte membranes. *Biochemistry.* 17:1216-1222.

Wieth, J. O. and P. J. Bjerrum. 1982. Titration of transport and modifier sites in the red cell anion transport system. *Journal of General Physiology.* 79:253–282.

Wieth, J. O., O. S. Andersen, J. Brahm, P. J. Bjerrum, and C. L. Borders, Jr. 1982*a*. Chloride-bicarbonate exchange in red blood cells. *Philosophical Transactions of the Royal Society of London, Series B.* 299:383–399.

Wieth, J. O., P. J. Bjerrum, and O. S. Andersen. 1982*b*. The anion transport protein of the red cell membrane. A zipper mechanism of anion exchange. *Tokai Journal of Experimental and Clinical Medicine.* 7:91–101.

Zaki, L. 1981. Inhibition of anion transport across red blood cells with 1,2-cyclohexanedione. *Biochemical and Biophysical Research Communications.* 99:243–251.

Chapter 16

Mechanisms of Anion Net Transport in the Human Erythrocyte

O. Fröhlich and P. A. King

Department of Physiology, Emory University School of Medicine, Atlanta, Georgia

Cell Physiology of Blood © 1988 by The Rockefeller University Press

Introduction

The human erythrocyte membrane has a much higher permeability (conductance) for anions than for cations, to the extent that the erythrocyte membrane potential is determined by the transmembrane Cl concentration gradient. An approximate value for the permeability ratio is $P_{Cl}/P_K = 100$ (Hoffman and Knauf, 1973). This anion selectivity can be demonstrated electrophysiologically in the case of amphibian erythrocytes, which are sufficiently large to tolerate microelectrode impalements. In the case of the smaller mammalian erythrocytes, it can be demonstrated with fluorescent, potential-sensitive dyes. In *Amphiuma* erythrocytes, the membrane potential was measured electrophysiologically as approximately -15 mV (Hoffman and Laris, 1974; Stoner and Kregenow, 1980), which can be compared with an estimated Cl equilibrium potential in the range of -17 to -20 mV (Hoffman and Laris, 1974). When measured by means of a potential-sensitive dye, the *Amphiuma* erythrocyte membrane potential was -15 mV. Human erythrocytes exhibit a slightly smaller membrane potential: -5 to -8 mV, as determined with fluorescent probes. Again, this is very similar to the estimated Cl equilibrium potential of -9 mV (Hoffman and Laris, 1974; Lassen, 1977).

The high selectivity for anions is due mainly to the very low cation permeability of the erythrocyte membrane since the anion conductance is relatively low itself. Electrophysiological measurements have provided a relatively high value for the electrical resistance of *Amphiuma* erythrocytes of ~ 2–$3 \times 10^6 \, \Omega \cdot cm^2$ (Lassen, 1977). A similar value can be obtained with mammalian cells by independent methods using either cationophores or the electrogenic action of the Na pump. Such measurements (at 37°C) with sheep erythrocytes yielded a membrane conductance of $\sim 10^{-6} \, S \cdot cm^{-2}$ (Tosteson et al., 1973), and measurements with human erythrocytes gave a value of $1.0 \times 10^{-5} \, S \cdot cm^{-2}$ (Hunter, 1977; Hoffman et al., 1980). Similar values can be calculated from net Cl fluxes at a fixed, known potential. A typical Cl net flux value of ~ 10 mmol $\cdot$ (kg cell solids $\cdot$ min)$^{-1}$, measured at -90 mV, corresponds to a conductance of 6 pS per cell or $4 \times 10^{-6} \, S \cdot cm^{-2}$ at 20°C. When expressed in units of a permeability, values of $\sim 2 \times 10^{-8} \, cm \cdot s^{-1}$ have been obtained (Hunter, 1977; Knauf et al., 1977).

The Cl permeability value of $2 \times 10^{-8} \, cm \cdot s^{-1}$ is the value of the (conductive) permeability that is typically used in the Goldman-Hodgkin-Katz equation to correlate ion concentration gradients, permeabilities, and membrane potential. It has long been known, however, that when measured in unidirectional tracer flux experiments, the Cl fluxes and the calculated permeability values were several orders of magnitude higher (Tosteson, 1959; Wieth et al., 1973; Gunn and Fröhlich, 1979). Initially, it was not known whether the low rate of net flux relative to the unidirectional flux rate was the action of an exchange carrier mechanism, or whether it was a consequence of the electrically restrictive effect of the low cation permeability on the net movements of anions. It took the development of cation-selective ionophores such as valinomycin and gramicidin to demonstrate that the unidirectional permeability was indeed higher than the net permeability (Harris and Pressman, 1967; Scarpa et al., 1970; Hunter, 1971). The anion-transport mechanism is not considered to be a very tightly coupling one-for-one exchanger of extracellular for intracellular anions. At 37°C, the rate constant of anion tracer equilibration is ~ 50 ms (Brahm, 1977), corresponding to a flux rate of 150 $\cdot$ (kg cell solids $\cdot$ min)$^{-1}$. From this, one can calculate a unidirectional permeability of $3 \times 10^{-4} \, cm \cdot s^{-1}$, compared to the net or conductive permeability of

2×10^{-8} cm·s^{-1} mentioned previously. Since exchange transport is the major feature of this transport mechanism, it is the transport mode that has been studied the most extensively (for reviews see Knauf, 1979; Passow, 1986).

Control of Membrane Potential in Net Flux Experiments

Since its small size has rendered the mammalian erythrocyte refractory to extensive electrophysiological experimentation, the ionophores have been the primary tool to study the erythrocyte anion conductance. The principle behind using these ionophores is simple: to introduce a cation-selective conductance pathway into the erythrocyte membrane, which changes the membrane selectivity from anion selective to cation selective. As a consequence, the membrane potential approaches that of the cation (usually K) equilibrium potential. This in turn enables one to control the membrane potential independently of the anion concentrations, by applying known cation concentration gradients. Ideally, one would want to use an ionophore with a selectivity for one particular cation, e.g., K, which would permit convenient substitution of K for an inert cation such as Na. It is for this reason that the K-selective ionophore valinomycin has been used in anion net flux experiments to introduce the cation permeability.

Unfortunately, valinomycin does not provide a sufficiently high K permeability to clamp the membrane potential under all experimental conditions. Valinomycin can raise the permeability ratio, P_K/P_{Cl}, to ~20 (Knauf et al., 1977; Hunter, 1977; Freedman and Hoffman, 1979). Raising the valinomycin concentration above 5 μM will not further increase the K permeability, probably because of the limited solubility of valinomycin and saturating partitioning into the erythrocyte membrane (Hunter, 1974; Fröhlich et al., 1983). The valinomycin-induced K conductance is sufficient within the limits of a small membrane potential and with slowly conducting anions, but the usefulness of valinomycin clearly reaches its limits under more extreme conditions: (*a*) with very low K concentrations on one side of the membrane to achieve a membrane potential of more than ~50 mV; (*b*) under conditions of low extracellular anion concentrations, where the anion conductance is nearly 10 times higher (Fröhlich et al., 1983; Fröhlich, 1984); and (*c*) with more rapidly permeating anions such as Br or NO$_3$ (see Fig. 4, below). These limitations of valinomycin in clamping the membrane potential have been recognized from the beginning. To circumvent them, most studies with valinomycin have relied on an analysis of the permeability ratio, using the Goldman or the Goldman-Hodgkin-Katz equation and determining independently the valinomycin-induced K permeability from the unidirectional tracer permeability. While this type of analysis is adequate for many studies, it has the limitation that the latter permeability may not be a good measure of the conductive permeability used in calculating the membrane potential. The kinetics of valinomycin-mediated transport can be complex: the carrier saturates at high K concentrations and the conductance depends on the membrane potential (Stark and Benz, 1971). This can lead to systematic errors that impede accurate measurement of the cation permeability. Recent studies by Bennekou and Christophersen (1986) have confirmed this concern by demonstrating the non-independence of the unidirectional K fluxes.

An alternative to valinomycin is gramidicin. Gramicidin induces a significantly higher K permeability, up to 100-fold higher than valinomycin (Fröhlich et al., 1983). This ensures that the membrane potential is closer to the K equilibrium potential, as long as impermeant cations such as choline or N-methyl-glucamine are used as a substitute for K. Although it still needs to be demonstrated, for example with membrane potential probes (Freedman et al., 1988), one would only expect minor deviations from the theoretical potential.

The Components of Anion Net Flux

When compared with the high rates of anion exchange, anion net transport appears to be only a minor transport mode of the transport mechanism. Nevertheless, it is sufficiently large to dominate the total membrane conductance of the intact cell and thus constitutes a functionally relevant feature. Cl net flux can readily be divided into two major components. This distinction is made by the sensitivity to the "classic" inhibitors of anion exchange, the stilbene disulfonate derivatives DIDS, SITS, and DNDS. There are a stilbene-inhibitable and a stilbene-insensitive component of anion net flux. Typically, under conditions of normal intracellular and extracellular Cl concentrations, ~60% of Cl net efflux is inhibited by these drugs while 40% remains unaffected (Knauf et al., 1977, 1983; Kaplan et al., 1983; Fröhlich et al., 1983). This stilbene sensitivity defines, at least as a first approximation, about two-thirds of anion net transport as mediated by band 3, the anion transport protein, and about one-third as nonspecific or unassigned. Next we will discuss the stilbene-sensitive component, and at the end we will return to the stilbene-insensitive component.

Initially, it had been presumed that Cl net transport by the band 3–specific pathway occurred by a classical carrier mechanism. In such a mechanism, the substrate anion is translocated by the transport site in one direction, followed by a return translocation of the unloaded transport site. Since net transport was so much slower than exchange transport, the term "slippage" has been used to describe the occasional translocation of the unloaded site, after the more general description by Vestergaard-Bogind and Lassen (1974) of the anion conductance as a "slip in the coupling mechanism."

The carrier-type or slippage mechanism, however, is not the sole contributor to the stilbene-sensitive net flux. It has been known for several years that Cl net efflux does not behave in the manner expected for a simple carrier mechanism. Kaplan et al. (1980, 1983) and Knauf and co-workers (Knauf and Law, 1980; Knauf et al., 1983) found that at equal extracellular and intracellular Cl concentrations, Cl net efflux increased with increasing Cl. The slippage mechanism cannot account for this concentration dependence since it predicts that with increasing substrate concentrations the fraction of unloaded sites decreases and with it the overall rate of translocation of unloaded sites. At physiological Cl concentrations (150 mM), therefore, another mechanism must be the major contributor to the anion conductance.

The same conclusion has been drawn for other Cl concentrations under conditions that should favor net transport by translocation of the unloaded transport site. The most favorable condition would be $Cl_o = 0$, when essentially all transport sites are unloaded and in the outward-facing state (Fröhlich and Gunn, 1986) and provide the maximal probability of anion net efflux via inward translocation of the unloaded site. Even then, however, slippage appears to provide only a small contribution to net efflux (Fröhlich et al., 1983). This was demonstrated in two ways. (*a*) Since translocation of the unloaded site is so slow, the overall rate of net transport by the slippage mechanism should be limited by this step and should therefore be independent of the chemical nature of the transported anion in the remainder of the transport cycle. The experiment showed considerable differences in the conductance for Cl, Br, and NO_3 at zero as well as at high extracellular anion concentrations (Fröhlich, 1984). (*b*) The slow rate of inward translocation of the unloaded site should be easily matched by the outward translocation of only a few rapidly transported, loaded sites, resulting in an apparent high transport affinity (low K_m) of Cl_i for net efflux. The experiments showed a nearly linear dependence on Cl_i up to at least 200 mM, which suggests the presence of a low-affinity pathway and an undetectably small contribution of slippage (Fröhlich, 1984).

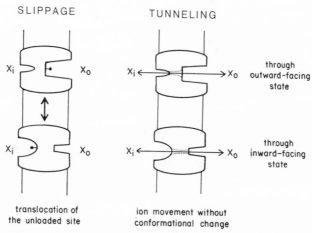

Figure 1. Schematic representation of the two band 3–specific anion-conductance mechanisms. On the left side, the slippage mode relies on the carrier-type conformational changes that reorient the translocation site between the outward-facing and inward-facing states. A net transport cycle is completed when an anion is bound on one side (e.g., on the inside to the inward-facing transport site), translocated to the extracellular space, and released, followed by a return translocation of the unloaded transport site. The scheme on the right side depicts the tunneling mechanism. In this mode, the anion moves through the transport protein as if through a channel; the conformational state of the transporter and the orientation of the transport sites remain the same. Consequently, tunneling can occur through either the outward-facing or the inward-facing state; the outward-facing state has a ~10-fold-higher tunneling rate than the inward-facing state (Fröhlich, 1984; see also Fig. 4 and Table II). Drawing adapted from Gunn et al. (1988).

What is the nature of this net transport pathway that does not follow carrier kinetics? The names suggested for this process vary among different laboratories, but the proposed processes are quite similar: during net transport, the anion moves through the transport protein without the carrier-type conformational changes but rather in the way envisioned for movement through a channel-like pathway (see Fig. 1). We have referred to this net transport mode as the tunneling mechanism, to indicate the channel-like nature of the process (Fröhlich et al., 1983; Fröhlich, 1984). In using the terms "channel" or "channel-like," one has to realize, however, that the calculated transport rates of the tunneling mode are considerably slower than those of true channels: at 20°C, the rate for Cl is only 3–30 s^{-1}, depending on the conformational state of the transporter (see Table II).

Kinetic Relationship Between the Tunneling and the Exchange Mechanisms

The fact that the anion conductance is largely mediated by the anion transport protein has been clearly documented by Knauf et al. (1977). They showed that the conductance was 60–70% inhibited after reaction with the irreversible stilbene disulfonate inhibitor DIDS. Furthermore, they showed that when the red cells were titrated with DIDS to different levels of saturation, there was a linear relationship between the degree of inhibition of Cl exchange and of Cl net efflux. This means that every time one DIDS molecule reacts with one band 3 molecule, one exchange transport unit and one net transport unit are eliminated. We examined the correlation between stilbene binding and net transport inhibition further, by testing whether the inhibitory action of the reversibly acting stilbene analogue DNDS occurred with the same affinity as exchange inhibition and DNDS binding. The latter two processes had been characterized previously in detail (Fröhlich, 1982), so all kinetic parameters are known, albeit at a different temperature (0°C vs. 20°C). We expect that in the absence of Cl_o, DNDS binds and inhibits with an affinity of 0.1 μM, and its apparent inhibitory constant

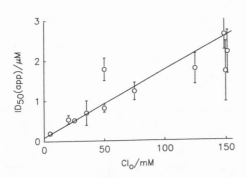

FIGURE 2. Interaction between Cl_o and the reversible transport inhibitor DNDS on the Cl net efflux mechanism. The half-inhibitory concentration of DNDS increases as Cl_o is increased, as expected for a competitive relationship between Cl_o and DNDS. The solid line represents the relationship between the ID_{50} of DNDS and Cl_o that had been determined in Cl exchange and DNDS binding experiments (Fröhlich, 1982).

increases to 2.6 μM as Cl_o is raised to 150 mM. Fig. 2 shows that the inhibitory potency of DNDS or Cl net efflux is influenced by Cl_o, in the manner expected for competition between DNDS and Cl_o. The simplest explanation for this competition is that Cl_o and DNDS compete for a common binding site; when DNDS binds, outward tunneling of intracellular Cl is blocked, and when Cl_o binds, it causes a shift into the inward-facing state to which DNDS does not bind. Whether Cl, when it is bound to the outward-facing transport site, also blocks outward tunneling in the same manner as DNDS, is not known.

Another inhibitor that can be used to compare the exchange and the tunneling pathway is phloretin. Phloretin is a mixed or noncompetitive inhibitor of Cl exchange with an inhibitory constant of ~1.5 μM (Fröhlich and Gunn, 1987). A noncompetitive inhibitor does not significantly interfere with the binding of the substrate. This means that phloretin must necessarily prevent the carrier-type conformational change of the Cl-loaded site that leads to Cl exchange. How does phloretin affect Cl net efflux? Fig. 3 shows that Cl net efflux is not affected by phloretin but is inhibited by DNDS. We have examined the effect of phloretin up to 100 μM on Cl net efflux into media containing 0 or 150 mM Cl and have found no detectable effect. These are conditions under which phloretin should cause significant inhibition of Cl exchange (Fröhlich, O., D. Bain, and L. Weimer, manuscript submitted for publication). While this discrepancy between Cl exchange and net transport may seem surprising, it is just what one would expect if net transport is not mediated by the conformational changes that inhibit Cl exchange. Net transport by tunneling is thought to be analogous to channel-like substrate movement and would not exhibit the same phloretin sensitivity as the carrier-type conformational changes.

To ascertain that phloretin actually binds to the anion transporter under these experimental conditions, we also examined the interaction of DNDS and phloretin on the anion transporter. We had shown before that DNDS and phloretin are mutually exclusive inhibitors of Cl exchange, which means that they compete in their binding to band 3 (Fröhlich and Gunn, 1987). If phloretin prevents DNDS from binding while by itself it does not affect Cl net efflux, it ought to protect net efflux against inhibition by DNDS. Indeed, we observed that increasing the phloretin concentration from 0 to 20

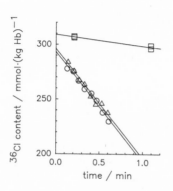

Figure 3. Effect of DNDS and phloretin on Cl net efflux from gramicidin-treated cells. Cl net efflux was measured as tracer Cl efflux into Cl-free media that had been carefully bubbled with N_2 to remove bicarbonate contaminations. Also, 100 μM acetazolamide was included to inhibit carbonic anhydrase and slow the interconversion between CO_2 and HCO_3 and thus stop any residual Cl-HCO_3 exchange. The efflux medium contained 4 mM K for a membrane potential of about -90 mV; 20°C, pH 7.0. Open circles: control efflux; triangles: in the presence of 20 μM phloretin; squares: in the presence of 50 μM DNDS.

μM in the presence of low micromolar concentrations of DNDS partially restored the DNDS-depressed net flux to uninhibited levels (Fröhlich et al., manuscript submitted for publication).

In conclusion, these experiments demonstrate the close kinetic relationship between Cl exchange and (DNDS-inhibitable) Cl net efflux, suggesting common structural elements for these two pathways. In particular, it is the quantitative similarity by which DNDS inhibits exchange and net efflux that suggests that there is one site to which DNDS binds and that upon binding it inhibits both transport modes.

Quantitative Evaluation of the Relative Contributions of Tunneling and Slippage to Chloride Net Efflux

Our previous experiments have demonstrated that most of Cl net efflux occurs by the tunneling mechanism. An estimate of the contribution of the slippage mechanism (net transport via translocation of the unloaded transport site) was only possible in the form of an upper limit, derived from estimates of the experimental errors in the measurements. Under optimal conditions for slippage, namely at $Cl_o = 0$, the upper-limit estimate is 5–10 mmol $\cdot$ (kg Hb $\cdot$ min)$^{-1}$ or 5–10% of the total Cl net efflux (Fröhlich, 1984; Fröhlich et al., manuscript submitted for publication).

To obtain a closer estimate of the translocation rate of the unloaded transport site, one needs to use an anion with a lower tunneling rate. Since the slippage rate is independent of the transported anion, a slower overall net transport rate means that the tunneling rate is slower. We therefore chose SO_4, whose conductive permeability (at isotonic intracellular and extracellular concentrations) has been reported to be approximately five times slower than that of Cl (Knauf et al., 1977).

We measured SO_4 net efflux into anion-free media (citrate/sucrose substitution) that had been carefully bubbled with N_2 to remove as much CO_2 as possible to minimize the possibility of SO_4/HCO_3 exchange. Indeed, we found that DNDS-sensitive SO_4 net efflux consisted of two components, a component that could be inhibited by phloretin and a phloretin-insensitive component (King and Fröhlich,

TABLE I
SO_4 Net Efflux from Erythrocyte Ghosts, Measured as [^{35}S]SO_4 Efflux into Citrate/Sucrose Media

$[SO_4]_i$	120 mM	250 mM	340 mM
Total SO_4 net efflux	1.46 ± 0.04	2.05 ± 0.18	2.27 ± 0.03
+100 μM phloretin	0.69 ± 0.10	1.30 ± 0.20	1.50 ± 0.17
+150 μM DNDS	0.30 ± 0.07	0.60 ± 0.10	0.69 ± 0.08
Phloretin-sensitive flux	0.77 ± 0.10	0.76 ± 0.26	0.77 ± 0.17
DNDS-sensitive flux	1.16 ± 0.08	1.44 ± 0.18	1.58 ± 0.08

The membrane potential was clamped near zero by keeping the intracellular and extracellular K concentrations the same and using gramicidin to induce a high cation conductance (data from King and Fröhlich, manuscript submitted for publication). 20°C, pH 7.0.

1988). These results (Table I) have two implications: they demonstrate the existence of the two predicted components of SO_4 net efflux, and they confirm their identity by characterizing their dependence on the intracellular SO_4 concentration. Let us consider first the phloretin-inhibitable component. If this component is due to slippage, it should depend on the intracellular SO_4 concentration with approximately the same apparent affinity as SO_4 exchange (~35 mM; Schnell and Besl, 1984) since activation of exchange and net efflux involve the same binding step. This means that over the concentration range of our study (120–340 mM), SO_4 net efflux should be nearly saturated and appear essentially independent of the intracellular SO_4 concentration. The phloretin-insensitive tunneling component, on the other hand, would vary with the intracellular concentration. If it behaves like Cl tunneling (Fröhlich, 1984), it should depend nearly linearly on the SO_4 concentration. These expectations are borne out by the data of Table I. It therefore appears that the phloretin-sensitive SO_4 net efflux component gives a closer estimate of the rate at which the unloaded anion transport site is translocated (to be precise, from the outward-facing to the inward-facing state). This permits us to compile a list of the relative rates of anion transport in the different transport modes and with different substrates (Table II).

On the Nature of the Stilbene-insensitive Anion Net Flux Component

It was mentioned above that at "normal" Cl concentrations, about two-thirds of Cl net efflux can be inhibited by the stilbene disulfonate inhibitors and that one-third is insensitive to these drugs. It is therefore almost by definition that the stilbene-sensitive component was assigned to the anion transporter and the insensitive component was labeled nonspecific (Knauf et al., 1977, 1983). Our experiments described above confirm the assignment of the stilbene-sensitive net flux component to the anion-exchange mechanism. However, the question of which protein mediates the stilbene-insensitive component remains.

TABLE II
Turnover Rates per Band 3 Molecule of the Different Transport Modes of the Red Cell Anion Transporter

	s^{-1}
Cl exchange*	13,000
SO$_4$ exchange[‡]	0.7
Inward translocation of unloaded site[‡§]	0.2
Outward Cl tunneling through the outward-facing state[‖]	30
Outward Cl tunneling through the inward-facing state[‖]	3
Outward SO$_4$ tunneling[‡§]	1

The turnover rates were calculated from flux rates by assuming 1×10^6 molecules band 3 · cell^{-1}: 0.3 mmol · (kg Hb · min)$^{-1}$ = 1 (band 3)$^{-1}$ · s^{-1}.
*From Brahm (1977).
[‡]From King and Fröhlich (manuscript submitted for publication).
[§]From Table I.
[‖]From Fröhlich et al. (manuscript submitted for publication).

Compared to the other known anion-transporting pathways of the erythrocyte membrane, namely the Na-K-Cl and the K-Cl cotransporters, this stilbene-insensitive conductance is overwhelming. While stilbene-insensitive Cl net efflux is ~5 mmol · (kg Hb · min)$^{-1}$ (at about -100 mV), the cotransporter-mediated fluxes are nearly two orders of magnitude slower, in the range of <1 mmol · (kg Hb · h)$^{-1}$ (Dunham and Benjamin, 1984). Furthermore, the stilbene-insensitive conductance exhibits an anion selectivity quite different from that known for the cotransport systems: the conductance rates have the sequence Cl < Br < NO$_3$, whereas the cotransport rates have the sequence NO$_3$ ≪ Br < Cl. However, it is notable that the anion selectivity is the same for the stilbene-inhibitable and the stilbene-insensitive net fluxes. This has been noted in the past (Knauf et al., 1979; Fröhlich, 1984). Fig. 4 shows the results of more recent experiments, where we measured net efflux for Cl, Br, I, NO$_3$, all under the same general experimental conditions. As observed previously, the selectivity pattern of the anion conductance in high-anion media is very similar in the presence and absence of DNDS. In addition, these data show that the selectivity is also the same for the two stilbene-sensitive net efflux components into zero- and high-anion media. This means

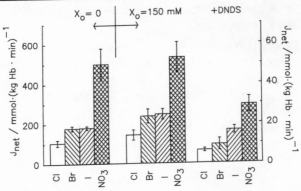

Figure 4. Comparison of the net efflux rates for different anions under three different conditions: (*a*) in the absence of extracellular exchangeable anions (substitution by isotonic citrate/sucrose), in which case the anion transporter is almost exclusively in the outward-facing state and the rate of tunneling through this state is determined; (*b*) in the presence of isotonic concentrations of the respective anions, in which case the anion transporters are mostly in the inward-facing state; (*c*) in the presence of 50–100 μM DNDS, in which case the anion transporters are forced into the outward-facing state, with their tunneling pathway blocked. Net efflux was determined as K net loss from gramicidin-treated cells under the same conditions as elsewhere: K_o = 4 mM, 20°C, pH 7.0 (Fröhlich et al., manuscript submitted for publication).

that the anion selectivity is the same for tunneling through the outward-facing and inward-facing conformations, respectively, of the anion transporter. This similarity does not exclude the possibility that an additional anion-conductance pathway exists in the erythrocyte membrane that is independent of the band 3 protein. Nevertheless, one is tempted to ask whether the stilbene-insensitive anion conductance is mediated by the anion-transport protein as well. If it is, this would have interesting structural implications for the anion-exchange and tunneling pathways in the band 3 protein. It could mean that the stilbenes, while binding to the anion-transport site, do not completely obstruct the tunneling pathway. Alternatively, it could mean that more than one tunneling pathway exists in the protein, only one of which is blocked by stilbene binding. Could these two pathways or portions thereof play a role in the alternating access of the anion-transport site that gives the anion exchanger the distinct carrier-type kinetics? Before such questions can be approached, however, more work is necessary to determine which protein is actually responsible for the stilbene-insensitive anion conductance.

Acknowledgments

We appreciate the excellent technical help of D. Bain in some of these experiments. This work has been supported in part by grant GM-31269 and post-doctoral fellowship HL-07342 from the National Institutes of Health.

References

Bennekou, P., and P. Christophersen. 1986. Flux ratio of valinomycin-mediated K+ fluxes across the human red cell membrane in the presence of the protonophore CCCP. *Journal of Membrane Biology.* 93:221–227.

Brahm, J. 1977. Temperature-dependent changes of chloride transport kinetics in human erythrocytes. *Journal of General Physiology.* 70:283–306.

Dunham, P. B., and M. A. Benjamin. 1984. Cl^--dependent cation transport in mammalian erythrocytes. *Federation Proceedings* 43:2476–2478.

Freedman, J. C., and J. F. Hoffman. 1979. The relationship between dicarbocyanine dye fluorescence and the membrane potential of human red blood cells set at varying Donnan equilibria. *Journal of General Physiology.* 74:187–212.

Freedman, J. C., E. M. Bifano, L. M. Crespo, P. R. Pratap, R. W. Walenga, R. E. Bailey, S. Zuk, and T. S. Novak. 1988. Membrane potential and the cytotoxic Ca cascade of human red blood cells. *In* Cell Physiology of Blood. R. B. Gunn and J. C. Parker, editors. The Rockefeller University Press, New York. *Society of General Physiologists Series.* 43:217–231.

Fröhlich, O. 1982. The external anion binding site of the human erythrocyte anion transporter: DNDS binding and competition with chloride. *Journal of Membrane Biology.* 65:111–123.

Fröhlich, O. 1984. Relative contributions of the slippage and tunneling mechanisms to anion net efflux from human erythrocytes. *Journal of General Physiology.* 84:877–893.

Fröhlich, O., and R. B. Gunn. 1986. Erythrocyte anion transport: the kinetics of a single-site obligatory exchange system. *Biochimica et Biophysica Acta.* 864:169–194.

Fröhlich, O., and R. B. Gunn. 1987. Interactions of inhibitors on the anion transporter of the human erythrocyte. *American Journal of Physiology.* 252:C153–C162.

Fröhlich, O., C. Leibson, and R. B. Gunn. 1983. Chloride net efflux from intact erythrocytes under slippage conditions. Evidence for a positive charge on the anion binding/transport site. *Journal of General Physiology.* 81:127–152.

Gunn, R. B., and O. Fröhlich. 1979. Asymmetry in the mechanism for anion exchange in human red blood cell membranes. Evidence for reciprocating sites that react with one transported anion at a time. *Journal of General Physiology.* 74:351–374.

Gunn, R. B., O. Fröhlich, P. A. King, and D. G. Shoemaker. 1988. Anion transport. *In* Red Blood Cell Membranes: Structure, Function and Clinical Implications. J. C. Parker and P. Agre, editors. Marcel Dekker, New York. In Press.

Harris, E. J., and B. C. Pressman. 1967. Obligate cation exchanges in red cells. *Nature.* 216:918–920.

Hoffman, J. F., and P. C. Laris. 1974. Determination of membrane potentials in human and *Amphiuma* red blood cells by means of a fluorescent probe. *Journal of Physiology.* 239:519–552.

Hoffman, J. F., J. H. Kaplan, T. J. Callahan, and J. C. Freedman. 1980. Electrical resistance of the red cell membrane and the relation between net anion transport and the anion exchange mechanism. *Annals of the New York Academy of Sciences.* 341:357–360.

Hoffman, J. F., and P. A. Knauf. 1973. The mechanism of the increased K transport induced by Ca in human red blood cells. *In* Erythrocytes, Thrombocytes, Leucocytes. E. Gerlach, K. Moser, E. Deutsch, and W. Williams, editors. Georg Thieme Verlag, Stuttgart. 66–70.

Hunter, M. J. 1971. A quantitative estimate of the non-exchange-restricted chloride permeability of the human red cell. *Journal of Physiology.* 218:P49–P50.

Hunter, M. J. 1974. The use of lipid bilayers as cell membrane models: an experimental test using the ionophore, valinomycin. *In* Drugs and Transport Processes. B. A. Callingham, editor. Macmillan, London. 227–240.

Hunter, M. J. 1977. Human erythrocyte anion permeabilities measured under conditions of net charge transfer. *Journal of Physiology.* 268:35–49.

Kaplan, J. H., M. Pring, and H. Passow. 1980. Concentration dependence of chloride movements that contribute to the conductance of the red cell membrane. *In* Membrane Transport in Erythrocytes. U. V. Lassen, H. H. Ussing, and J. O. Wieth, editors. Munksgaard, Copenhagen. 494–497.

Kaplan, J. H., M. Pring, and H. Passow. 1983. Band-3 protein-mediated anion conductance of the red cell membrane. Slippage vs. ionic diffusion. *FEBS Letters*. 156:175–179.

Knauf, P. A. 1979. Erythrocyte anion exchange and the band 3 protein; transport kinetics and molecular structure. *Current Topics in Membranes and Transport*. 12:249–363.

Knauf, P. A., G. F. Fuhrmann, S. Rothstein, and A. Rothstein. 1977. The relationship between exchange and net anion flow across the human red blood cell membrane. *Journal of General Physiology*. 69:363–386.

Knauf, P. A., and F.-Y. Law. 1980. Relationship of net anion flow to the anion exchange system. *In* Membrane Transport in Erythrocytes. (Alfred Benzon Symposium 14), U. V. Lassen, H. H. Ussing, and J. O. Wieth, editors. Munksgaard, Copenhagen. 488–493.

Knauf, P. A., F.-Y. Law, and P. J. Marchant. 1983. Relationship of net chloride flow across the human erythrocyte membrane to the anion exchange mechanism. *Journal of General Physiology*. 81:95–126.

Lassen, U. V. 1977. Electrical potential and conductance of the red cell membrane. *In* Membrane Transport in Red Cells. J. C. Ellory and V. L. Lew, editors. Academic Press, Inc., New York. 137–172.

Passow, H. 1986. Molecular aspects of band 3 protein-mediated anion transport across the red blood cell membrane. *Reviews of Physiology, Biochemistry and Pharmacology*. 103:61–223.

Scarpa, A., A. Cecchetto, and G. F. Azzone. 1970. The mechanism of anion translocation and pH equilibration in erythrocytes. *Biochimica et Biophysica Acta*. 219:179–188.

Schnell, K. F., and E. Besl. 1984. Concentration dependence of the unidirectional sulfate and phosphate flux in human red cell ghosts under selfexchange and under homoexchange conditions. *Pflügers Archiv*. 402:197–206.

Stark, G., and R. Benz. 1971. The transport of potassium through lipid bilayer membranes by the neutral carriers valinomycin and monactin. Experimental studies to a previously proposed model. *Journal of Membrane Biology*. 5:133–153.

Stoner, L. C., and F. M. Kregenow. 1980. A single-cell technique for the measurement of membrane potential, membrane conductance, and the efflux of rapidly penetrating solutes in *Amphiuma* erythrocytes. *Journal of General Physiology*. 76:455–478.

Tosteson, D. C. 1959. Halide transport in red blood cells. *Acta Physiologica Scandinavica*. 46:19–41.

Tosteson, D. C., R. B. Gunn, and J. O. Wieth. 1973. Chloride and hydroxyl ion conductance of sheep red cell membrane. *In* Erythrocytes, Thrombocytes, Leucocytes. E. Gerlach, K. Moser, E. Deutsch, and W. Williams, editors. Georg Thieme, Stuttgart. 62–66.

Vestergaard-Bogind, B., and U. V. Lassen. 1974. Membrane potential of *Amphiuma* red cells: hyperpolarizing effect of phloretin. *In* Comparative Biochemistry and Physiology of Transport. L. Bolis, K. Bloch, S. E. Luria, and F. Lynen, editors. Elsevier/North-Holland. 346–353.

Wieth, J. O., M. Dalmark, R. B. Gunn, and D. C. Tosteson. 1973. The transfer of monovalent inorganic anions through the red cell membrane. *In* Erythrocytes, Thrombocytes, Leucocytes. E. Gerlach, K. Moser, E. Deutsch, and W. Williams, editors. Georg Thieme, Stuttgart. 71–76.

Chapter 17

Properties of the Principal Anion-Exchange Mechanism in Human Neutrophils

Louis Simchowitz

*Department of Medicine, the John Cochran Veterans
Administration Medical Center, and the Departments of Medicine
and of Cell Biology and Physiology, Washington University School
of Medicine, St. Louis, Missouri*

Cell Physiology of Blood © 1988 by The Rockefeller University Press

Introduction

The nature of the classic inorganic anion-exchange mechanism of red blood cells has been the subject of active investigation for the better part of three decades (for reviews, see Sachs et al., 1975; Gunn, 1979; Knauf, 1979; Lowe and Lambert, 1983). However, it is only over the past few years that interest in a number of different systems for anion transport in a variety of other cell types has come about. This chapter is devoted to a discussion of what is currently known about the kinetic properties of the principal anion-exchange carrier in circulating human polymorphonuclear leukocytes, or neutrophils, derived from whole blood.

In our original reports on Cl^- movements in human neutrophils (Simchowitz and De Weer, 1986; Simchowitz et al., 1986), we identified three distinct pathways by which Cl ions cross the plasma membrane of resting cells: anion exchange, active transport, and electrodiffusion. The bulk ($\sim70\%$) of the total steady state $^{36}Cl^-$ influx and efflux (~1.5 meq/liter of cell water $\cdot$ min) represents an electroneutral anion exchange, a discussion of which forms the basis of this chapter. The carrier is relatively insensitive to disulfonic stilbenes and to the loop diuretics furosemide and ethacrynic acid, commonly used transport inhibitors of the classic inorganic anion-exchange mechanism of red blood cells. However, the anion countertransport system of neutrophils can be competitively inhibited by α-cyano-4-hydroxycinnamate (CHC), a well-known inhibitor of monocarboxylate transport in mitochondria and of anion exchange in erythrocytes (Halestrap and Denton, 1975; Halestrap, 1976; Deuticke, 1982). Unfortunately, in neutrophils, the affinity of the carrier for CHC is rather low, with an apparent K_i for the drug of ~9 mM in 148 mM Cl^- medium.

In related studies on intracellular pH (pH_i) regulation (Simchowitz and Roos, 1985), we demonstrated that this anion-exchange mechanism functions physiologically as a Cl^-/HCO_3^- exchanger in pH_i recovery from intracellular alkalinization: restoration of pH_i to its normal resting value (~7.25 at an extracellular pH [pH_o] of 7.40) following imposed alkalinization required external Cl^-, was dependent on HCO_3^-, and could be blocked by CHC.

The present work arises out of our continuing interest in examining the factors that control steady state pH_i in human neutrophils and from our efforts toward a detailed biochemical characterization of the kinetic parameters of the principal anion-exchange carrier. In this chapter, I will briefly review some of the important features of this system and provide new information concerning the kinetics of anion exchange and the identification of a number of inhibitors with enhanced activity against the carrier. These findings are then summarized and contrasted with the properties of the classic anion exchanger of red blood cells.

Characterization of Anion Exchange

As in all prior work originating from this laboratory, the studies reported here were performed on suspensions of purified human neutrophils. The cells were isolated from the blood of normal donors by sequential dextran sedimentation and Ficoll-Hypaque gradient centrifugation as previously described (Simchowitz and De Weer, 1986). The suspensions contained $\sim98\%$ neutrophils, all of which were viable.

The standard medium had the following composition: 140 mM NaCl, 5 mM KCl, 1 mM $CaCl_2$, 0.5 mM $MgCl_2$, and 5 mM HEPES buffer, pH 7.40; the solutions were supplemented with 1 mg/ml crystalline bovine serum albumin. As indicated, K^+

replaced Na^+, and glucuronate or various other anions were used as substitutes for Cl^-. Although nominally free of CO_2 and HCO_3^-, the solutions were routinely gassed with N_2 to further reduce the HCO_3^- content of the media (see below). For the most part, then, the various drugs were tested for their effects on Cl^-/Cl^- self-exchange.

One-way $^{36}Cl^-$ influx and efflux were measured using a silicone oil centrifugation technique that allows separation of the cells from the suspending medium in <5 s. The details of this procedure have been described in two recent articles (Simchowitz and

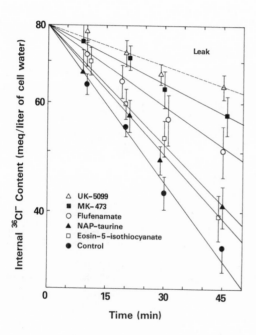

Figure 1. Time course of $^{36}Cl^-$ loss from neutrophils: effect of drugs. The medium contained 148 mM Cl^-, pH_o 7.40. At zero time, aliquots of the neutrophil suspension were added to tubes containing the standard medium alone (control), 0.4 mM UK-5099, 1 mM MK-473, 1 mM flufenamate, 1 mM eosin-5-isothiocyanate, or 1 mM NAP-taurine. At stated times, samples were taken and the cell pellets were isolated and counted for radioactivity. The lines represent least-squares fits to single exponentials. The efflux rates were calculated from the rate coefficients and the assumption that $[Cl^-]_i = 80$ meq/liter cell water. Results are from three to four separate experiments for each condition. The dashed line labeled "leak" denotes passive electrodiffusive $^{36}Cl^-$ efflux through permeability channels: its rate is 0.40 meq/liter · min (Simchowitz and De Weer, 1986). The efflux rates for control, 0.4 mM UK-5099, 1 mM MK-473, 1 mM flufenamate, 1 mM eosin-5-isothiocyanate, and 1 mM NAP-taurine were 1.56 ± 0.09, 0.42 ± 0.07, 0.58 ± 0.09, 0.86 ± 0.12, 1.18 ± 0.10, and 1.28 ± 0.13 meq/liter · min, respectively.

Roos, 1985; Simchowitz and De Weer, 1986). Efflux rates were calculated as the product of the rate coefficient of $^{36}Cl^-$ loss from the cells and an initial intracellular Cl^- content of 80 meq/liter cell water (Simchowitz and De Weer, 1986).

The time course of $^{36}Cl^-$ efflux from human neutrophils into a 148 mM Cl^- medium in the presence of a number of different drugs is shown in Fig. 1. All effluxes were first-order: the rate of $^{36}Cl^-$ loss under control conditions was 1.56 ± 0.09 meq/liter · min. As previously reported (Simchowitz and De Weer, 1986), this efflux was not appreciably reduced by the disulfonic stilbenes SITS (1 mM) or DIDS (0.1

mM) or by the "loop" diuretics furosemide or ethacrynic acid (each at 1 mM), all of which have been shown to inhibit anion-exchange fluxes in human red blood cells (for reviews, see Sachs et al., 1975; Gunn, 1979; Knauf, 1979). Table I lists the apparent K_i values for inhibition of steady state $^{36}Cl^-$ efflux into 148 mM Cl^- medium for a number of different test compounds. The apparent K_i values were calculated by least-squares fitting a Michaelis-Menten inhibition equation to the individual set of data points for each drug.

TABLE I
Apparent K_i Values for Inhibition of Anion Exchange by Various Drugs

Compound name	Apparent K_i*
Cinnamate derivatives	μM
UK-5099	29
α-Phenylcinnamate	371
α-Methylcinnamate	3,140
CHC	9,400
Diuretic agents	
MK-473	146
(+)5c	184
(+)DCPIB	207
MK-935	≫1,000
MK-447	≫1,000
Miscellaneous drugs	
Diazepam	176
Flufenamate	376
Niflumate	452
SITS	3,470
NAP-taurine	4,800
Eosin-5-isothiocyanate	>1,000
Aldrithiol	≫1,000
2-Methoxy-5-nitrotropone	≫1,000
Anthracene-9-carboxylate	≫1,000
Salicylate	—‡
Picrylsulfonate	—
Maleic anhydride	—
5,5′-dithio-bis(2-nitrobenzoate)	—

*The inhibition constants were derived as in Fig. 2 for $^{36}Cl^-$ efflux into 148 mM Cl^- medium. Results represent the means of three to four experiments.
‡No inhibition was observed at a drug concentration of 1 mM.

The data of Fig. 1 and Table I indicate that several other compounds known to inhibit anion-exchange fluxes in erythrocytes are for all practical purposes virtually inactive in blocking $^{36}Cl^-$ fluxes in human neutrophils in 148 mM Cl^- medium. These include NAP-taurine and eosin-5-isothiocyanate (Fig. 1), which caused only modest inhibition at 1 mM, and agents such as aldrithiol, 2-methoxy-5-nitrotropone, salicylate, picrylsulfonate, maleic anhydride, 5,5′-dithio-bis(2-nitrobenzoate), and anthracene-9-carboxylate, which had little or no detectable effect at this concentration

(Table I). In contrast, the rate of $^{36}Cl^-$ efflux was reduced considerably in the presence of 1 mM flufenamate, 1 mM MK-473 (a loop diuretic), or 0.4 mM UK-5099 (a CHC analogue), with rates of 0.86 ± 0.12, 0.58 ± 0.09, and 0.42 ± 0.07 meq/liter · min, respectively (Fig. 1). Relative to the total $^{36}Cl^-$ efflux rate of 1.56 meq/liter · min from control cells, these values represent inhibitions of 45, 63, and 73%. The actual extent of inhibition of anion exchange is much greater since ~30% of the total one-way $^{36}Cl^-$ efflux (rate ~0.40 meq/liter · min; dashed line of Fig. 1) represents not anion countertransport, but rather electrodiffusion through ion permeability channels (Simchowitz and De Weer, 1986). In fact, in the presence of 0.4 mM UK-5099, efflux was reduced to about this level, which implies that the drug totally abolishes the anion-exchange component of total efflux. In contrast to $^{36}Cl^-$ movements through the exchange carrier, which are sensitive to inhibition by CHC, the passive leak flux represents a UK-5099-and CHC-resistant background. None of the drugs tested in the present studies appeared to have any additional effect in reducing $^{36}Cl^-$ efflux in the presence of near-saturating concentrations (40 mM) of CHC, conditions that effectively measure only electrodiffusive $^{36}Cl^-$ efflux. Thus, it would appear that none of the agents tested suppressed passive $^{36}Cl^-$ efflux. For convenience, in order to display the effects of the various drugs on the exchange carrier alone, the remainder of the results in this section relate exclusively to anion exchange. This was accomplished by subtracting a passive efflux rate of 0.40 meq/liter · min from the measurements of total efflux.

CHC Analogues

Data for a few structural analogues of CHC, all cinnamic (phenylpropenoic) acid derivatives, are shown in Fig. 2. In these experiments, the dose dependence of the various drugs on $^{36}Cl^-$ efflux into 148 mM Cl^- medium was assessed. As previously reported (Simchowitz et al., 1986), under these conditions, the parent compound CHC

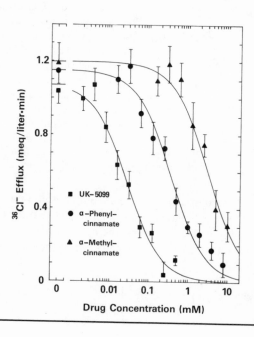

Figure 2. Dose dependence of inhibition of $^{36}Cl^-$ efflux from human neutrophils by analogues of CHC. Neutrophils were suspended in 148 mM Cl^- medium, pH_o 7.40, and exposed to varying concentrations of UK-5099 (0–0.5 mM), α-phenylcinnamate (0–8 mM), or α-methylcinnamate (0–10 mM). The efflux rates were calculated as in Fig. 1 and plotted against the prevailing drug concentrations in the bathing solutions. Results are from three to four experiments. The data points were fitted to Michaelis-Menten inhibition equations, which yielded apparent K_i values for UK-5099, α-phenylcinnamate, and α-methylcinnamate of 28.6 ± 5.8, 371 ± 70, and 3,140 ± 932 μM, respectively.

is a relatively weak inhibitor with an apparent K_i of $\simeq 9.4$ mM. Removal of the hydroxyl group at the 4-carbon position of the benzene ring to form α-cyanocinnamate had no significant effect on activity (not shown). However, substitution of a methyl or phenyl group for the cyanogen moiety on the α-carbon of the cinnamate backbone to form the α-methyl– and α-phenylcinnamate derivatives led to a substantial increase in potency: the compounds exhibited apparent K_i values of 3,140 $\pm$ 932 and 371 $\pm$ 70 μM, respectively, and thus were ~3- and ~25-fold more active than CHC.

The most potent compound tested was α-cyano-β-(1-phenylindol-3-yl)acrylate (designated UK-5099), which displayed an apparent K_i of 28.6 $\pm$ 5.8 μM, ~300-fold

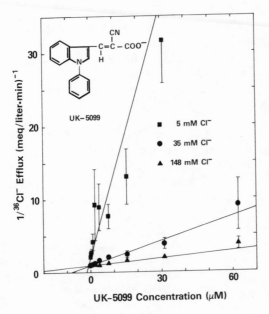

Figure 3. Effect of external Cl^- on the dose dependence of inhibition of $^{36}Cl^-$ efflux by UK-5099. The efflux of $^{36}Cl^-$ from human neutrophils was measured in the presence of varying concentrations of UK-5099 (0–125 μM). The effect of the drug was also evaluated as a function of the extracellular Cl^- concentration of the bathing medium where $[Cl^-]_o$ = 5, 35, and 148 mM (balance glucuronate), all at pH_o 7.40. The data derived at 148 mM Cl^- are the same as those shown in Fig. 2. Experiments at 5 and 35 mM Cl^- were performed in the presence of 85 mM K^+ and 2 μM nigericin to "pH-clamp" the cells (see text for explanation). The $^{36}Cl^-$ efflux rates were calculated as in Fig. 1 and plotted against the added drug concentration. Results represent three to four experiments. The data points were fitted to Michaelis-Menten inhibition equations and graphed in the form of Dixon plots. Fits to the individual sets of data yielded apparent K_i values for UK-5099 of 1.63 $\pm$ 0.60, 8.68 $\pm$ 1.94, and 28.6 $\pm$ 5.4 μM at 5, 35, and 148 mM Cl^-, respectively. A simultaneous fit of all three data sets (not shown) gave a true K_i for UK-5099 of 0.963 $\pm$ 0.130 μM.

lower than that of CHC. As with the latter compound, the nature of the inhibition was strictly competitive, as shown in Fig. 3. The rationale is based on the fact that for two competing substrates, S_1 and S_2 (one may be an inhibitor), the relationship between the true and apparent kinetic constants and the external concentration of the reference substrate S_2, $[S_2]_o$, is given by:

$$K_m^{app}(S_1) = K_m^{true}(S_1) \cdot \left[1 + \frac{[S_2]_o}{K_m^{true}(S_2)} \right]. \quad (1)$$

For these studies, the dose dependence of UK-5099 (i.e., S_1) was assessed in the presence of three different external Cl^- (i.e., S_2) concentrations: 5, 35, and 148 mM, multiples of 1, 7, and 30 times $K_m(Cl^-)$ (5.0 mM; Simchowitz et al., 1986). The results in 148 mM Cl^- have been taken from Fig. 2. The data sets have been plotted according to the method of Dixon (1964) and indicate that lowering the extracellular Cl^- concentration caused an increase in the apparent affinity for UK-5099: as $[Cl^-]_o$ was reduced from 148 to 35 to 5 mM, the apparent K_i for UK-5099 inhibition of $^{36}Cl^-$ efflux fell from 28.6 ± 5.8 to 8.68 ± 1.94 to 1.63 ± 0.60 μM, respectively. Considering, for example, that at 5 mM in the bathing solutions, Cl^- was present at a concentration equal to its own K_m, it follows from Eq. 1 that the true K_i value for UK-5099 is ~1.0 μM, ~300-fold lower than for CHC (true K_i, ~270 μM; Simchowitz et al., 1986). A simultaneous fit of all three data sets (not shown) yielded a true K_i value for UK-5099 of 0.963 ± 0.130 μM. It should be pointed out that a more complicated analysis of CHC indicated that it, too, behaved as a competitive inhibitor of Cl^-, with a true K_i of ~270 μM (Simchowitz et al., 1986). Presumably, this is true of all related compounds in this series.

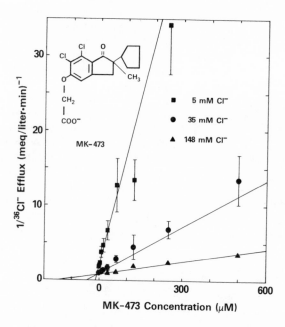

Figure 4. Effect of external Cl^- on the dose dependence of inhibition of $^{36}Cl^-$ efflux by MK-473. Conditions were the same as in Fig. 3, except that experiments were conducted in the presence of varying concentrations of MK-473 (0–1,000 μM). Fits to the individual sets of data yielded apparent K_i values for MK-473 of 11.8 ± 2.2, 45.8 ± 12.7, and 146 ± 31 μM at 5, 35, and 148 mM Cl^-, respectively. A simultaneous fit of all three data sets (not shown) gave a true K_i for MK-473 of 5.37 ± 0.70 μM.

Miscellaneous Drugs

The inhibitory effect of several other agents, whether competitive, noncompetitive, or mixed, was also investigated in a similar manner. As with UK-5099, there was an increase in the apparent affinity for the diuretic MK-473, an (indanyloxy)acetate derivative (Cragoe, 1983), as the concentration of the competing substrate (i.e., Cl^-) was reduced (Fig. 4). At $[Cl^-]_o = 5$, 35, and 148 mM, apparent K_i values of 146 ± 31, 45.8 ± 12.7, and 11.8 ± 2.2 μM, respectively, were observed, consistent with a true K_i for MK-473 of 5.37 ± 0.70 μM.

Several other diuretics from different chemical classes were also screened for activity against anion exchange in neutrophils. These included MK-935, an (acylvinylphenoxy)acetate derivative, and MK-447, an aminomethylphenol (Cragoe, 1983). These compounds exhibited only slight inhibition at a concentration of 1 mM in 148 mM Cl⁻ medium. Other agents, structurally similar to MK-473, such as (+)DCPIB (in the notation of Cragoe et al., 1982) and (+)5c (in the notation of Cragoe et al., 1986), were of particular interest since each had been shown to be a potent inactivator of anion exchange in human erythrocytes, of HCO_3^--stimulated $Na^+ + Cl^-$ cotransport in cat astrocytes, and of cell swelling in cerebrocortical slices with apparent K_i values in the nanomolar range (Cragoe et al., 1982, 1986; Garay et al., 1986). However, these drugs were only about as potent as MK-473 toward anion exchange in human neutrophils (Table I).

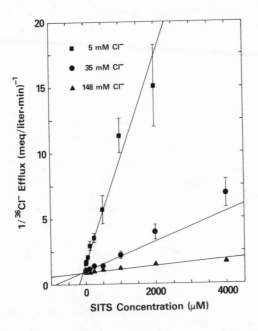

Figure 5. Effect of external Cl⁻ on the dose dependence of inhibition of ³⁶Cl⁻ efflux by SITS. Conditions were the same as in Fig. 3, except that experiments were conducted in the presence of varying concentrations of SITS (0–4,000 μM). Fits to the individual sets of data yielded apparent K_i values for SITS of 195 ± 33, 904 ± 172, and 3,466 ± 896 μM at 5, 35, and 148 mM Cl⁻, respectively. A simultaneous fit of all three data sets (not shown) gave a true K_i for SITS of 110 ± 13 μM.

In our original study of Cl⁻ movements (Simchowitz and De Weer, 1986), we reported that 1 mM SITS had no significant effect on steady state ³⁶Cl⁻ efflux into a 148 mM Cl⁻ medium. This, together with the lack of effect of 0.1 mM DIDS, prompted us to conclude that the anion-exchange carrier of human neutrophils is insensitive to disulfonic stilbenes. While it is certainly true that the exchanger is relatively resistant to the actions of these stilbene derivatives, inhibition of anion exchange by SITS can nonetheless be demonstrated under special conditions, namely low external Cl⁻. These studies (Fig. 5) indicate that SITS is actually a weak competitive inhibitor of Cl⁻. As will become evident from a consideration of the data of Fig. 5, the degree of inhibition of ³⁶Cl⁻ efflux by SITS can be increased dramatically by lowering [Cl⁻]ₒ, thereby reducing competition by another substrate that presumably binds to the same external translocation site on the carrier. As [Cl⁻]ₒ was reduced from 148 to 35 to 5 mM, the apparent K_i for SITS fell progressively from 3,466 ± 896 to 904 ± 172 to 195 ± 33 μM, respectively. Taking true $K_m(Cl^-)$ as 5.0 mM

(Simchowitz et al., 1986), a derived value of 110 ± 13 μM for true K_i(SITS) could be obtained. These findings indicate that the principal anion-exchange mechanism of human neutrophils probably bears some resemblance to that of the classic inorganic anion-exchange system in erythrocytes. This may explain why antibodies directed against the band 3 protein of red blood cell membranes also seem to cross-react with determinants on the surface of neutrophils (Kay et al., 1983).

Finally, a number of miscellaneous agents, most with activity against the exchange carrier of erythrocytes (for reviews, see Sachs et al., 1975; Gunn, 1979; Knauf, 1979; Lowe and Lambert, 1983), were also tested for their effects on ^{36}Cl$^-$ efflux in neutrophils (Table I). NAP-taurine, with an apparent K_i of 4.80 ± 1.43 mM in 148 mM Cl$^-$, could be shown to be a weak competitive inhibitor of Cl$^-$ with a true K_i of 137 ± 19 μM. Niflumate and flufenamate, with apparent K_i values of 452 ± 97 and 376 ± 89 μM (true $K_i = 11.0 \pm 1.9$ μM) in 148 mM Cl$^-$, also interacted with Cl$^-$ in a strictly competitive manner. On the other hand, diazepam exhibited an apparent K_i of 176 ± 40 μM under standard (148 mM Cl$^-$) conditions. In this case, however, limited studies suggest the nature of the inhibition to be either noncompetitive or mixed. The inhibition by each of the agents discussed above, including CHC, UK-5099, MK-473, and SITS, was readily reversible upon washing the cells in drug-free medium.

Effect of pH$_o$

The activity of the principal anion-exchange mechanism of human neutrophils is very sensitive to changes in pH$_o$. In Fig. 6, the rate of ^{36}Cl$^-$ efflux from labeled cells was measured as a function of pH$_o$ in the presence of 148 mM Cl$^-$. The results demonstrate that, relative to steady state conditions at pH$_o$ 7.40, the rate of Cl$^-$/Cl$^-$ exchange rises dramatically with extracellular alkalinization and falls with acidification. For example, compared with an exchange rate of 1.11 meq/liter $\cdot$ min at pH$_o$ 7.40, ^{36}Cl$^-$ efflux was enhanced ~6-fold at pH$_o$ 8.40 and reduced ~12-fold at pH$_o$ 6.40. The data could

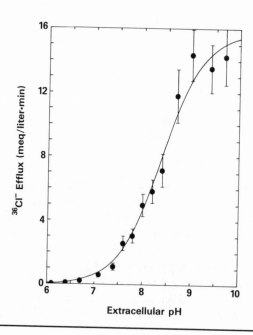

Figure 6. Effect of extracellular pH on the rate of ^{36}Cl$^-$ efflux into 148 mM Cl$^-$ media. The pH$_o$ of the bathing solutions was varied between 6.1 and 9.7 and the rate of ^{36}Cl$^-$ loss from the cells was measured as described in Fig. 1. Results have been taken from four separate experiments. The data points were fitted to a titration curve with an inflection point corresponding to an apparent pK at 8.38 ± 0.07.

be fitted to a titration curve with an inflection point corresponding to an apparent pK of 8.38 ± 0.07. The stimulation of anion exchange by extracellular alkalinization is probably analogous to the activation by internal HCO_3 (or OH) ions that occurs during pH_i recovery after imposed alkaline loads when the carrier mediates net external Cl^-/internal HCO_3^- exchange.

Several lines of evidence support the contention of a direct effect of pH (i.e., H^+ or OH^- concentrations) on the maximal transport rate of the carrier rather than an effect on either the concentration or the affinity constant of one or more of the ions participating in the exchange as, for example, HCO_3^-. (*a*) Since the K_m for external Cl^- is ~5.0 mM at pH_o 7.40 (Simchowitz et al., 1986), the carrier is already >95% saturated with Cl^- when $[Cl^-]_o = 148$ mM. Thus, an increase in affinity for Cl^- with alkalinization cannot by itself explain the ~12-fold enhancement of $^{36}Cl^-$ efflux rate in going from pH_o 7.40 to 9.40. (*b*) An increase in HCO_3^- concentration with rising pH_o could provide an additional exchange partner for internal Cl^- in the countertransport reaction. This could potentially explain the observed results if the V_{max} for HCO_3^- were considerably greater than that for Cl^-. This does not seem to be the case, however, since Cl^- and HCO_3^- exhibit very similar maximal transport rates (unpublished observations). Moreover, if the effect of pH_o were simply explainable on the basis of $[HCO_3^-]_o$, then the data of Fig. 6 should exhibit a pK of ~6.1, the "overall" first pK of carbonic acid. This value is sufficiently far removed from the value of ~8.4 determined experimentally as to largely exclude this possibility. Finally, the experiments presented here were conducted under CO_2- and HCO_3^--free conditions in that the media had been extensively bubbled with N_2 in order to effectively rid solutions of all but trace amounts of HCO_3^-. Collectively, these arguments suggest that the most plausible interpretation for the data of Fig. 6 is a direct effect of pH on the maximal transport rate of the anion-exchange carrier. This hypothesis implies that under physiological conditions (i.e., in the presence of CO_2/HCO_3^-), as in the process of pH_i recovery from intracellular alkalinization, pH_i (either directly or via H^+, OH^-, or HCO_3^-) exerts a regulatory role on carrier activity over and above any effects on the concentration of one or more of the primary reactants in the exchange (e.g., HCO_3^-).

These studies provide the rationale for our use of the high-K^+–nigericin technique to "pH-clamp" the cells at their normal resting pH of ~7.25 (pH_o 7.40) in this and in previous articles (Simchowitz and Roos, 1985; Simchowitz et al., 1986). This was done for experiments conducted in low-Cl^- media where Cl^- was replaced by glucuronate, a nominally inert anion. As will become evident in the next section, the affinity for HCO_3^- is high. Since it is technically difficult to rid solutions completely of HCO_3^-, attempts to measure $^{36}Cl^-$ efflux into low-Cl^- media (equimolar substitution of glucuronate) resulted in an exchange of internal Cl^- for the small amounts of HCO_3^- usually present in the bathing media and led to an intracellular alkalinization. This pH change would then have the secondary effect of dramatically enhancing the V_{max} of the exchange carrier as shown in Fig. 6. In order to avoid these complications from even trace amounts of HCO_3^- (derived perhaps from cell metabolism), we took the added precaution of pH-clamping the neutrophils at their normal pH_i of ~7.25 using a high-K^+–nigericin technique (Simchowitz and Roos, 1985; Simchowitz et al., 1986). The cells are bathed in medium of pH_o 7.40 as before, but now containing 85 mM K^+ and 2 μM nigericin, a K^+/H^+-exchanging ionophore (Pressman, 1969). Under these conditions ($[K^+]_i \simeq 120$ mM, $pH_o = 7.40$), $[H^+]_i/[H^+]_o \simeq [K^+]_i/[K^+]_o$. The resulting depolarization with 85 mM K^+ has no practical effect on the interpretation of the data since membrane potential causes little or no alteration in the flux through the exchange carrier (Simchowitz and De Weer, 1986). The experiments presented in Figs. 3–5 were in fact generated in the presence of 85 mM K$^+$ and 2 μM nigericin.

Anion Selectivity

In our previous reports on Cl$^-$ movements in human neutrophils (Simchowitz and De Weer, 1986; Simchowitz et al., 1986), we noted that the anion-exchange carrier is not perfectly selective for Cl$^-$ in that it also binds and transports other ions, such as

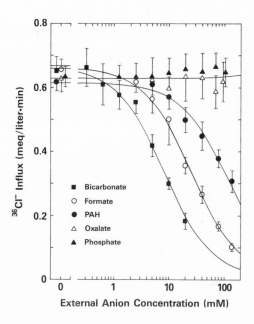

Figure 7. Substrate competition: effect of different anions on the initial rate of ^{36}Cl$^-$ influx from 5 mM Cl$^-$ media. The bathing solutions contained 85 mM K$^+$ and 2 μM nigericin, together with 1 mM 2-deoxy-D-glucose to block active Cl$^-$ uptake (Simchowitz and De Weer, 1986). The concentrations of HCO$_3^-$, formate, PAH, oxalate, and phosphate were varied between 0 and 140 mM by replacement of glucuronate. The influx of ^{36}Cl$^-$ was assessed at 5 and 10 min and the initial influx rates were calculated by fitting the data points to a single-exponential equation. Results are from three to four experiments. The curves for HCO$_3^-$, formate, and PAH follow Michaelis-Menten inhibition equations superimposed on a constant background of 0.0073 meq/liter·min, representing passive ^{36}Cl$^-$ influx at 5 mM Cl$^-$. The apparent K_i values for HCO$_3^-$, formate, and PAH are 8.27 ± 1.74, 26.0 ± 5.0, and 122 ± 25 mM, respectively. The combined data sets for oxalate and phosphate were fitted to a straight line with a slope of 0.000058 ± 0.000391, which could not be distinguished from zero.

p-aminohippurate (PAH), albeit with much lower affinity. The relative lack of specificity is further shown in Fig. 7, which displays the ability of HCO$_3^-$, formate, PAH, oxalate, and phosphate to compete with Cl$^-$ at the carrier's external binding site and thus to inhibit the influx of ^{36}Cl$^-$ from a 5 mM Cl$^-$ medium. The extracellular concentration of each of the anions was varied between 0 and 140 mM by equimolar replacement of glucuronate, for which the carrier seems to be devoid of affinity. Raising the concentrations of HCO$_3^-$, formate, and PAH over this range caused a progressive reduction in the rate of ^{36}Cl$^-$ influx into the cells. In each case, inhibition followed simple Michaelis-Menten kinetics with apparent K_i values of 8.27 ± 1.74, 26.0 ± 5.0, and 122 ± 25 mM, respectively. Since Cl^- was present in the bathing media at a concentration equal to its own true K_m (5.0 mM), it follows from Eq. 1 that the measured apparent K_i values are actually twice their respective true Michaelis constants for binding to the external translocation site of the exchange carrier. The value of 60.8 mM for true K_m(PAH) thus derived from this single set of experiments compares quite favorably with the value of 50.3 mM that was obtained using a much larger number of data sets (Simchowitz et al., 1986). It should be emphasized that the rate of $^{36}Cl^-$ influx was completely unaffected by the divalent anions oxalate and phosphate over the range 0–95 mM. A similar lack of effect was observed with SO_4^{2-} (not shown).

TABLE II
Kinetic Constants for the External Binding Sites of the Anion-Exchange Carrier of Human Neutrophils

Ligands	Michaelis constants
Monovalent anions	*mM*
Bicarbonate	4.1 ± 0.9
Chloride	5.0 ± 0.8*
Nitrate	8.4 ± 1.3[‡]
Bromide	9.4 ± 1.0[‡]
Formate	13.0 ± 2.5
Fluoride	23.2 ± 2.9[‡]
Iodide	44.2 ± 7.5[‡]
p-Aminohippurate	50.3 ± 14.9*
Thiocyanate	55.6 ± 18.2[‡]
Glucuronate	—*,[§]
Divalent anions	
Sulfate	—[‡],[§]
Oxalate	—[§]
Phosphate	—[§]

*From Simchowitz et al. (1986).
[‡]From Simchowitz (1988).
[§]No apparent affinity.

The ability of other halide ions and several lyotropic anions to compete with external Cl^- for binding to the exchange carrier was also investigated; the results are detailed in another report on anion selectivity (Simchowitz, 1988). The major findings are summarized in Table II, where the true K_m values for all of the anions tested have been listed.

The topic of anion selectivity in physical and biological systems has been the subject of several in-depth reviews (Diamond and Wright, 1969; Wright and Diamond, 1977) since the original report of Eisenman (1961). He proposed a theory to account for the natural occurrence of only 11 of the possible 120 sequences of alkali metal cation selectivity based on the anionic field strength of the membrane. Since then, selectivities for the halide series have also been calculated according to this theory

(Diamond and Wright, 1969; Wright and Diamond, 1977) and related to the sequences observed in nature (7 out of a possible 24).

The neutrophil anion-exchange carrier binds anions in a series of decreasing affinities: $HCO_3^- \sim Cl^- > Br^- \sim NO_3^- >$ formate $> F^- > I^- \sim SCN^- \sim PAH$, which corresponds to sequence 5 for the halides. This order is similar to that reported by Aickin and Brading (1985) for the Cl^-/HCO_3^- exchanger of smooth muscle cells of guinea pig vas deferens. In the best-studied system, the ability of the inorganic anion-exchange carrier of red blood cells to handle other anions has come under intense scrutiny (Tosteson, 1959; Wieth, 1970; Dalmark and Wieth, 1972; Dalmark, 1976; Lambert and Lowe, 1978; Lowe and Lambert, 1983). Most investigators find a higher affinity for HCO_3^- over Cl^- by a factor of ~ 2. For the halides, results generally fall within sequence 4 ($Cl^- > Br^- > I^- > F^-$) or 5 ($Cl^- > Br^- > F^- > I^-$).

It should be pointed out that the anion-exchange mechanism of human neutrophils, while exhibiting broad selectivity among monovalent anions, appears to be devoid of affinity for divalent anions such as SO_4^{2-}, oxalate, and phosphate. This contrasts dramatically with the situation in red cells whose anion-exchanger binds and transports SO_4^{2-} and oxalate (Gunn, 1978; Knauf, 1979). There, for instance, Cl^- and SO_4^{2-} possess comparably high affinities, although the transport rate for Cl^- is three orders of magnitude greater than that for SO_4^{2-} (Schnell et al., 1977; Barzilay and Cabantchick, 1979; Milanick and Gunn, 1982, 1984). These findings, together with the lack of sensitivity to various drugs, emphasize that major differences exist between the anion-exchange carrier of leukocytes and that of red blood cells.

Summary and Conclusions

As part of our ongoing studies aimed at understanding the physiology of human neutrophil function, we have been investigating the nature of the principal anion-exchange mechanism in these cells. This countertransport system functions physiologically as a Cl^-/HCO_3^- exchanger in intracellular pH regulation from alkalinization (Simchowitz and Roos, 1985). In summarizing the results of our work, it is worthwhile to keep in mind the well-defined properties of the classic inorganic anion-exchange carrier of red blood cells, as important differences between the two are evident.

Inhibitor Profile

In our original articles on Cl^- movements (Simchowitz and De Weer, 1986; Simchowitz et al., 1986), we had identified CHC, which suppresses monocarboxylate transport in mitochondria (Halestrap and Denton, 1975), as a relatively weak inhibitor. Several structural analogues, including α-methylcinnamate and α-phenylcinnamate, exhibited enhanced activity against anion exchange. The most potent compound thus far identified was UK-5099, which was ~ 300-fold more active than CHC. The inhibitory effect of each member of this series of compounds was strictly competitive in nature: both the drug and Cl ions appeared to bind to the same external translocation site on the exchange carrier. The inhibition by several other unrelated agents (e.g., MK-473, niflumate, flufenamate, and NAP-taurine), all weak organic acids, was also of the competitive type, which suggests that these compounds bind to the carrier by virtue of their negatively charged carboxyl or sulfonic groups.

In marked contrast to erythrocytes, the anion-exchange carrier of neutrophils is rather insensitive to the disulfonic stilbenes SITS and DIDS (Simchowitz and De Weer, 1986): little or no effect can be detected in 148 mM Cl^-. However, under conditions of low external Cl^- (5–35 mM), SITS can in fact be shown to interact weakly with the anion binding site of the exchange carrier. In a comparable vein, several other drugs, all of which have been reported to suppress Cl^- fluxes in red cells (for reviews, see Sachs et al., 1975; Gunn, 1979; Knauf, 1979; Lowe and Lambert, 1983), were either completely inactive against anion exchange in neutrophils at a concentration of 1 mM or only marginally effective at doses several orders of magnitude greater than those required in red blood cells. This list included furosemide, picrylsulfonate, maleic anhydride, 5,5′-dithio-bis(2-nitrobenzoate), salicylate, 2-methoxy-5-nitrotropone, and aldrithiol.

The external translocation site of the neutrophil carrier binds anions in a sequence of decreasing affinities: $HCO_3^- \sim Cl^- > Br^- \sim NO_3^- >$ formate $> F^- > I^- \sim SCN^- \sim$ PAH. It should be pointed out that while displaying broad specificity for a number of monovalent anions, the carrier lacks any measurable affinity for divalent anions, such as sulfate, oxalate, and phosphate. This is in contradistinction to the behavior of the red cell exchanger, which, for example, binds Cl^- and SO_4^{2-} with similarly high affinities (Schnell et al., 1977; Barzilay and Cabantchick, 1979; Milanick and Gunn, 1982, 1984).

Finally, it should also be remembered that the steady state flux through the anion (i.e., Cl^-/Cl^-) exchange mechanism in resting neutrophils at 37°C amounts to ~1 meq/liter cell water·min, a value roughly three to four orders of magnitude smaller than that in erythrocytes when measured under comparable conditions. Thus, while knowledge concerning a precise kinetic characterization of anion countertransport in leukocytes is in its infancy, the sum total of these early observations makes it clear that major differences exist between the predominant anion-exchange carriers of neutrophils and red cells.

Acknowledgments

I would like to acknowledge the expert technical assistance of Arabella R. Kizzart, Ajuah O. Davis, Margaret A. Foy, and Jacquelyn T. Engle, and the secretarial skills of Teri L. Becker. I am especially grateful to Dr. Paul De Weer for his helpful discussions while the work was in progress and for supplying the least-squares programs for the computer fits. I also wish to thank Dr. Edward J. Cragoe, Jr., of the Merck, Sharp & Dohme Research Laboratories, West Point, PA, for his generous gifts of the various diuretic agents and also Pfizer Central Research Laboratories of Sandwich, Kent, England for providing the UK-5099.

This work was supported by the Veterans Administration and by U.S. Public Health Service grant GM-38094.

References

Aickin, C. C., and A. F. Brading. 1985. The effects of bicarbonate and foreign anions on chloride transport in smooth muscle of the guinea-pig vas deferens. *Journal of Physiology.* 366:267–280.

Barzilay, M., and Z. I. Cabantchik. 1979. Anion transport in red blood cells. II. Kinetics of reversible inhibition by nitroaromatic sulfonic acids. *Membrane Biochemistry.* 2:255–281.

Cragoe, E. J., Jr. 1983. The (aryloxy)acetic acid family of diuretics. *In* Diuretics: Chemistry, Pharmacology, and Medicine. E. J. Cragoe, Jr., editor. John Wiley & Sons, New York, NY. 201–266.

Cragoe, E. J., Jr., N. P. Gould, O. W. Woltersdorf, Jr., C. Ziegler, R. S. Bourke, L. R. Nelson, H. K. Kimelberg, J. B. Waldman, A. J. Popp, and N. Sedransk. 1982. Agents for the treatment of brain injury. 1. (Aryloxy)alkanoic acids. *Journal of Medicinal Chemistry.* 25:567–579.

Cragoe, E. J., Jr., O. W. Woltersdorf, Jr., N. P. Gould, A. M. Pietruszkiewicz, C. Ziegler, Y. Sakurai, G. E. Stokker, P. S. Anderson, R. S. Bourke, H. K. Kimelberg, L. R. Nelson, K. D. Barron, J. R. Rose, D. Szarowski, A. J. Popp, and J. B. Waldman. 1986. Agents for the treatment of brain edema. 2. [(2,3,9,9a-tetrahydro-3-oxo-9a-substituted-1H-fluoren-7-yl)oxy]alkanoic acids and some of their analogues. *Journal of Medicinal Chemistry.* 29:825–841.

Dalmark, M. 1976. Effects of halides and bicarbonate on chloride transport in human red blood cells. *Journal of General Physiology.* 67:223–234.

Dalmark, M., and J. O. Wieth. 1972. Temperature dependence of chloride, bromide, iodide, thiocyanate and salicylate transport in human red cells. *Journal of Physiology.* 224:583–610.

Deuticke, B. 1982. Monocarboxylate transport in erythrocytes. *Journal of Membrane Biology.* 70:89–103.

Diamond, J. M., and E. M. Wright. 1969. Biological membranes: the physical basis of ion and nonelectrolyte selectivity. *Annual Review of Physiology.* 31:581–646.

Dixon, M., and E. C. Webb. 1964. Enzymes. 2nd Ed. Academic Press, Inc., New York, NY.

Eisenman, G. 1961. On the elementary atomic origin of equilibrium ionic specificity. *In* Symposium on Membrane Transport and Metabolism. A. Kleinzeller and A. Kotyk, editors. Academic Press, Inc., New York, NY. 163–179.

Garay, R. P., P. A. Hannaert, C. Nazaret, and E. J. Cragoe, Jr. 1986. The significance of the relative effects of loop diuretics and anti-brain edema agents on the Na^+, K^+, Cl^--cotransport system and the $Cl^-/NaCO_3^-$ anion exchanger. *Naunyn-Schmiedeberg's Archives of Pharmacology.* 334:202–219.

Gunn, R. B. 1978. Considerations of the titratable carrier model for sulfate transport in human red blood cells. *In* Membrane Transport Processes. J. F. Hoffman, editor. Raven Press, New York, NY. 1:61–77.

Gunn, R. B. 1979. Transport of anions across red cell membranes. *In* Membrane Transport in Biology. Vol. II: Transport across Single Biological Membranes. G. Giebisch, D. C. Tosteson, and H. H. Ussing, editors. Springer Verlag, Berlin. 59–80.

Halestrap, A. P. 1976. Transport of pyruvate and lactate into human erythrocytes. Evidence for the involvement of the chloride carrier and a chloride-independent carrier. *Biochemical Journal.* 156:193–207.

Halestrap, A. P., and R. M. Denton. 1975. The specificity and metabolic implications of the inhibition of pyruvate transport in isolated mitochondria and intact tissue preparations by α-cyano-4-hydroxycinnamate and related compounds. *Biochemical Journal.* 148:97–106.

Kay, M. M. B., C. M. Tracey, J. R. Goodman, J. C. Cone, and P. S. Bassel. 1983. Polypeptides immunologically related to band 3 are present in nucleated somatic cells. *Proceedings of the National Academy of Sciences.* 80:6882–6886.

Knauf, P. A. 1979. Erythrocyte anion exchange and the band 3 protein: transport kinetics and molecular structure. *Current Topics in Membrane Transport.* 12:249–363.

Lambert, A., and A. G. Lowe. 1978. Chloride/bicarbonate exchange in human erythrocytes. *Journal of Physiology.* 275:51–63.

Lowe, A. G., and A. Lambert. 1983. Chloride-bicarbonate exchange and related transport processes. *Biochimica et Biophysica Acta.* 694:353–374.

Milanick, M. A., and R. B. Gunn. 1982. Proton-sulfate cotransport: H^+ and sulfate addition to the chloride transporter of human red blood cells. *Journal of General Physiology.* 79:87–113.

Milanick, M. A., and R. B. Gunn. 1984. Proton-sulfate cotransport: external proton activation of sulfate influx into human red blood cells. *American Journal of Physiology.* 247:C247–C259.

Pressman, B. C. 1969. Mechanism of action of transport-mediating antibiotics. *Annals of the New York Academy of Sciences.* 147:829–841.

Sachs, J. R., P. A. Knauf, and P. B. Dunham. 1975. Transport through red cell membranes. *In* The Red Blood Cell. 2nd edition. D. M. Surgenor, editor. Academic Press, Inc., New York, NY. 2:613–703.

Schnell, K. F., S. Gerhardt, and A. Schöppe-Fredenburg. 1977. Kinetic characteristics of the sulfate self-exchange in human red blood cells and red blood cell ghosts. *Journal of Membrane Biology.* 30:319–350.

Simchowitz, L. 1988. Interactions of bromide, iodide, and fluoride with the pathways of chloride transport and diffusion in human neutrophils. *Journal of General Physiology.* 91:835–860.

Simchowitz, L., and P. De Weer. 1986. Chloride movements in human neutrophils. Diffusion, exchange, and active transport. *Journal of General Physiology.* 88:167–194.

Simchowitz, L., R. Ratzlaff, and P. De Weer. 1986. Anion/anion exchange in human neutrophils. *Journal of General Physiology.* 88:195–217.

Simchowitz, L., and A. Roos. 1985. Regulation of intracellular pH in human neutrophils. *Journal of General Physiology.* 85:443–470.

Tosteson, D. C. 1959. Halide transport in red cells. *Acta Physiologica Scandinavica.* 46:19–41.

Wieth, J. O. 1970. Effect of some monovalent anions on chloride and sulfate permeability of human red cells. *Journal of Physiology.* 207:581–609.

Wright, E. M., and J. M. Diamond. 1977. Anion selectivity in biological systems. *Physiological Reviews.* 57:109–156.

Chapter 18

Mechanisms of Membrane Potential Probes

Alan S. Waggoner

Department of Biological Sciences, Carnegie Mellon University, Pittsburgh, Pennsylvania

Cell Physiology of Blood © 1988 by The Rockefeller University Press

Introduction

Three groups of dyes have given sufficiently large signals with blood cells to justify their use. These are the membrane-permeant cations, the membrane-permeant anions, and the membrane-impermeant oxonols (Fig. 1). Additional information on the mechanisms and applications can be found in reviews (Freedman and Novak, 1988; Freedman and Laris 1988; Waggoner, 1985; Cohen and Hoffman, 1982; Bashford and Smith, 1979).

diS-C$_3$-(5)

RHODAMINE 123

diBA-C$_4$-(5)

OX-VI

WW781

RH 160

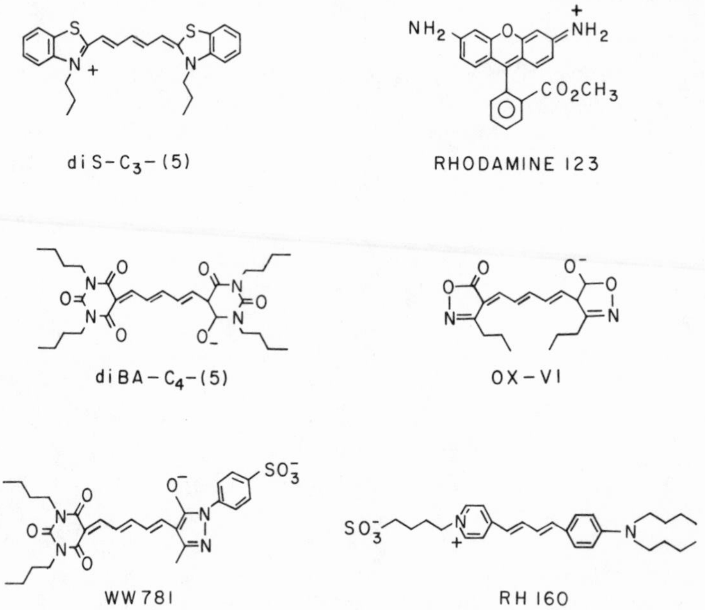

Figure 1. Potential-sensitive dyes used for blood cell studies. (Top row) Permeant cations. Cyanine analogues other than diS-C$_3$-(5) have been used. Changes in the number of methine carbons between the two heterocyclic nuclei and in the heteroatoms in the nuclei, e.g., oxygen or carbon instead of sulfur, control the color of the probe and its chemical stability. The length of the alkyl chain attached to the nitrogens modulates the lipid affinity of the dye. The nomenclature used was developed by Sims et al. (1974). The "mitochondrial probe" rhodamine 123 was introduced by Johnson et al. (1981). (Middle row) Permeant oxonols. Like the cyanines, the heteroatoms, methine carbons, and the alkyl groups determine the color and lipid affinity, respectively, of the oxonols (Freedman and Novak, 1988; Bashford and Smith, 1979). (Bottom row) Impermeant voltage probes. WW781 is one of a large group of impermant oxonols used for red blood cell studies by Freedman and Novak (1983). The mechanism of potential sensitivity of the impermant oxonols was elucidated by George et al. (1988). RH 160 was developed by Grinvald et al. (1982) and, along with other impermeant styryl dyes of similar structure, has been evaluated as a red cell potential probe (Freedman and Novak, 1988).

Membrane-permeant Cations

The cyanine dyes have been used extensively for blood cell studies and their behavior during potential changes is the best understood of the permeant cations. It is likely that the mechanism of the other cationic dyes shown in Fig. 1 is similar. The cyanine dyes are accumulated by cells with large negative internal membrane potentials. They are further accumulated within intracellular compartments that are negatively polarized. The redistribution of cyanine dye molecules between compartments is usually complete within a few seconds to tens of seconds following a potential change. While the Nernst equation describes the potential-dependent distribution of soluble monomeric dye molecules across membranes, it does not quantitatively relate the total amount of dye within a compartment to the concentration of dye bathing the compartment because the permeant cations are lipophilic and interact with membranes and with hydrophobic regions of proteins.

An additional complication precludes a direct determination of membrane potential from the probe fluorescence signal. The sign and magnitude of a potential-dependent fluorescence change depends on whether the measurement is made on an individual cell or instead is made on a suspension of cells. In the latter case, fluorescence from dye in the suspension medium is detected in addition to cellular fluorescence. Surprisingly, cell suspensions containing moderate to high concentrations (10^{-6}–10^{-7} M) of most cyanines lose fluorescence when cells, organelles, or vesicles in the suspension become hyperpolarized (more negative inside). Dye monomers, which were fluorescent in the bathing solution, tend to form nonfluorescent aggregates within the cells when the dye is accumulated. Also, at high internal concentrations, collisional quenching of fluorescence contributes to the reduced fluorescence of internalized dyes, even when the dye molecules do not form stable aggregates. On the other hand, when the total concentration of dye in the medium is very low (10^{-7}–10^{-9} M), and when the particular dye used has little tendency to aggregate, the suspension fluorescence increases upon cell hyperpolarization. In this case, accumulated dye molecules find nonpolar, high-viscosity intracellular membrane and protein binding sites in which cyanine dye molecules are more fluorescent than aqueous monomers.

The fluorescence from individual cells studied by flow cytometry or microscope imaging must be considered differently. These instruments are very sensitive and the dye concentration can be kept low (10^{-7}–10^{-9} M). The dyes are less toxic to mitochondria at low concentrations. Under these conditions the total cell fluorescence increases as the cytoplasm or the mitochondrial matrix space is hyperpolarized relative to the medium. Individual cell analysis has other advantages that are discussed by Seligmann, Simons et al., Wilson et al., and Ransom and Cambier in this volume.

A variety of cyanines can be used (Freedman and Novak, 1988). The length of the alkyl groups attached to the chromophore control the speed of the fluorescence response, the water solubility, and the lipid binding of the dyes (Sims et al., 1974). The heteroatoms of the two ring systems affect the color of the dye as well as the sensitivity of the dye to oxidative destruction. For example, diS-C$_3$-(5) fluorescence is lowered by H_2O_2 released from stimulated neutrophils (Whitin et al., 1981). Although this effect reduces the quantitative utility of this dye for measuring neutrophil membrane potentials, Simons et al. (this volume) find the quenching to be a good qualitative indicator for the respiratory burst, and the presence of a membrane potential response can qualitatively be detected simultaneously. As the number of methine carbon atoms between the nitrogen-containing ring systems is increased from three to seven, the absorption wavelength moves from 500 to 750 nm. The emission maxima are ~20 nm to longer wavelengths from the absorption maxima.

The total amount of cyanine dye associated with a cell is dependent on the plasma membrane potential, the mitochondrial membrane potential, the number of mitochondria, and the extent of nonspecific binding to cell components. Bright mitochondrial fluorescence can be especially striking in cultured cells stained with cyanine dyes and with the permeant cationic dye rhodamine 123, which is a potential-sensitive dye, sometimes called a "mitochondrial-specific dye" because of the strong signals from mitochondria (Johnson et al., 1981). Brand and Felber (1984) have quantitatively described the potential-dependent distribution of permeant triphenylmethylphosphonium (TPMP) cations in compartmentalized systems. Relating cyanine dye fluorescence to membrane potential is much more difficult, however, because of dye binding to cell constituents and because of the strong effects of dye binding and dye aggregation on fluorescence. Nevertheless, we can get a quick sense of the contribution of the mitochondria to the total fluorescence of the cell by using a very simple model. Assume that cyanine dye molecules do not interact with cell substituents and that we can describe the concentrations of dye in the various compartments with the Boltzman equation. The dye concentrations in the bathing medium, cytoplasm, and mitochondrial matrix (D_w, D_c, and D_m) are determined by the plasma membrane potential (V_{pm}) and the mitochondrial membrane potential (V_m). In the equation that relates these quantities, $\gamma = ZF/RT$, where Z is the charge on the dye, F is Faraday's constant, R is the gas constant, and T is the temperature (Kelvin):

$$D_c = D_w \, e^{\gamma V_{pm}}$$

$$D_m = D_w \, e^{\gamma(V_{pm} + V_m)}.$$

Assuming further that the fluorescence of the dye is independent of its location, we can write an expression for the total cell fluorescence (F_{cell}) in terms of the fractional cellular volume occupied by the mitochondria (f_m) and the cytoplasm (f_c):

$$F_{cell} = f_c \, e^{\gamma V_{pm}} + f_m \, e^{\gamma(V_{pm} + V_m)}.$$

Crude but reasonable estimates of the membrane potentials ($V_{pm} = -60$ mV; $V_m = -180$ mV) and fractional volumes ($f_m = 0.01$; $f_c = 0.99$) indicate that the amount of total cell fluorescence contributed by the mitochondria can be significant, 50% for this example. Therefore, it is essential that experimenters who work with blood cells in suspension or measure the total fluorescence of individual cells address the possibility that they are detecting the effects of both mitochondrial and plasma membrane changes. If mitochondria are absent from the cell (red cells) or if the mitochondria have been inhibited or poisoned, the interpretation of fluorescence changes may be much simpler. Fluorescence image analysis, where mitochondrial and cytoplasmic fluorescence can both be quantified, may offer the opportunity to measure simultaneously both V_{pm} and V_m.

Permeant Anions

The oxonol dyes shown in Fig. 1 are the only membrane-permeant oxonol dyes that have been exploited as membrane potential probes. They are negatively charged and are ejected from cellular compartments that become more hyperpolarized. Therefore, fluorescence signals from individual oxonol-stained cells change in the opposite direction as compared with signals from cyanine-stained cells. Although the exclusion of oxonol dyes from negatively charged blood cells leads to lower fluorescence and possible difficulties in detection, the oxonols have the advantage that they tend not to be accumulated by mitochondria and can be used to estimate plasma membrane potentials directly (Seligmann, and Wilson et al., this volume). As with the cyanines, it is important to select an oxonol with the right balance of lipophilicity. In order to be sufficiently membrane permeant for a quick response to potential changes, the dye molecules must have an affinity for membranes. However, if the affinity is too high, most of the cell fluorescence is due to membrane-bound material that will not respond to a potential change. Two to four carbon alkyl groups on the oxonol structure seem to be optimal.

Calibration of cyanine fluorescence signals is usually accomplished by setting the membrane potential with valinomycin or by using proton ionophores (Freedman and Novak, 1988). More caution is needed in calibrating the hydrophobic oxonol anions because they interact with hydrophobic cation-ionophore complexes and fluorescence is reduced.

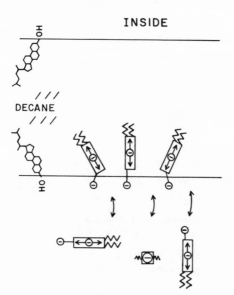

Figure 2. "On-off" mechanism for WW781 and other impermeant oxonol analogues (George et al., 1988). This illustration shows three probe molecules oriented more or less perpendicularly to the plane of the membrane. When the inside of the bilayer (top of figure) moves to a negative potential, probe molecules are driven off the membrane and assume random orientations. The center of the bilayer contains an undetermined amount of decane solvent.

Impermeant Probes

The impermeant oxonols were first used to follow action potential changes in nerve and muscle fibers. These dyes have a submillisecond response to nerve cell potential changes but their fluorescence or light transmission signals ($\Delta F/F$ or $\Delta I/I$) are usually small (~0.1–10%/100 mV) relative to the fluorescence changes of the cyanines and oxonols in blood cells ($\Delta F/F = >80\%/100$ mV). Freedman and Novak (1983) have used the impermeant oxonol analogue WW781 to quantify red cell potential changes. Fluorescence changes for this system are in the range of 10–20%/100 mV.

The sulfonic acid group prevents the oxonol chromophore, which by itself would be membrane permeant, from crossing membranes at a detectable rate. It is now known that this class of potential probes operates by an "on-off" mechanism, illustrated in Fig. 2 (George et al., 1988). Electric field changes that occur within the membrane during hyperpolarization or depolarization drive the hydrophobic anionic chromophore between binding sites ~10 Å deep in the membrane and the aqueous bathing medium near the surface of the membrane. The optical signal occurs because the absorption spectrum of the membrane-associated dye is at longer wavelengths than the spectrum of aqueous dye. Also, membrane-bound dye is more fluorescent. The "on-off" mechanism is consistent with potential-dependent spectral shifts and light transmission changes from lipid bilayer membranes bathed in impermeant oxonol solutions. The loss of dye from hyperpolarized red blood cell membranes confirms this mechanism.

The stryryl dye RH246 (an analogue of RH 160 in Fig. 1, but with a pentylsulfonate instead of a butylsulfonate) has been shown to respond to red cell potential changes (Freedman and Novak, 1988), but it has not generally found its way into blood studies.

Potential-sensitive dyes can be obtained from a number of sources, but not all sources can provide all dyes. Molecular Probes (Eugene, OR) and Eastman Kodak (Rochester, NY) are the commercial sources that should be contacted first. Amiram Grinvald (The Rockefeller University, New York, NY), Leslie Loew (University of Connecticut, CT), and Alan Waggoner (Carnegie Mellon, Pittsburgh, PA) have in the past provided small samples to investigators.

References

Bashford, C. L., and J. C. Smith. 1979. The use of optical probes to monitor membrane potential. *Methods in Enzymology.* 55:569–586.

Brand, M. D., and S. M. Felber. 1984. Membrane potential of mitochondria in intact lymphocytes during early mitogenic stimulation. *Biochemical Journal.* 217:453–459.

Cohen, L. B., and J. F. Hoffman. 1982. Optical measurements of membrane potential. *Techniques in Cell Physiology.* 118:1–13.

Freedman, J. C., and P. C. Laris. 1988. Optical potentiometric indicators of nonexcitable cells. *In* Spectroscopic Membrane Probes, L. Loew, editor. CRC Press. In press.

Freedman, J. C., and T. S. Novak. 1983. Membrane potentials associated with Ca-induced K conductance in human red blood cells: studies with a fluorescent oxonol dye, WW 781. *Journal of Membrane Biology.* 72:59–74.

Freedman, J. C., and T. S. Novak. 1988. Optical measurement of membrane potential in cells, organelles, and vesicles. *Methods in Enzymology.* In press.

George, E. B., P. Nyirjesy, M. Basson, L. E. Ernst, P. R. Pratap, J. C. Freedman, and A. S. Waggoner. 1988. Membrane potential sensitivity of the impermeant oxonol dyes. I. Wavelength and polarization dependence of the light absorption signal. *Journal of Membrane Biology.* In press.

Grinvald, A., R. Hildsheim, I. C. Farber, and L. Anglister. 1982. Improved fluorescent probes for the measurement of rapid changes in membrane potential. *Biophysical Journal.* 39:301–308.

Johnson, L. V., M. L. Walsh, B. J. Bocus, and L. B. Chen. 1981. Monitoring of relative mitochondrial membrane potential in living cells by fluorescence microscopy. *Journal of Cell Biology.* 88:526–535.

Sims, P. J., A. S. Waggoner, C. H. Wang, and J. F. Hoffman. 1974. Studies on the mechanism by which cyanine dyes measure membrane potential in red blood cells and phosphatidylcholine vesicles. *Biochemistry.* 13:3315–3323.

Waggoner, A. S. 1985. Dye probes of cell, organelle, and vesicle membrane potentials. *In* The Enzymes of Biological Membranes. Vol. 1. 2nd ed. A. Martonosi, editor. 313–331.

Whitin, J. C., R. A. Clark, E. R. Simons, and J. J. Cohen. 1981. Effects of the myeloperoxidase system on fluorescent probes of granulocyte membrane potential. *Journal of Biological Chemistry.* 256:8904–8906.

Dye Indicators of
Membrane Potentials
in Blood Cells

Chapter 19

Membrane Potential and the Cytotoxic Ca Cascade of Human Red Blood Cells

Jeffrey C. Freedman, Ellen M. Bifano, Lynn M. Crespo, Promod R. Pratap, Ronald Walenga, R. Eugene Bailey, Susan Zuk, and Terri S. Novak

Departments of Physiology, Pediatrics, and Pharmacology, State University of New York Health Science Center at Syracuse, Syracuse, New York

Cell Physiology of Blood © 1988 by The Rockefeller University Press

Introduction

Since the first application of fluorescent cyanine dyes to measure and monitor the transmembrane electrical potential (E_m) of human red blood cells (Hoffman and Laris, 1974; Sims et al., 1974), numerous cellular electrophysiological studies have used this technique with red cells, with other blood cells, and with suspensions of many other types of cells, isolated organelles, and vesicles. Recent studies using optical potentiometric dyes with cell and organelle suspensions have considered the role of E_m in stimulus-response systems such as secretion, mitogenesis, chemotaxis, and viral infection. Other problems studied include the electrogenicity associated with active transport mediated by the Na,K pump, the Ca pump, and the proton pump, and with symports, antiports, and conductive fluxes (see Freedman and Laris, 1988, for review). With blood cells, optical potentiometric indicators have been used for current-voltage analysis of red cells and for studies of stimulus-response coupling in neutrophils, lymphocytes, platelets, macrophages, and cultured cells. Background information on the major classes of indicators, a summary of current methodology, with specific protocols for using slow and fast dyes with red cells, methods for calibrating optical potentiometric signals to millivolts, and a discussion of precautions and controls for interpreting dye responses can be found elsewhere (Freedman and Novak, 1988; see also Cohen and Hoffman, 1982; Laris and Hoffman, 1986). Tabulations identifying which particular dyes have been used with which types of cells by which investigators are also available (Freedman and Laris, 1981, 1988). While many studies have used permeant cyanine and oxonol dyes, this chapter summarizes results obtained in our laboratory with human red blood cells using an impermeant oxonol dye, WW781.

Electrophysiological studies of red cells using optical potentiometric indicators have focused on the relationship between cyanine dye fluorescence and E_m (Hladky and Rink, 1976; Tsien and Hladky, 1978; Freedman and Hoffman, 1979a, b), electrogenicity associated with Na pumping (Hoffman et al., 1979, 1980; Dissing and Hoffman, 1983; Hoffman and Laris, 1984), effects of Ca on E_m (Freedman and Novak, 1983), the relationship between red cell shape and E_m (Bifano et al., 1984), and determinations of Cl conductance and its voltage dependence (Freedman and Novak, 1984, 1987; Freedman and Miller, 1984). In certain red cell ion-transport experiments, including the case of cells treated with valinomycin (Hunter, 1977; Knauf et al., 1977), it is critical to measure the magnitude of E_m so as to avoid making arbitrary assumptions about the relationship between voltage and fluxes. This chapter summarizes recent progress concerning the measurement of E_m and Cl conductance, and discusses the possible relationship of E_m to a set of Ca-induced reactions, which we refer to collectively as the "cytotoxic Ca cascade."

WW781: An Impermeant Potentiometric Indicator

WW781 tracks the propagation of nerve and cardiac action potentials (Gupta et al., 1981; Dillon and Morad, 1981) and also gives signals associated with excitation-contraction coupling in skeletal muscle (Baylor et al., 1984). With red cell suspensions, the fluorescent potentiometric signals of WW781 have proved useful in the study of Ca-activated K conductance (Freedman and Novak, 1983) and of Cl conductance (Freedman and Novak, 1984, 1987). This divalent, anionic blue oxonol dye is thought to be impermeant because it contains a sulfonate group, or "anchor charge," which was designed by A. S. Waggoner to keep one end of the dye in the external aqueous medium. A second negative charge is delocalized. Typical kinetic responses to hyperpolarization of human red cells by addition of valinomycin or the Ca ionophore A23187 in the presence of 1 mM extracellular Ca (Ca_o) are shown in Fig. 1. Inset *A* shows that WW781 responds within the mixing time of magnetically stirred suspensions in cuvettes, ~1–2 s. Inset *B* shows identical null points with valinomycin and with Ca plus A23187. WW781 has a sensitivity to voltage of 0.13 %ΔF/mV between 0 and − 120 mV; this calibration is based on simultaneous measurements of extracellular pH

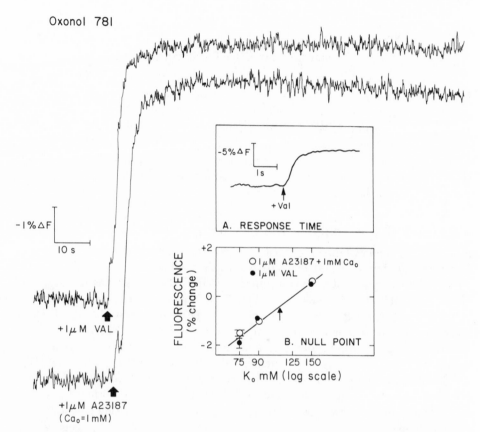

Figure 1. Fluorescent potentiometric responses of WW781. The traces above show decreases in dye fluorescence after adding valinomycin (upper trace) or A23187 (lower trace) to suspensions of human red blood cells. Upward deflections correspond to decreased fluorescence. Inset *A*, at a higher chart speed, shows the response to be complete within 1 s. Inset *B* shows identical null points obtained with valinomycin and with Ca plus A23187. The arrow designating the null point is displaced to −0.4%ΔF to correct for the effect of ethanol. (Adapted with permission from Freedman and Novak, 1983.)

(pH_o) and fluorescence in unbuffered suspensions of DIDS-treated red cells (Freedman and Novak, 1983, 1984, 1988). Optimal sensitivity depends on the proper choice of cell and dye concentrations (Freedman and Novak, 1988). In a series of alcohols of progressively lowered dielectric constant, the fluorescence intensity of WW781 increases by 15 times; it also increases 20 times upon equilibration with red cells (Freedman and Novak, 1983). Binding studies with intact red cells by P. R. Pratap

(George et al., 1988) indicate that WW781 responds by an on-off mechanism; the anionic dye partitions out of the membrane upon hyperpolarization, thus decreasing the total fluorescence of the suspension.

WW781 was chosen over 3,3'-dipropylthiadicarbocyanine [diS-C$_3$(5)] for the study of red cell Cl conductance because the range of its linear response to voltage is at least twice as great as that of the cyanine dye (Freedman and Novak, 1983). This large dynamic range permits measurements of ΔE_m of ~100 mV, as occurs after adding valinomycin to cells whose Cl conductance is inhibited by DIDS. A survey of 20 structural analogues of WW781 (all generously provided by A. S. Waggoner) indicated that WW781 had the best sensitivity of the series and the least interaction with valinomycin. Such interfering reagent interactions are detected either by spectral shifts in suspensions of white ghost membranes or by an offset of the null point. The agreement between the null points determined with WW781 and with diS-C$_3$(5) (Freedman and Novak, 1983), and between those determined with valinomycin and with Ca plus A23187 (Fig. 1, inset *B*), indicates an absence of significant interference. Despite its relative utility, WW781 does stimulate the DIDS-sensitive Cl conductance (but not the DIDS-insensitive Cl conductance), and also gives some signals unrelated to E_m at high levels of cytosolic Ca (Ca$_c$) (Freedman and Novak, 1983). Additional studies with other impermeant dyes are needed for further quantitative characterization of red cell conductive transport systems.

Current-Voltage Analysis and Cl Conductance

Studies of net K and Cl fluxes induced either with valinomycin or gramicidin at varied outward K gradients suggest that DIDS-insensitive Cl conductance is dependent on voltage. This property of Cl conductance was first noted in studies using WW781 (Freedman and Novak, 1984) (see Fig. 2, left panel, filled squares), and was subsequently confirmed in the absence of the dye (Freedman and Novak, 1987). The basis for this inference is that maximal inhibition by DIDS or SITS is only 65% at hyperpolarizing potentials, but increases and becomes virtually complete as E_m approaches the normal resting potential of -9 mV. To the extent that DIDS specifically reacts with capnophorin, this unexpected result suggests that Cl conductance may be entirely mediated by the anion-exchange protein at normal E_m, and that a DIDS-insensitive pathway becomes appreciable at hyperpolarizing potentials below about -40 mV.

Changes in WW781 fluorescence induced by valinomycin or gramicidin were calibrated to millivolt values in unbuffered, DIDS-treated suspensions by measuring changes in pH$_o$ after addition of FCCP (Freedman and Novak, 1983, 1984, 1987, 1988). This method of dye calibration is based on studies of E_m estimated from pH$_o$ by Macey et al. (1978) and Wieth et al. (1980). Characteristic plots of E_m vs. K$_o$ for valinomycin-treated cells obtained with WW781 are well described by the constant-field equation with single values of $P_{K \cdot VAL}/P_{Cl}$ either in the presence or absence of DIDS (Freedman and Novak, 1984; cf. Bennekou, 1984; Bennekou and Christophersen, 1986). Valinomycin-treated red cells can be considered to be electrically equivalent to two parallel resistors, each in series with a battery representing the concentration gradients (Tosteson et al., 1973). Accordingly, current-voltage plots for Cl and for K·valinomycin were constructed from measured net K and Cl effluxes (converted to currents by the Faraday) and from calibrated fluorescence measurements of voltages

(Fig. 2, left panel). The results indicate that the DIDS-insensitive Cl conductance becomes appreciable below -40 mV (filled squares).

For cells not treated with DIDS, the voltage changes induced by valinomycin at varied K_o, as obtained by the dye method (calibrated with DIDS-treated cells), are as much as 30 mV smaller than the voltage changes measured in the same suspensions, as estimated by the pH method of Macey et al. (1978). This substantial difference indicates a serious problem with at least one of the methods. Despite this quantitative discrepancy between the dye method and the pH method in the absence of DIDS, both methods indicate the same general pattern for K·valinomycin conductance (Fig. 2, open symbols) and for Cl conductance (Fig. 2, filled symbols). To attempt to resolve the quantitative discrepancy between the dye and pH methods, studies are in progress using the distribution of the radioactively labeled lipophilic cations tetraphenylphos-

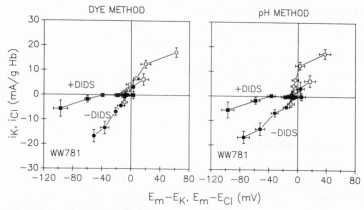

Figure 2. Current-voltage plots for valinomycin-treated human red blood cells both treated (squares) and not treated (circles) with DIDS. For each K_o, net K and Cl fluxes were measured and converted to currents, and were then plotted vs. the driving forces, $E_m - E_K$ and $E_m - E_{Cl}$. E_m was determined with WW781 in parallel suspensions of the same cells. The slopes indicate the conductances, $g_{K·Val}$ (open symbols) and g_{Cl} (filled symbols). The points with error bars represent the means $\pm$ 1 SD from six experiments in which E_m was measured. In four of the experiments, net K and Cl fluxes were also determined. For the results in both panels, voltages in the presence of DIDS were calibrated by the ΔpH method. For the left panel, voltages in the absence of DIDS were determined from WW781 fluorescence, as calibrated in the presence of DIDS. For the right panel, voltages in the absence of DIDS were determined by the ΔpH method (see Freedman and Novak, 1984, 1988).

phonium (TPP) and triphenylmethylphosphonium (TPMP) as an independent method of measuring the diffusion potentials (see Deutsch et al., 1979; Cheng et al., 1980).

Given the tendency of anionic oxonols to interact with valinomycin, it is possible that reagent interactions might underlie the 30-mV discrepancy between the dye method and the pH method. Such interactions would have to be dependent on voltage or K in order to account for the fact that the null point is not displaced with WW781. Another possibility is that the pH method might overestimate the voltages in the absence of DIDS. After addition of valinomycin, the cells shrink because of KCl efflux. As the hemoglobin concentration rises, its increased osmotic coefficient results in sufficient water retention to lower the intracellular concentrations of K and Cl

(Freedman and Novak, 1987). This effect results in an increased inward Cl gradient. By using a pH-stat apparatus, significant valinomycin-induced, DIDS-sensitive proton influxes were measured at varied K_o (Freedman and Novak, 1987). Such proton fluxes could either be conductive or be mediated as inward HCl cotransport, as first postulated for other experimental circumstances by Jennings (1978). Proton influxes coupled to the Cl gradient would be in the right direction to cause the pH method to overestimate the valinomycin-induced voltages in the absence of DIDS. With three methods of measuring E_m in red cells—fluorescent dyes, unbuffered pH changes, lipophilic ion distribution—at least two must agree before confidence can be placed in the quantitative results.

The Cytotoxic Ca Cascade

Elevated Ca_c in human red blood cells, as in many other cells, stimulates a cytotoxic cascade of pathophysiological and biochemical changes (Weed et al., 1969; see Wiley and McCulloch, 1982, for review). Some of these events are depicted in Fig. 3. This

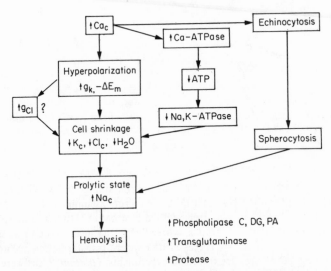

Figure 3. The cytotoxic Ca cascade of human red blood cells. The scheme illustrates the effects of elevated intracellular Ca in human red blood cells. The relationship of the activated enzymatic activities to the permeability and morphology changes is not understood.

scheme proposes that the rise in K permeability, first attributed to Ca by Gardos (1959), is an early event in a cytotoxic Ca cascade. KCl loss leads directly to cell shrinkage, which in turn leads to the prolytic state and to eventual hemolysis. Elevated Ca_c also stimulates the Ca-ATPase, causing depletion of ATP and inhibition of the volume-regulatory function of the Na,K-ATPase. ATP depletion is prevented by inhibiting the ATPases with vanadate, but echinocytosis still develops (Bifano et al., 1985).

Elevated Ca_c induces different cellular reactions with widely varying potencies. Addition of A23187 stimulates KCl loss at 0.1 μM free external Ca (Ca_f); the membrane hyperpolarizes within 2 s, as detected by WW781 (Fig. 1, bottom trace; Fig. 4, top; Fig. 5, dotted line). The fluorescence changes of WW781 induced by

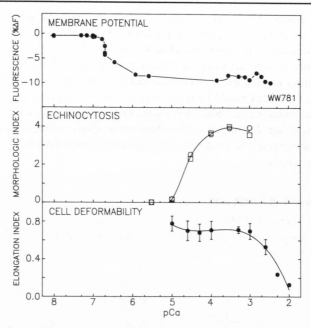

Figure 4. The dose-response curves for Ca activation of hyperpolarization (top), echinocytosis (center), and deformability (bottom). For all experiments, fresh human red blood cells were washed as described previously (Bifano et al., 1984) and used for experiments on the same day. (*Top*) Red cells were suspended to 1.2% hematocrit (HCT) in media containing 1 mM KCl, 7 mM EGTA, 5 mM HEPES (pH 7.4), and $(145 - 1.5x)$ NaCl, where $x = 0$–10 mM $CaCl_2$, and 3.5 μM WW781. The fluorescence of the potentiometric oxonol dye WW781 was monitored at 37°C before and after addition of 1 μM A23187, as described elsewhere (Freedman and Novak, 1988). Free Ca_o was measured with a Ca electrode. A representative experiment is shown. (*Center*) Cells were suspended at 2% HCT in media containing 90 mM KCl, 60 mM NaCl, 5 mM HEPES (pH 7.4), 25 μM EGTA, 0.01% bovine serum albumin, and 0–1 mM Ca_f at 25°C. Photomicrographs were taken 5 min after addition of 4 μM A23187. The morphologic index, MI, was assessed from photomicrographs, where MI is the mean score of at least 200 cells (scored 0–5) on the basis of the six stages of disk-sphere transformation according to Bessis (1972). The circles and squares are from two experiments. (*Bottom*) Cells were suspended to 0.125% HCT in media containing 90 mM KCl, $(33.5 - 1.5x)$ NaCl, where $x = 0$–10 mM Ca_o, 25 μM EGTA, 6.3 mM Na_2HPO_4, 23% dextran, pH 7.4 at 25°C, with the osmolarity adjusted with NaCl to 300 mosmol. 60 min after addition of 2 μM A23187, the elongation index, or the axial ratio of the laser diffraction pattern, was recorded at 256 dyne/cm^2 shear stress using the Ektacytometer (method of Heath et al., 1982). The points represent means $\pm$ 1 SD of three donors.

A23187 in the presence of Ca_o are in qualitative accord with many known properties of Ca-induced K fluxes in red cells (Freedman and Novak, 1983). The threshold for Ca-induced echinocytosis (assessed by the morphologic index) is two orders of magnitude greater at 10 μM Ca_f (Fig. 4, middle). Thus, it is possible to set Ca_o at a level sufficient to induce hyperpolarization, but below the threshold for echinocytosis (Fig. 5, half-filled and open symbols; from Bifano et al., 1984). Under similar conditions, the Ca-induced decrease in cell deformability, assessed by a decreased elongation index during flow at set shear stress (measured by ektacytometry, the method of Health et al., 1982), occurs at 1 mM Ca_f (Fig. 4, bottom). The separate apparent potencies of Ca, spanning five orders of magnitude, for inducing hyperpolarization, echinocytosis, and reduced deformability are consistent with the hypothesis of independent sites of action for these elements of the cytotoxic Ca cascade.

In order to link hyperpolarization with subsequent events in the cascade, it is desirable to measure Ca_c in the physiological range. The fluorescent Ca indicator quin2 was evaluated with intact red cells and resealed ghosts. Absorption by hemoglobin interfered with quin2 fluorescence in intact cells, but with resealed ghosts both active efflux and passive influx of Ca could be monitored (James-Kracke and Freedman, 1985).

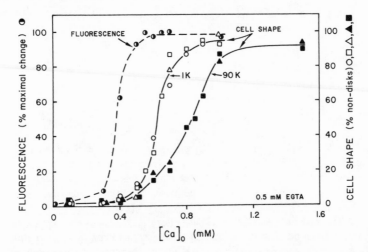

Figure 5. The effect of Ca on the change in diS-C₃(5) fluorescence (dotted line, left ordinate) and on the shape of human red blood cells (solid lines, right ordinate). Cell suspensions were prepared in media containing 0–1.5 mM $CaCl_2$, x mM KCl ($x = 1$ or 90), 0.5 mM EGTA, 5 mM HEPES (pH 7.35 at 25°C), 0.01% bovine serum albumin, and sufficient NaCl ($150 - x - 3/2$ Ca) so that each solution was 300 ideal milliosmolar (imOsm). 15 min after the addition of 0.8 μM A23187, at least 200 cells of each suspension were observed, using Nomarski interference contrast microscopy, photographed, and counted, and the percentage of nondiscoid cells was recorded. The dotted line indicates the percentage changes of fluorescence (%ΔF) of diS-C₃(5), as determined at 1 mM K_o and varied Ca_o under conditions equivalent to those used to observe the changes in cell shape. (From Bifano et al., 1984, with permission).

Upon treatment with A23187, red cells can be hyperpolarized but they remain discoidal (Fig. 5). However, E_m, as set by K_o, modulates the threshold for Ca-induced echinocytosis (Fig. 5, solid lines). This effect occurs because the altered E_m affects the influx of Ca and the resultant Ca_c concentration (Bifano et al., 1984). If a similar effect occurs in the absence of A23187, positive feedback would be present in the cytotoxic Ca cascade.

As illustrated by the scheme drawn in Fig. 3, the relationships of the activation by Ca of a transglutaminase (Lorand et al., 1979), a phospholipase C (Allan et al., 1976; Moore and Appel, 1984), and a protease (Lorand et al., 1983) to the permeability, cell volume, and morphology changes are unknown. While micromolar Ca dramatically affects the viscosity of complexes of spectrin, actin, and band 4.1 (Fowler and Taylor, 1980), the relative contributions of lipid and cytoskeletal alterations induced by Ca to the morphological changes is similarly unclear. With red cells exposed to A23187 and either 100 μM or 1 mM Ca_o, the morphologic index reaches 4–5 (sphero-echinocytes), but thin-layer chromatography of extracted membrane lipids performed by R. E. Bailey and R. Walenga (Table I) revealed no increase in lysolecithin. Other lipid alterations, including phosphoinositide hydrolysis and phosphatidic acid production, occur at 1–10 μM Ca (Allan et al., 1976; Moore and Appel, 1983).

The final consequence of increased Ca_c in vitro, after prolonged cell shrinkage caused by loss of KCl, is the development of prolytic cells leaky to Na, which then undergo colloid osmotic hemolysis. Experiments by Crespo et al. (1987) suggest that cell shrinkage, rather than Ca itself, results in the gain of Na. The observation that comparable Na gains occur in cells simply shrunken by sucrose, as upon exposure to

TABLE I
Lack of Lysophosphatidylcholine Accumulation after Ca-induced Echinocytosis

Ca_o	Time	Lysophosphatidylcholine
mM	*min*	*%*
0.0	0	1.2, 1.3
0.0	60	1.4, 1.2
0.1	60	0.9, 0.8
1.0	60	0.9, 0.9

Fresh washed human red blood cells were suspended at 10% HCT in media containing 60 mM NaCl, 90 mM KCl, 25 μM EGTA, 5 mM HEPES (pH 7.4), and either 0.1 or 1.0 mM $CaCl_2$ plus 20 μM A23187. Control suspensions lacked both Ca and A23187. Portions of the suspensions were removed at time zero or 60 min after addition of A23187, and white ghosts were prepared (Dodge et al., 1963). The phospholipids were extracted (Bligh and Dyer, 1959), separated by thin-layer chromatography, and identified by means of lipid standards. The spots were scraped and phosphorous assays were performed (Ames, 1966) to determine lysophosphatidylcholine as a percentage of total phospholipid. Data are from two experiments.

valinomycin or to Ca plus A23187 at low K_o (Fig. 6, left), suggests that Ca is not essential in causing elevated Na_c. Other flux studies and assessment of relative cell volume distributions (Fig. 6, right) indicate that a subpopulation of cells become highly permeable to Na after shrinkage, induced either osmotically or by Ca-induced KCl loss. An apparent steady state level of elevated Na_c is attained because a fraction of the cells become prolytic while another fraction hemolyzes. An interesting question is what determines which cells enter the prolytic state. The older, denser cells, isolated by ultracentrifugation, demonstrated the same Na uptake in response to Ca stress as the less dense, younger cells. These results reconciled previous reports by others in that where Ca entry had resulted in elevated Na_c, cell shrinkage was also present; in other reports where Ca had no effect on cell Na, shrinkage was not present.

In order to determine the function of the cytotoxic Ca cascade, a potent inhibitor would be helpful. Charybdotoxin (CTX; purified by and obtained from C. Miller; see Smith et al., 1986) is just such a potent inhibitor of Ca-activated K conductance in human red blood cells (Castle and Strong, 1986; Cecchi et al., 1987). The results in Fig. 7 indicate that half-maximal inhibition of Ca-induced ^{86}Rb efflux occurs at nanomolar toxin concentrations. The changes of WW781 fluorescence associated with Ca-activated K conductance are also inhibited by CTX (not shown).

Conclusions

The features of the cytotoxic Ca cascade suggest that it may have a physiological or pathophysiological function, as yet unknown. While WW781 has been useful in

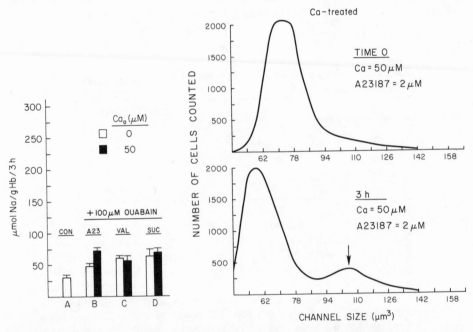

Figure 6. Prolytic state of human red blood cells is induced by cell shrinkage. The bar graphs show the effects of no additions (*A*), Ca plus 2 μM A12187 (*B*), 1 μM valinomycin (*C*), and 0.5 M sucrose (*D*) on the net gain of cell Na after incubation for 3 h. Cells were incubated as described in the presence (filled bars) or absence (open bars) of 50 μM Ca$_o$, and with 100 μM ouabain (*B–D*). Direct evidence for a subpopulation of swollen, prolytic cells after prolonged shrinkage is revealed by the bimodal relative cell volume distribution determined in a Coulter Channelyzer. (Adapted from Crespo et al., 1987, with permission.)

characterizing Ca-activated K conductance and Cl conductance in human red blood cells, a nontoxic dye is needed for further quantitative studies. Moreover, quantitative agreement between at least two of the three methods for measuring voltages is needed before reliable quantitative inferences can be drawn. Despite these problems, recent results both in the presence and absence of WW781 suggest that voltage either affects DIDS-insensitive Cl conductance or, alternatively, the ability of DIDS to inhibit Cl conductance. If hyperpolarization increases Cl conductance in the absence of DIDS, then cell shrinkage after initiation of the Gardos effect would be facilitated.

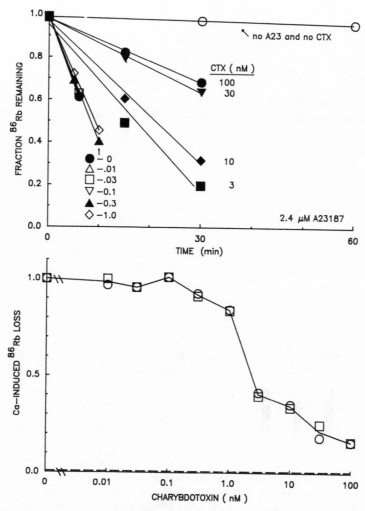

Figure 7. Effect of CTX on Ca-induced ^{86}Rb efflux. Fresh washed cells were loaded overnight with tracer, washed five times in ice-cold tracer-free media containing 1 mM KCl, 149 mM NaCl, and 5 mM HEPES buffer (pH 7.4 at 25°C), and then added to flux media containing 1 mM KCl, 141.5 mM NaCl, 5 mM HEPES buffer, 3 mM H-EDTA, and 1 mM CaCl$_2$ (pH 7.2 at 37°C). After 1 min, CTX was added, and after another 5 min the flux was initiated by adding 2.4 μM A23187. At desired times using longer times for slower fluxes (top panel), each suspension was sampled, and the supernatants and hemolysates were counted to determine the fraction of ^{86}Rb lost from the cells. The data shown in the bottom panel are from two experiments (circles and squares). The effluxes were normalized with respect to the flux in the absence of inhibitor. Other methods were as described previously (Freedman and Novak, 1983).

Acknowledgments

This research was supported by grant GM-28839 from the National Institutes of Health.

References

Allan, D., R. Watts, and R. H. Michell. 1976. Production of 1,2-diacylglycerol and phosphatidate in human erythrocytes treated with calcium ions and ionophore A23187. *Biochemical Journal.* 156:225–232.

Ames, B. N. 1966. Assay of inorganic phosphate, total phosphate and phosphatases. *Methods in Enzymology.* 8:115–118.

Baylor, S. M., W. K. Chandler, and M. W. Marshall. 1984. Calcium release and sarcoplasmic reticulum membrane potential in frog skeletal muscle fibres. *Journal of Physiology.* 348:209–238.

Bennekou, P. 1984. K^+-valinomycin and chloride conductance of the human red cell membrane. Influence of the membrane protonophore carbonylcyanide *m*-chlorophenylhydrazone. *Biochimica et Biophysica Acta.* 776:1–9.

Bennekou, P., and P. Christophersen. 1986. Flux ratio of valinomycin-mediated K^+ fluxes across the human red cell membrane in the presence of the protonophore CCCP. *Journal of Membrane Biology.* 93:221–227.

Bessis, M. 1972. Red cell shapes. An illustrated classification and its rationale. *In* Red Cell Shape. M. Bessis, R. I. Weed, and P. F. Leblond, editors. Springer-Verlag, New York, NY. 1–24.

Bifano, E. M., S. M. Becker, and J. C. Freedman. 1985. Calcium-induced echinocytosis in human red blood cells from neonates and adults. *Pediatric Research.* 19:258*A*. (Abstr.)

Bifano, E. M., T. S. Novak, and J. C. Freedman. 1984. The relationship between the shape and the membrane potential of human red blood cells. *Journal of Membrane Biology.* 82:1–13.

Bligh, E. G., and W. J. Dyer. 1959. A rapid method of total lipid extraction and purification. *Canadian Journal of Biochemistry and Physiology.* 37:911–917.

Castle, N. A., and P. N. Strong. 1986. Identification of two toxins from scorpion (*Leiurus quinquestriatus*) venom which block distinct classes of calcium-activated potassium channel. *FEBS Letters.* 209:117–121.

Cecchi, X., D. Wolff, and M. Canessa. 1987. Charybdotoxin blocks the Ca-activated K channel in Hb AA and Hb SS human red cells. *Journal of General Physiology.* 90:12*a*. (Abstr.)

Cheng, K., J. C. Haspel, M. L. Vallano, B. Osotimehin, and M. Sonenberg. 1980. Measurement of membrane potentials (Ψ) of erythrocytes and white adipocytes by the accumulation of triphenylmethylphosphonium cation. *Journal of Membrane Biology.* 56:191–201.

Cohen, L. B., and J. F. Hoffman. 1982. Optical measurements of membrane potential. *In* Techniques in Cellular Physiology. Elsevier/North-Holland, New York, NY. P118:1–13.

Crespo, L. M., T. S. Novak, and J. C. Freedman. 1987. Calcium, cell shrinkage, and prolytic state of human red blood cells. *American Journal of Physiology.* 252:C123–C152.

Deutsch, C. J., A. Holian, S. K. Holian, R. P. Daniele, and D. F. Wilson. 1979. Transmembrane electrical and pH gradients across human erythrocytes and human peripheral lymphocytes. *Journal of Cell Physiology.* 99:79–94.

Dillon, S., and M. Morad. 1981. A new laser scanning system for measuring action potential propagation in the heart. *Science.* 214:453–456.

Dissing, S., and J. F. Hoffman. 1983. Anion-coupled Na efflux mediated by the Na/K pump in human red blood cells. *Current Topics in Membranes and Transport.* 19:693–695.

Dodge, J. T., C. Mitchell, and D. J. Hanahan. 1963. The preparation and chemical characteristics of hemoglobin-free ghosts of human erythrocytes. *Archives of Biochemistry and Biophysics.* 100:119–130.

Fowler, V., and D. L. Taylor. 1980. Spectrin plus band 4.1 cross-link actin. Regulation by micromolar calcium. *Journal of Cell Biology.* 85:361–376.

Freedman, J. C., and J. F. Hoffman. 1979a. Ionic and osmotic equilibria of human red blood cells treated with nystatin. *Journal of General Physiology.* 74:157–185.

Freedman, J. C., and J. F. Hoffman. 1979b. The relation between dicarbocyanine dye fluorescence and the membrane potential of human red blood cells set at varying Donnan equilibria. *Journal of General Physiology.* 74:187–212.

Freedman, J. C., and P. C. Laris. 1981. Electrophysiology of cells and organelles: studies with optical potentiometric indicators. *International Review of Cytology. Suppl.* 12:177–246.

Freedman, J. C., and P. C. Laris. 1988. Optical measurements of membrane potential in cell and organelle suspensions. *In* Spectroscopic Membrane Probes. Vol. 3. L. Loew, editor. CRC Press, Boca Raton, FL. pp. 1–49.

Freedman, J. C., and C. Miller. 1984. Membrane vesicles from human red blood cells in planar lipid bilayers. *Annals of the New York Academy of Sciences.* 435:541–544.

Freedman, J. C., and T. S. Novak. 1983. Membrane potentials associated with Ca-induced K conductance in human red blood cells. Studies with a fluorescent oxonol dye, WW781. *Journal of Membrane Biology.* 72:59–74.

Freedman, J. C., and T. S. Novak. 1984. K and Cl conductance of valinomycin-treated human red blood cells, as determined with the fluorescent potentiometric indicator WW781. *Journal of General Physiology.* 84:18a (Abstr.)

Freedman, J. C., and T. S. Novak. 1987. Chloride conductance of human red blood cells at varied E_K. *Biophysical Journal.* 51:565a. (Abstr.)

Freedman, J. C., and T. S. Novak. 1988. Optical measurement of membrane potential in cells, organelles, and vesicles. *Methods in Enzymology.* 172: In press.

Gardos, G. 1959. The role of calcium in the potassium permeability of human erythrocytes. *Acta Physiologica Academiae Scientarium Hungaricae.* 15:121–125.

George, E. B., P. Nyirjesy, M. Basson, L. E. Ernst, A. S. Waggoner, P. R. Pratap, and J. C. Freedman. 1988. Impermeant potential-sensitive oxonol dyes. I. Evidence for an on-off mechanism. *Journal of Membrane Biology.* 103:245–253.

Gupta, R. K., B. M. Salzberg, A. Grinvald, L. B. Cohen, K. Kamino, S. Lesher, M. B. Boyle, A. S. Waggoner, and C. H. Wang. 1981. Improvements in optical methods for measuring rapid changes in membrane potential. *Journal of Membrane Biology.* 58:123–138.

Heath, B. P., N. Mohandas, J. L. Wyatt, and S. B. Shohet. 1982. Deformability of isolated red blood cell membranes. *Biochimica et Biophysica Acta.* 69:211–219.

Hladky, S. B., and T. J. Rink. 1976. Potential difference and the distribution of ions across the human red blood cell membrane: a study of the mechanism by which the fluorescent cation diS-C_3(5) reports membrane potential. *Journal of Physiology.* 263:287–319.

Hoffman, J. F., J. H. Kaplan, and T. J. Callahan. 1979. The Na:K pump in red cells is electrogenic. *Federation Proceedings.* 38:2440–2441.

Hoffman, J. F., J. H. Kaplan, T. J. Callahan, and J. C. Freedman. 1980. Electrical resistance of the red cell membrane and the relation between net anion transport and the anion exchange mechanism. *Annals of the New York Academy of Sciences.* 341:357–360.

Hoffman, J. F., and P. C. Laris. 1974. Determination of membrane potential in human and *Amphiuma* red blood cells by means of a fluorescent probe. *Journal of Physiology.* 239:519–552.

Hoffman, J. F., and P. C. Laris. 1984. Membrane electrical parameters of normal human red blood cells. *In* Electrogenic Transport: Fundamental Principles and Physiological Implications. M. P. Blaustein and M. Lieberman, editors. Raven Press, New York, NY. 287–293.

Hunter, M. J. 1977. Human erythrocyte anion permeabilities measured under conditions of net charge transfer. *Journal of Physiology.* 268:35–49.

James-Kracke, M. R., and J. C. Freedman. 1985. Ca transport monitored by quin-2 fluorescence in human red blood cell ghosts. *Annals of the New York Academy of Sciences.* 463:389–391.

Jennings, M. L. 1978. Characteristics of CO_2-independent pH equilibration in human red blood cells. *Journal of Membrane Biology.* 40:365–391.

Knauf, P. A., G. F. Fuhrmann, S. Rothstein, and A. Rothstein. 1977. The relationship between anion exchange and net anion flow across the human red blood cell membrane. *Journal of General Physiology.* 69:363–386.

Laris, P. C., and J. F. Hoffman. 1986. Optical determination of electrical properties of red blood cell and Ehrlich ascites tumor cell membranes with fluorescent dyes. *In* Optical Methods in Cell Physiology. P. De Weer and B. M. Salzberg, editors. John Wiley & Sons, New York, NY. *Society of General Physiologists Series.* 40:199–210.

Lorand, L., O. J. Bjerrum, M. Hawkins, L. Lowe-Krentz, and G. E. Siefring. 1983. Degradation of transmembrane proteins in Ca^{2+}-enriched human erythrocytes. *Journal of Biological Chemistry.* 258:5300–5305.

Lorand, L., G. E. Siefring, and L. Lowe-Krentz. 1979. Enzymatic basis of membrane stiffening in human erythrocytes. *Seminars in Hematology.* 16:65–74.

Macey, R. I., J. S. Adorante, and F. W. Orme. 1978. Erythrocyte membrane potentials determined by hydrogen ion distribution. *Biochimica et Biophysica Acta.* 512:284–295.

Moore, R. B., and S. H. Appel. 1984. Calcium-dependent hydrolyses of polyphosphoinositides in human erythrocyte membranes. *Canadian Journal of Biochemistry and Cell Biology.* 62:363–368.

Sims, P. J., A. S. Waggoner, C. Wang, and J. F. Hoffman. 1974. Studies on the mechanism by which cyanine dyes measure membrane potential in red blood cells and phosphatidylcholine vesicles. *Biochemistry.* 13:3315–3330.

Smith, C., M. Phillips, and C. Miller. 1986. Purification of charybdotoxin, a specific inhibitor of the high-conductance Ca^{2+}-activated K^+ channel. *Journal of Biological Chemistry.* 261:14607–14613.

Tosteson, D. C., R. B. Gunn, and J. O. Wieth. 1973. Chloride and hydroxyl ion conductance of sheep red cell membrane. *In:* Erythrocytes, Thrombocytes, Leukocytes. E. Gerlach, K. Moser, E. Deutsch, and W. Wilmans, editors. G. Thieme, Stuttgart. 62–66.

Tsien, R. Y., and S. B. Hladky. 1978. A quantitative resolution of the spectra of a membrane potential indicator, diS-C_3(5), bound to cell components and to red blood cells. *Journal of Membrane Biology.* 38:73–97.

Weed, R. I., P. L. LaCelle, and E. W. Merrill. 1969. Metabolic dependence of red cell deformability. *Journal of Clinical Investigation.* 48:795–809.

Wieth, J. O., J. Brahm, and J. Funder. 1980. Transport and interactions of anions and protons in the red blood cell membrane. *Annals of the New York Academy of Sciences.* 341:394–418.

Wiley, J. S., and K. E. McCulloch. 1982. Calcium ions, drug action and the red cell membrane. *Pharmacology & Therapeutics.* 18:271–292.

Chapter 20

Measurement of Membrane Potential Responses Elicited from Blood Cells: Effect of the Dye/Cell Ratio and the Presence of an Intracellular Calcium Probe

Bruce Seligmann, William O. Haston, James S. Wasvary, and John R. Rediske

Research Department, Pharmaceutical Division, Ciba-Geigy Corp., Summit, New Jersey

Cell Physiology of Blood © 1988 by The Rockefeller University Press

Introduction

The use of charged fluorescent dyes to monitor membrane potential has become quite common over the past ten years. The use of other fluorescent probes monitoring other aspects of cell physiology has also become quite widespread, with the fluorescent calcium-chelating probes (quin2, fura-2, and indo-1) being perhaps the most frequently used. While indirect fluorescent probes clearly offer distinct advantages over other methods for measuring membrane potential and intracellular ions and permitting new experiments, numerous concerns must be addressed before the results can be reliably interpreted.

The cyanine dyes used to monitor membrane potential have a number of recognized attributes that are not generally desirable. For example, there is a membrane potential–insensitive component to the total cell fluorescence that can lead to artifacts. Also, since the dyes are positively charged, there is a considerable accumulation of dye in subcellular organelles such as mitochondria, which have a negative potential relative to the cytosol. The total cell fluorescence signal may thus contain a greater component owing to the membrane potential of mitochondria than from the plasma membrane trans-membrane potential, confusing the interpretation of data (Wilson et al., 1985; Chused et al., 1988). The fluorescence of thiol-containing cyanine dyes (such as 3,3'-dipentylthiodicarbocyanine [diS-C$_3$(5)]) is destroyed by reactive oxygen products. The cyanine dyes that are not sensitive to reactive oxygen products include 3,3'-dipentyloxacarbocyanine [diO-C$_5$(3)] and 3,3'-dipentylindocarbocyanine [diI-C$_5$(3)] (Seligmann and Gallin, 1983; Whitin et al., 1980, 1981).

The relationship of cyanine fluorescence to membrane potential depends on the dye concentration and structure. Cyanine dyes self-aggregate at high concentrations, forming nonfluorescent aggregates at a critical concentration that is characteristic of each dye structure. The dyes diO-C$_5$(3) and diI-C$_5$(3) are less prone to form aggregates than diS-C$_3$(5) (Seligmann and Gallin, 1980, 1983). Since the cyanine dyes are positively charged, they reach a higher concentration in the cytosol than in the extracellular medium. This results in a quenching of cell fluorescence at high dye concentrations, where dye aggregation occurs as it is accumulated by cells, and enhancement of fluorescence owing to solvent effects at low dye concentrations, where dye aggregation does not occur as dye is accumulated by the cells.

Alternatives to the cyanine dyes include the negatively charged oxonol dyes. The one advantage of oxonol dyes over cyanine dyes is that they are not accumulated in negatively charged subcellular organelles (Wilson and Chused, 1985). Thus, oxonol dyes are preferred in studies of cells containing a large number of mitochondria. Generally, the fluorescence yield and low uptake of oxonol dyes make their use in a spectrofluorometer difficult. Oxonol dyes are more suited for instrumentation where the measured extracellular volume is reduced (such as in a flow cytometer or microscope). However, we have selected one that gives satisfactory results when used in a spectrofluorometer assay, bis-[1,3-diisopropyl-2-barbiturate-(5)]-trimethin oxonol [di-isoC$_3$-BA(3)].

Using human neutrophils, we investigated the use of both cyanine and oxonol dyes alone and in conjunction with the negatively charged fluorescent probe of intracellular calcium, fura-2. The sensitivity of cyanine and oxonol dye fluorescence to membrane potential was assayed while both the dye and cell concentrations were varied. We observed that dye sensitivity, particularly to elicited changes in membrane potential, changed with the experimental conditions. The conclusion from these studies is that both cyanine dyes and oxonol dyes can be used reliably, but great care must be taken to use conditions such that fluorescence remains sensitive to membrane potential.

Materials and Methods

Preparation of Cells

Human blood neutrophils were purified by sedimentation in Hespan, followed by a modification of the centrifugal elutriation procedure (Berkow et al., 1983). The purified cells were kept on ice in 10 mM HEPES-buffered Hanks' solution in the absence of calcium and magnesium. In experiments involving fura-2, the cells were incubated at a concentration of 2×10^6/ml for 30 min at 37°C in Hanks' buffer containing calcium and magnesium and 0.1 μM of the acetoxymethyl ester of fura-2 (fura-2/AM), centrifuged, resuspended in fresh medium without calcium and magnesium, and stored on ice until use. 15 min before assay, the cells were suspended in the assay medium and incubated at 37°C to permit equilibration of the cells with the dye, calcium, and magnesium.

Experimental Solutions and Assay

Assays were conducted in modified Hanks' media at a low potassium concentration, denoted as 5 K (5 mM KCl, 125 mM NaCl, 1 mM $CaCl_2$, 1 mM $MgCl_2$, 11 mM glucose, 1.3 mM NaH_2PO_4, 10 mM HEPES), or at a high potassium concentration, denoted as either 120 K (120 mM KCl, 10 mM NaCl, 1 mM $CaCl_2$, 1 mM $MgCl_2$, 11 mM glucose, 1.3 mM NaH_2PO_4, 10 mM HEPES) or 110 K (110 mM KCl, 20 mM NaCl, 1 mM $CaCl_2$, 1 mM $MgCl_2$, 11 mM glucose, 1.3 mM NaH_2PO_4, 10 mM HEPES). The stimulus used was the chemotactic hexapeptide N-formyl-norleucyl-leucyl-phenylalanyl-norleucyl-tyrosyl-lysine (CHP). Assays were conducted at 37°C in disposable glass cuvettes with a 1.5-ml assay volume. The cells were maintained in suspension using a disposable magnetic stir bar, and the stimulus (15 μl volume) was injected directly into the cuvette through an injection port using a disposable-tip pipetter. Mixing time was <1 s. The instrument used was an SLM 8000 fluorometer (SLM-Aminco Urbana, IL). The membrane potential probe fluorescence was recorded with an excitation of 480 nm and an emission of 510 nm. The fura-2 excitation ratio of (343 nm)/(390 nm) was measured at 510 nm emission.

The cyanine dye diO-C_5(3) and oxonol dye di-isoC$_3$-BA(3) were obtained from Dr. Alan Waggoner (Carnegie-Mellon, Pittsburgh, PA). CHP and fura-2/AM were purchased from Sigma Chemical Co. (St. Louis, MO). All of these reagents were dissolved in dimethylsulfoxide (DMSO). Dilute working stock solutions were made in 50% DMSO/buffer at a 100-fold-higher concentration than the final assay concentration.

Results

Effect of Cell and Dye Concentration on Oxonol Dye Fluorescence

Fig. 1 shows a series of representative experiments using the oxonol dye di-isoC$_3$-BA(3), in which both the cell concentration and cyanine dye concentration were varied while the sensitivity of fluorescence to changes in membrane potential was assessed. The cells were suspended in either 5 mM K or 120 mM K buffer and the baseline fluorescence was recorded for 40 s before stimulation. Each panel represents a different dye concentration: 5, 1, and 0.5 μM. The upper tracings were obtained with 4×10^6 PMN/ml in either 5 K or 120 K buffer, and the lower tracings were obtained at cell concentrations that produced membrane potential–dependent fluorescence changes upon stimulation with the chemoattractant peptide CHP (0.1 μM). Resting fluorescence appeared to be dependent upon membrane potential in all cases since the baseline was substantially higher in 120 K than in 5 K. At each dye concentration, the increase in baseline fluorescence caused by increasing the concentration of cells suspended in 120 K was not linear, but reached a plateau around the 4×10^6 PMN/ml concentration, which indicates that at this cell concentration most if not all of the extracellular oxonol dye had been accumulated by the cells before stimulation. Thus, it was surprising that upon stimulation there was such a large increase in fluorescence (see particularly panels *A* and *B*, 120 K tracing), as though there were accumulation of

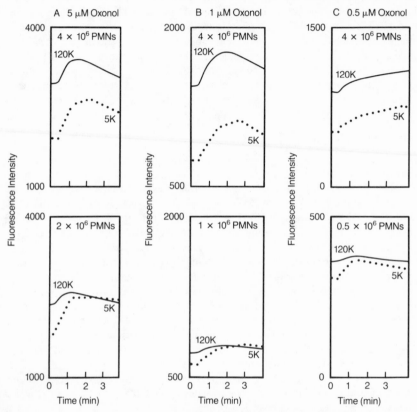

Figure 1. Effect of varying the dye and cell concentration on the fluorescence sensitivity of oxonol dye to membrane potential. The fluorescence intensity of cells depolarized by suspension in 120 mM K (solid line tracings labeled 120 K) and cells suspended in 5 mM K (dotted line tracings labeled 5 K) is shown as a function of time (min). A 40-s baseline of fluorescence was recorded before stimulation with 0.1 μM CHP. (*A*) Data obtained using 5 μM oxonol dye and either 4×10^6 PMN/ml or 2×10^6 PMN/ml, as labeled. (*B*) Data obtained using 1 μM oxonol dye and either 4×10^6 PMN/ml or 1×10^6 PMN/ml. (*C*) Data obtained using 0.5 μM oxonol dye and either 4×10^6 PMN/ml or 0.5×10^6 PMN/ml.

more dye after stimulation. The failure of high potassium to ablate the response demonstrated that at high cell concentrations the responses were not dependent on membrane potential. This is consistent with the observation that there was not sufficient dye remaining extracellularly to support such a large change in fluorescence owing to influx of additional dye. Only at lower cell concentrations were the responses dependent on membrane potential.

Effect of Cell and Dye Concentration on Cyanine Dye Fluorescence

Similar problems were documented using cyanine dyes (Fig. 2). The results obtained using two different dye concentrations, 100 nM (*A*) and 50 nM (*B*), are shown. The top tracings represent the fluorescence using 2×10^6 PMN/ml. The lower tracings were obtained using cell concentrations producing membrane potential–dependent fluorescence responses to CHP. Note the similarities to the oxonol dye studies. Increasing the cell concentration from 1 to 2×10^6 PMN/ml in the presence of 100 nM diO-C_5(3) did not result in a higher fluorescence signal in the 5 K solution, which

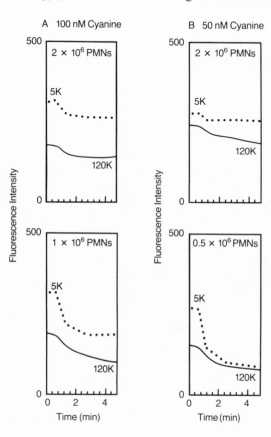

Figure 2. Effect of varying the dye and cell concentration on the fluorescence sensitivity of cyanine dye to membrane potential. The fluorescence intensity of cells depolarized by suspension in 120 mM K (solid line tracings labeled 120 K) and cells suspended in 5 mM K (dotted line tracings labeled 5 K) is shown as a function of time (min). A 40-s baseline of fluorescence was recorded before stimulation with 0.1 μM CHP. (*A*) Data obtained using 100 nM cyanine dye and either 2×10^6 PMN/ml or 1×10^6 PMN/ml, as labeled. (*B*) Data obtained using 50 nM cyanine dye and either 2×10^6 PMN/ml or 0.5×10^6 PMN/ml.

indicates that all of the dye was already cell associated. Similar results are evident with 50 nM dye comparing the 5 K baseline fluorescence with 0.5 ± 10^6 PMN/ml to that obtained with 2×10^6 PMN/ml. The stimulated responses were sensitive to membrane potential only at the lower cell concentrations.

Interaction of Fura-2 with Cyanine Dye

Fig. 3 demonstrates the effect on cyanine dye fluorescence of loading cells with the negatively charged molecule fura-2. The tracings of cyanine dye fluorescence are displayed in *A*. The intensity of baseline diO-C_5(3) fluorescence of fura-2–loaded cells in 5 K buffer (labeled 5K + F) was over twice that of cells not loaded with fura-2. The response to CHP was comparably greater, but sensitivity to membrane potential was maintained. Tracings of the fura-2 fluorescence ratio are displayed in *B*, which demonstrate that the calcium rise was not substantially affected by the extracellular potassium concentration, though the duration of the calcium rise was less in high potassium medium, consistent with the report by Di Virgilio et al. (1987).

Discussion

Oxonol Dye Sensitivity to Membrane Potential

The results presented in Figs. 1 and 2 clearly demonstrate that only with specific dye and cell concentrations is cell fluorescence truly sensitive to membrane potential changes. The nonideal behavior can be explained by considering the dynamics of dye

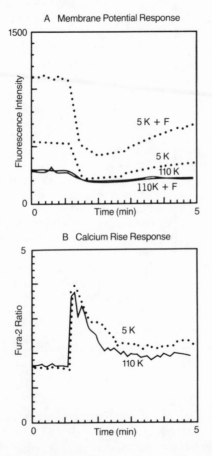

Figure 3. Effect of loading cells with fura-2 on cyanine dye fluorescence and sensitivity to membrane potential. (*A*) The fluorescence intensity of cells depolarized by suspension in 110 mM K (dotted line tracing) and cells suspended in 5 mM K (solid line tracing) is shown as a function of time (min). A 1-min baseline of fluorescence was recorded before stimulation with 0.1 μM CHP. The tracings labeled with a K concentration + F were obtained from cells loaded with fura-2. (*B*) the fura-2 fluorescence ratio of cells depolarized by suspension in 110 mM K (solid line tracing) and cells suspended in 5 mM K (dotted line tracing) is shown as a function of time (min). As in *A*, stimulus was added after 1 min.

equilibration. The driving force of membrane potential in these experiments depends only on the extracellular K concentration or cell activation. The driving force for dye accumulation was thus fixed at a physiologic potential in 5 K buffer, or at a depolarized potential in 120 K buffer, and was independent of the two variables of cell concentration and dye concentration. Oxonol dyes are negatively charged, and thus under ideal conditions maintain a steady state where the ratio of dye concentration outside the cells to the concentration inside cells is proportional to the membrane potential. As the cell concentration is increased while keeping the total dye concentration constant, the extracellular dye concentration becomes vanishingly small, an effect that is particularly acute under conditions of depolarization (120 K). Conditions are reached where most of the dye is cell associated in the resting state, and insufficient dye remains outside the cells to account for the large change in fluorescence observed upon stimulation. It is probable that the membrane potential–sensitive component of intracellular dye is simply too small to give a measurable signal when the external dye concentration is very low. Regardless, the response that is obtained does not result from the accumulation of more dye due to depolarization, but instead is an artifact due to a change in the fluorescence of dye already accumulated by the cells, possibly reflecting intracellular redistribution.

Cyanine Dye Sensitivity to Membrane Potential

The results with the cyanine dye are consistent with this line of reasoning. Since this positively charged dye is accumulated to a greater extent by cells in the resting, hyperpolarized state of 5 mM K, fluorescence in the 5 mM K solutions was higher than when cells were depolarized by suspension in 120 mM K (Fig. 2), giving the appearance that membrane potential was accurately reflected by the difference in fluorescence under all the conditions. However, as shown in Fig. 2, increasing the cell concentration from 1 to 2×10^6 PMN/ml in the presence of 100 nM diO-C_5(3) did not result in a higher fluorescence signal in the 5 mM K solution, which indicates that virtually all the dye was cell associated at the lower cell concentration. Similar results are evident with 50 nM dye comparing the 5 mM K baseline fluorescence with 0.5×10^6 PMN/ml to that obtained with 2×10^6 PMN/ml. In fact, even when depolarized in 120 mM K the baseline fluorescence with 2×10^6 PMN/ml indicated nearly all the dye was cell associated. Under the conditions of high cell concentration and insufficient cyanine dye, the stimulated responses, while suggestive of a depolarization, were abrogated. This result was consistent with the expectation that there was insufficient dye accumulated in the resting state for there to be a significant membrane potential–dependent component lost due to depolarization.

Interaction of Cyanine Dye with Fura-2

The effect of a second charged probe on dye fluorescence and sensitivity to membrane potential was also assessed. The negatively charged probe fura-2 was chosen because of the importance of intracellular Ca. As expected, there was no effect on the negatively charged oxonol dye di-isoC₃-BA(3) fluorescence or responsiveness (results not shown). However, there was increased uptake of the positively charged cyanine dye diO-C_5(3). As shown in Fig. 3, the diO-C_5(3) fluorescence remained sensitive to membrane potential even in the presence of fura-2. We observed (not shown) that the point at which all the extracellular dye was accumulated by cells was reached at a lower cell concentration with fura-2–loaded cells than with unloaded cells. This was a consequence of the fura-2–loaded cells accumulating more cyanine dye than fura-2–unloaded cells. The enhancing effect of fura-2 was greater at lower cyanine dye concentrations.

Conclusions

Taken together, these results demonstrate that great caution must be exercised when the assay conditions of the dye concentration, cell concentration, and cell type were varied, or when other chemical species that are accumulated by cells were incorporated. However, if the appropriate control experiments are performed to demonstrate the membrane potential dependence of both baseline fluorescence and stimulated responses, satisfactory results can be obtained.

References

Derkow, R. I., D. Y. Tzeng, L. V. Williams, and R. L. Baehener. 1983. The comparative responses of human polymorphonuclear leukocytes obtained by counterflow centrifugal elutriation and Ficoll-Hypaque density centrifugation: resting volume, stimulus induced superoxide production, and primary and specific granule release. *Journal of Laboratory and Clinical Medicine.* 102:732–742.

Chused, T. M., H. A. Wilson, B. Seligmann, and R. Y. Tsien. 1988. Probes for use in the study of leukocyte physiology by flow cytometry. *Fluorescence in Biomedical Research.* In press.

Di Virgilio, F., P. D. Lew, T. Anderson, and T. Pozzan. 1987. Plasma membrane potential modulates chemotactic peptide-stimulated cytosolic free Ca^{+2} changes in human neutrophils. *Journal of Biological Chemistry.* 262:4574–4579.

Seligmann, B., and J. I. Gallin. 1980. Neutrophil activation studied using two indirect probes of membrane potential which respond by different fluorescence mechanisms. *Advances in Experimental Medicine and Biology.* 141:335–349.

Seligmann, B., and J. I. Gallin. 1983. Comparison of indirect probes of membrane potential utilized in studies of human neutrophils. *Journal of Cellular Physiology.* 115:105–115.

Whitin, J. L., C. B. Chapman, E. R. Simons, M. Choaniec, and H. J. Cohen. 1980. Correlation between membrane potential changes and superoxide production in human granulocytes stimulated by phorbol myristate acetate. *Journal of Biological Chemistry.* 255:1874–1878.

Whitin, J. L., R. A. Clark, E. R. Simons, and H. J. Cohen. 1981. Effects of the myeloperoxidase system on fluorescent probes of granulocyte membrane potential. *Journal of Biological Chemistry.* 265:8904–8906.

Wilson, H. A., B. Seligmann, and T. M. Chused. 1985. Voltage-sensitive cyanine dye fluorescence signals in lymphocytes: plasma membrane and mitochondrial components. *Journal of Cellular Physiology.* 125:61–71.

Wilson H. A., and T. M. Chused. 1985. Lymphocyte membrane potential and Ca^{2+}-sensitive potassium channels described by oxonol dye fluorescence measurements. *Journal of Cellular Physiology.* 125:72–81.

Chapter 21

The Dynamics and Relationship of K$^+$ Efflux and Ca^{++} Influx in B Lymphocytes after Antigen-Receptor Cross-Linking

John T. Ransom and John C. Cambier

Department of Immunology, DNAX Research Institute, Palo Alto, California, and Department of Pediatrics, National Jewish Center for Immunology and Respiratory Medicine, Denver, Colorado

Cell Physiology of Blood © 1988 by The Rockefeller University Press

Introduction

The molecular basis of transmembrane signaling by B lymphocyte receptors for antigen has recently been the subject of intense study. Cross-linking of membrane immunoglobulin (mIg) causes the rapid hydrolysis (within 30 s) of members of a specific class of plasma membrane phospholipids, phosphatidylinositol (PtdIns), phosphatidylinositol 4-phosphate (PtdInsP), and phosphatidylinositol 4,5-bisphosphate (PtdInsP$_2$), by phospholipase C (PLC), generating inositol 1-phosphate (InsP), inositol 1,4-bisphosphate (InsP$_2$), inositol 1,4,5-trisphosphate (InsP$_3$), and diacylglycerol (DG) (Coggeshall and Cambier, 1985; Bijsterbosch et al., 1985; Ransom et al., 1986). In B cells, InsP3 mediates the release (30–60 s) of Ca^{++} from the endoplasmic reticulum (Ransom et al., 1986). It has been suggested that this release is accompanied by an influx of extracellular Ca^{++} via a non–voltage-activated Ca channel (Bijsterbosch et al., 1986). The elevated intracellular Ca^{++} appears to act in concert with DG to mediate the translocation (1–2 min) of protein kinase C to the plasma membrane (Chen et al., 1986), where it is bound and activated by a complex of Ca^{++}, DG, and phosphatidylserine (Hannum et al., 1985). Phosphorylation of a number of substrates follows (Coggeshall et al., 1985; Hornbeck and Paul, 1986), and membrane depolarization occurs somewhat later (3–90 min; Monroe and Cambier, 1983). Within 15 min of stimulation by anti-IgM or anti-IgD, increased transcription of the c-*fos* oncogene is detectable (Snow et al., 1986; Klemsz, M., E. Palmer, and J. C. Cambier, manuscript submitted for publication). In the succeeding 2 h, increased transcription of c-*myc* and Ia genes is detectable, as is stabilization of c-*myc* message (Klemsz, M., E. Palmer, and J. C. Cambier, manuscript submitted for publication). The events that intervene in the cascade between protein kinase C activation and increased gene transcription and stabilization are as yet undefined.

The occurrence of these events in a cascade is supported by the findings that exogenous PLC, phorbol myristate acetate (PMA), DG, and Ca ionophores induce the events at and distal to their predicted sites of action, including c-*myc, c-fos,* and Ia expression (Coggeshall and Cambier, 1985; Ransom and Cambier, 1986; Klemsz, M., E. Palmer, and J. C. Cambier, manuscript submitted for publication). Recent reports indicate that Ca^{++} ionophore plus PMA provide a complete signal for B cell proliferation (Klaus et al., 1986), which indicates that, under appropriate circumstances, activation of this cascade is sufficient to drive proliferation. The hypothesis that this cascade is responsible for the biologic response to anti-Ig is supported by findings that inhibitors of phosphoinositide hydrolysis in B cells, including TMB8, neomycin (Ransom et al., 1988), and dbcAMP (Coggeshall and Cambier, 1985), block this cascade as well as hyperexpression of Ia in response to anti-Ig. Finally, recent studies have demonstrated that thymus-dependent antigens stimulate isolated, antigen-specific, normal B cells to increase inositol lipid metabolism (Grupp et al., 1986) and c-*myc* expression (Snow et al., 1986). Therefore, both antigen and surrogate antireceptor antibodies activate this cascade.

Thus, many of the principal elements of the signaling cascade used by mIg have been defined. However, the precise relationship between the activation of phosphoinositide hydrolysis and Ca^{++} influx remains an enigma. Here we report recent studies that suggest that the occurrence of a K$^+$ efflux appears to be necessary for subsequent Ca^{++} influx, which follows mIg cross-linking.

Methods

B Cell Preparation

Splenic B cells from 6–8-wk-old mice (BDF_1) were prepared as previously described (Ransom and Cambier, 1986). All cell preparations were suspended (5×10^6 cells/ml) in balanced salts solution (BSS) supplemented with 5% fetal calf serum (FCS) until use.

Relative Membrane Potential Measurements

Flow cytometric measurements of the relative membrane potential were performed with the carbocyanine dye $DiO\text{-}C_5(3)$ as previously described (Ransom and Cambier, 1986). Separate experiments have shown that anti-Ig–stimulated depolarization measured by this technique is partially inhibited by azide (1 mM), which suggests some contribution of mitochondrial potential to the measurement. Further experiments with valinomycin, gramicidin D, and manipulations of the extracellular K^+ concentration ($[K^+]_o$) indicate that the technique accurately assesses the polarity of shifts in B cell membrane potential. Fluorimetric analyses of the relative membrane potential were performed at 37°C with the oxonol dye bis-(1,3-diethylthiobarbiturate)-trimethineoxonol (bis-oxonol; Molecular Probes, Inc., Eugene, OR), as described by Rink et al. (1980), except that the bis-oxonol (final concentration, 150 nM) was diluted directly into the cell suspension. A stable baseline fluorescence was determined after repeated brief mixings with a magnetic stirrer. Reagents or stimuli were subsequently added with a minimum of mixing to avoid artifactual changes in fluorescence intensity. Gramicidin D (20 nM; Sigma Chemical Co., St. Louis, MO) was added at the end of each recording to verify the potentiometric measurements.

Flow Cytometric Analyses of Intracellular Ca^{++} Concentration

Single cell analyses of intracellular Ca^{++} concentration ($[Ca^{++}]_i$) were performed using indo-1–loaded B cells as described previously by Rabinovitch et al. (1986). The cells were analyzed at a flow rate of ~1,000/s at 37°C in the simplified saline described by Tsien et al. (1982). Fluorescence histograms were collected every 9 s and displayed in a continuous profile over a 4–9-min period. Breaks in the display represent time points where the analysis was stopped and then resumed after addition of reagents.

Analysis of $^{86}Rb^+$ Content

Determination of the $^{86}Rb^+$ distribution of anti-Ig–stimulated B cells was performed as described elsewhere (Ransom, J. T., and J. C. Cambier, manuscript in preparation). Briefly, the technique uses the distribution ratios between the cell pellet and supernatant samples of $[^{14}C]$sucrose (r_S, 1 μCi/ml) and $^{86}Rb^+$ (r_{Rb}, 5 μCi/ml), as normalized to the total water space of each sample, which is determined using ^3H_2O (1 μCi/ml). This technique permits the total water space, the cellular water space, and the cellular $^{86}Rb^+$ content to be determined from one sample. The accumulation factor for $^{86}Rb^+$ (F_{Rb}), which indicates the ratio of intracellular to extracellular concentration, was calculated from the following relationship:

$$F_{Rb} = (r_{Rb} - r_S)/(1 - r_S).$$

Immunofluorescence Analysis of Surface I-A Expression

B cells were cultured for 12 h, harvested, and stained with directly fluoresceinated anti-I-A^{b,d} monoclonal antibody (D3.137) as previously described (Ransom and Cambier, 1986). All reagents were mixed into the cultures from stock solutions (nifedipine, <0.1% dimethyl sulfoxide) immediately before addition of anti-κ stimulus (1 μg/mL).

Results

Membrane Depolarization Is Dependent on Ca^{++} Influx

Since anti-Ig–stimulated membrane depolarization occurs after DG release and Ca^{++} mobilization, we sought to determine whether the increase in [Ca^{++}]$_i$ was necessary for membrane depolarization. Table I shows the percentage of cells that become depolarized after stimulation with suboptimal doses of PMA and the electroneutral

TABLE I
A23817 and PMA Synergize to Induce Membrane Depolarization

	Depolarized cells (%)			
	PMA (ng/ml)			
A23187	0.4	0.2	0.1	0
nM				
80	48	41	40	19
40	38	29	21	8
20	32	15	3	0
0	15	4	4	0

B cells were cultured in the presence of the indicated concentrations of A23187 and PMA for 90 min and the percentage of depolarized cells was determined as described using DiO-C$_5$(3).

Ca^{++} ionophore A23187. Although A23187 and PMA alone caused some cells to depolarize, combinations of the two reagents induced the depolarization of a greater number of cells than either reagent alone. This suggests that the increased cytosolic free Ca^{++} concentration and the liberated DG act synergistically to mediate plasma membrane depolarization. B cell plasma membrane depolarization has been demonstrated to be coupled to the activation of protein kinase C (Monroe et al., 1984). These findings are consistent with in vitro evidence that binding of protein kinase C to red cell membranes is dependent upon DG and is enhanced by elevated free Ca^{++} (Wolf et al., 1985). The role of Ca^{++} is supported by further evidence that anti-Ig–stimulated depolarization is dependent upon [Ca^{++}]$_o$. The data in Fig. 1 show the effect of lowering [Ca^{++}]$_o$ on the anti-Ig–induced changes in membrane potential. At all subphysiologic [Ca^{++}]$_o$, the control cells become increasingly depolarized. Furthermore, increased heterogeneity of dye accumulation by the cells was noted (not shown). However, as [Ca^{++}]$_o$ was lowered from 1 mM to <100 μM, the membrane potential response shifted from a depolarization to a hyperpolarization. This result correlates with observations that the Ca^{++} influx that occurs during anti-Ig stimulation (see below) is minimal at $[Ca^{++}]_o \leq 100\ \mu M$ (Bijsterbosch et al., 1986; Ransom et al., 1988). Further, they suggest that, during a response, conductive Ca^{++} influx may normally be balanced electrically.

The Ca^{++} Influx Pathway Is Coupled to K^+ Efflux

Previous experiments using EGTA buffers indicated that Ca^{++} mobilization was dependent upon $[Ca^{++}]_o$ and suggested that Ca^{++} influx, as well as a release of sequestered Ca^{++} from the endoplasmic reticulum (ER), may occur (Ransom et al., 1988). However, such experiments do not account for the possibility that a reduction of $[Ca^{++}]_o$ with EGTA may cause redistribution of intracellular Ca^{++} stores and a diminution of the releasable store in the ER. If this were to occur, then a reduction of the Ca^{++} response using EGTA could reflect a loss of the releasable store from the ER and not the elimination of a true Ca^{++} influx pathway. To test more rigorously for the activation of a Ca^{++} influx pathway, we first determined whether the small but significant Ca^{++} response observed when $[Ca^{++}]_o \leq 10\ \mu M$ was due to the release of intracellular stores (Ransom et al., 1986; Ransom et al., 1988). We determined that the response was insensitive to EGTA and competitive cations such as Ba^{++} and Sr^{++}

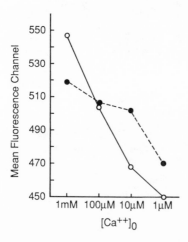

Figure 1. Plasma membrane depolarization in anti-Ig–stimulated B cells is dependent on extracellular Ca^{++}. Mouse B cells were cultured in simplified saline containing the indicated concentrations of Ca^{++} with (●) or without (○) anti-κ monoclonal antibody 187.1 (1 $\mu g/ml$). The cells were harvested after 60 min, analyzed by flow microfluorimetry using the carbocyanine dye diO-C_5(3), and the mean channel number of the fluorescence histogram were determined to assess the relative membrane potential of the population.

(Ransom, J., D. DiGiusto, V. Sandoval, D. Thorpe, J. Pasternak, and J. C. Cambier, manuscript in preparation). When $[Ca^{++}]_o$ was set to <10 μM, the $[Ca^{++}]_i$ of stimulated cells was found to rise to ~200 nM and then return to the initial resting level (100 nM) within 3 min (Fig. 2 B). If $[Ca^{++}]_o$ was elevated to 5 mM 3 min after stimulation, $[Ca^{++}]_i$ in all cells rapidly but transiently increased to an average value of ~500 nM or greater. A similar $[Ca^{++}]_o$ depletion/repletion protocol was used with unstimulated cells, and we found that $[Ca^{++}]_i$ increased to only 200 nM after repletion (Fig. 2 A). These results indicate that anti-Ig stimulation caused the plasma membrane to become more permeable to Ca ions than the membranes of resting cells.

Previous results indicated that isotonic replacement of NaCl with KCl or the addition of 50 mM KCl to the cell suspension significantly inhibits the magnitude of the Ca^{++} response. Using the $[Ca^{++}]_o$ depletion/repletion protocol in the presence of 50 mM KCl, it can be shown that the elevated $[K^+]_o$ inhibits Ca^{++} influx while only marginally affecting the intracellular release (Fig. 2 C). This suggests that Ca^{++} influx is dependent either upon a normally hyperpolarized plasma membrane or upon a low $[K^+]_o$. To assess the effect of membrane depolarization or hyperpolarization on

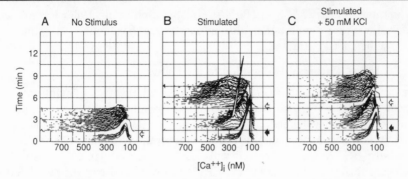

Figure 2. Ca^{++} influx is blocked by increased $[K^+]_o$. Indo-1–loaded mouse B cells were suspended in simplified saline containing 10 μM Ca^{++}, analyzed briefly by flow microfluorimetry, and then either stimulated with anti-κ at the time indicated by the solid arrow (1 μg/ml; *B* and *C*) or not stimulated (*A*) and the analyses were continued. 3 min after stimulation (*B* and *C*), or after the initial baseline analysis of unstimulated cells (*A*), $[Ca^{++}]_o$ was increased to 5 mM (indicated by the open arrow) and the analyses continued further. The results presented in *C* are from cells stimulated in the presence of 50 mM KCl.

Ca^{++} influx, electrogenic ion-selective ionophores were used to depolarize (gramicidin) or hyperpolarize (valinomycin) the plasma membrane. Although the valinomycin results were difficult to interpret because of an ionophore-induced increase in $[Ca^{++}]_i$, even in low-$[Ca^{++}]_o$ media, the results with gramicidin indicated that membrane depolarization did not inhibit Ca^{++} influx or affect the release of Ca^{++} from the endoplasmic reticulum (Ransom, J., D. DiGiusto, V. Sandoval, D. Thorpe, J. Pasternak, and J. C. Cambier, manuscript in preparation). Furthermore, isotonic replacement of NaCl with choline chloride had no effect on the Ca^{++} response, which suggests that Ca^{++} influx is independent of $[Na^+]_o$.

Thus, it was apparent that the inhibition of Ca^{++} influx by elevated $[K^+]_o$ was a

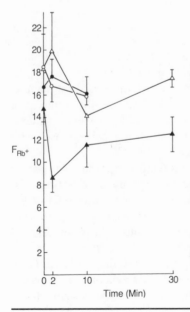

Figure 3. Anti-Ig stimulation causes a decrease of B cell $[^{86}Rb^+]_i$, which is blocked by channel blockers. Mouse B cells were pre-equilibrated with $^{86}Rb^+$ and ^3H_2O as described and then stimulated with anti-κ alone (▲), stimulated in the presence of 10 mM 4-AP (○) or 50 mM nifedipine (●), or not stimulated (△). At the indicated times, aliquots were removed, mixed with $[^{14}C]$sucrose, and transferred to sealed pipette tips and centrifuged to pellet the cells. Samples of the supernatant and portions of the cell pellet were analyzed by ion-exchange chromatography and the relative accumulation of $^{86}Rb^+$ (corrected for the intracellular water space) was calculated as described.

direct effect of the K ions on the Ca^{++} permeability and was not due to the depolarization caused by the elevated $[K^+]_o$. Since it has been shown that K channel blockers and extraordinarily high concentrations of voltage-gated Ca channel blockers can block the voltage-gated K^+ conductance in human T cells (described by Chandy et al., 1984) and the voltage-gated Ca^{++} conductance in murine B cell myelomas (Fukushima and Hagiwara, 1983), it seemed possible that similar conductances might be present and functionally coupled in the murine B cell. Ca^{++} mobilization measurements with quin-2–loaded B cells indicated that Ca^{++} mobilization is mostly but not completely blocked by 50 μM nifedipine, diltiazem, or verapamil and by 10 mM

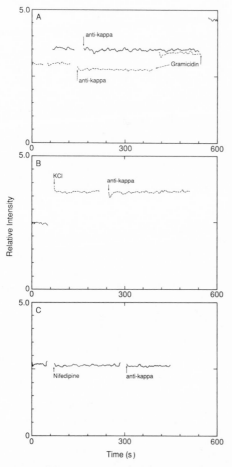

Figure 4. In the absence of Ca^{++} influx anti-Ig stimulation causes B cell hyperpolarization, which is blocked by channel blockers. (*A*) B cells were suspended in simplified saline containing either 1 mM (solid line) or 1 μM (dotted line) $[Ca^{++}]_o$ and transferred to a quartz cuvette for fluorimetric analysis. The bis-oxonol dye was added and a baseline fluorescence was established. The cells were then stimulated with anti-κ and gently stirred and the analysis was continued. Gramicidin D was added at the time points shown to depolarize the cells and confirm the shifts in membrane potential indicated by the dye. (*B*) B cells were suspended in 1 μM $[Ca^{++}]_o$, oxonol dye was added, and the baseline fluorescence was established. KCl (50 mM) and anti-κ were added at the times indicated. (*C*), B cells in 1 μM $[Ca^{++}]_o$ were stimulated with anti-κ after the addition of nifedipine (50 μM). The gramicidin controls were eliminated from *B* and *C* for clarity.

4-aminopyridine (4-AP; Ransom, J., D. DiGiusto, V. Sandoval, D. Thorpe, J. Pasternak, and J. C. Cambier, manuscript in preparation). Further analyses of indo-1–loaded B cells by the $[Ca^{++}]_o$ depletion/repletion protocol have indicated that Ca^{++} release from the endoplasmic reticulum is not affected by these reagents but that the majority of Ca^{++} influx is blocked by 20–50 μM of the organic Ca channel blockers or by 10 mM 4-AP. These drug concentrations coincide with the concentrations that effectively blocked K^+ conductance in human T cells (Chandy et al., 1984). A recent description of a voltage-gated K^+ conductance in mouse splenic B cells is consistent with the hypothesis that a K^+ permeability is operative in the anti-Ig–stimulated B cell

(Choquet et al., 1987). Hence, the inhibition of Ca^{++} influx by the channel blockers and by $[K^+]_o$ suggests that Ca^{++} influx may be coupled to the efflux of K^+ from the cell.

Anti-Ig Stimulation Induces a Loss of B Cell K^+

To establish whether anti-Ig stimulation causes a rapid K^+ efflux from the cell, B cells were pre-equilibrated with $^{86}Rb^+$, an isotopic analogue for $^{42}K^+$ that has been shown to be transported by K^+ transport mechanisms similar to $^{42}K^+$. The cells were also equilibrated with ^3H_2O to determine the total water space of each sample and $[^{14}C]$sucrose was added just before initiation of the experiment to determine the extracellular water space. After equilibration, the cells for each experiment were aliquoted into single tubes and, before and after addition of anti-Ig, samples were

TABLE II
Channel Blockers Inhibit Anti-Ig–stimulated Membrane Depolarization

Culture conditions	Percentage of cells depolarized
Control	33
KCl (50 mM)	59
Anti-κ	64
4-AP	20
Anti-κ + 4-AP	39
Anti-κ, 4-AP‡	52
Nifedipine	30
Anti-κ + nifedipine	32

B cells were cultured with the indicated reagents for 90 min and the number of depolarized cells was determined using diO-C_5(3). The concentrations of 4-AP and nifedipine used were 10 mM and 50 μM, respectively.
‡Cells were stimulated with anti-κ for 90 min. 4-AP was added at the end of the experiment with diO-C_5(3).

withdrawn for each time point. The cellular $^{86}Rb^+$ accumulation factor (F_{Rb}) was determined for each time point. As shown in Fig. 3, 2 min after stimulation, the F_{Rb} of the cells had decreased significantly compared with the control cells. Within 10 min, there was no statistically significant difference between the two conditions, which suggests that the stimulated cells have begun to restore $[^{86}Rb^+]_i$ toward resting levels. Measurement of the intracellular water space in the samples (i.e., relative cell volume) also indicate that the cells swell within 2 min and remain enlarged for at least 60 min (not shown). When these measurements were made in the presence of either 4-AP (10 mM) or nifedipine (50 μM), both the decrease of cell $[^{86}Rb^+]_i$ and cell enlargement were blocked, which suggests that these reagents block K^+ efflux from anti-Ig–stimulated B cells.

An alternative way to investigate whether anti-Ig causes K^+ efflux as well as Ca^{++} influx is to monitor the membrane potential of stimulated cells when Ca^{++} influx is minimized. B cells were quickly suspended in simplified saline with 1 μM or 1 mM

[Ca^{++}]$_o$ and the relative membrane potential was monitored in a fluorimeter using an anionic oxonol dye (Rink et al., 1980). When [Ca^{++}]$_o$ was 1 mM, addition of anti-κ to the cell suspension did not induce any significant change in membrane potential for at least 6 min (Fig. 4 *A*). However, in medium where [Ca^{++}]$_o$ was 1 μM, stimulation caused the cells to immediately become hyperpolarized (Fig. 4 *A*). This hyperpolarization was blocked by 50 mM KCl (Fig. 4 *B*), which indicates that hyperpolarization is mediated by the efflux of K$^+$ from the cell. When the potential was monitored in the presence of 50 μM nifedipine, the anti-Ig–mediated hyperpolarization was also blocked, which suggests that this reagent also blocked K$^+$ efflux (Fig. 4 *C*). It was not possible to assess the effect of 4-AP on this response because of the large sample dilution required, which significantly alters the oxonol fluorescence intensity. Thus, measurements of [^{86}Rb$^+$]$_i$ and indirect measurements of K$^+$ efflux by analysis of membrane potential indicate that anti-Ig stimulation causes a transient efflux of cell

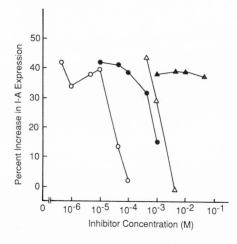

Figure 5. Channel blockers inhibit anti-Ig–stimulated increased I-A expression. B cells were cultured for 12 h in the presence of anti-κ and the indicated concentrations of nifedipine (O), quinine (●), 4-AP ($\triangle$), or TEA (▲). The cells were then harvested, directly stained with fluorescein-conjugated anti-I-A^{b,d} monoclonal antibody, and analyzed by flow microfluorimetry for surface I-A expression.

K$^+$ as well as a transient influx of Ca^{++}. The evidence that K$^+$ channel blockers and high concentrations of organic Ca channel blockers as well as an elevated [K$^+$]$_o$ can block both Ca^{++} influx and K$^+$ efflux suggests that these two events are coupled. More specifically, since K$^+$ efflux apparently occurs in the absence of Ca^{++} influx (Fig. 4 *A*), but Ca^{++} influx is blocked when K$^+$ efflux is blocked (Fig. 2 *C*), the results suggest that anti-Ig–stimulated Ca^{++} influx is dependent upon K$^+$ efflux.

Anti-Ig-stimulated Membrane Depolarization and Increased I-A Expression Require Ca^{++} Influx

The data in Fig. 1 suggested that the membrane depolarization determined 60 min after anti-Ig stimulation is dependent upon extracellular Ca^{++}. To more firmly demonstrate that this dependence is due to a requirement for Ca^{++} influx, the relative membrane potential of cells stimulated in the presence of nifedipine or 4-AP was measured with the carbocyanine dye. Nifedipine (50 μM) completely blocked anti-Ig–stimulated depolarization without affecting the membrane potential of the control cells (Table II). When present throughout the assay, 4-AP (10 mM) apparently caused a slight hyperpolarization of the resting cells but also almost completely blocked the depolarization response. As a control to show that this apparent inhibition was not due to the direct hyperpolarization caused by 4-AP alone, the K channel blocker was added at the end of the assay along with the indicator dye. In this case, 4-AP also caused some apparent hyperpolarization of the cells but had no effect on the net depolarization caused by anti-Ig stimulation. Since the channel blockers inhibit Ca^{++} influx but do not affect Ca^{++} release from the ER, these results indicate that anti-Ig–stimulated membrane depolarization is dependent upon Ca^{++} influx.

It has been previously suggested that the anti-Ig–stimulated increase in I-A expression is also dependent upon an increase in $[Ca^{++}]_i$ (Klaus et al., 1986). Again, to determine whether increased I-A expression is dependent upon Ca^{++} influx, the cells were cultured with the channel blockers and stimulus for 12 h. As shown in Fig. 5, both nifedipine and 4-AP block increased I-A expression at concentrations that block Ca^{++} influx. Quinine, which could not be assayed for its ability to inhibit Ca^{++} mobilization, also inhibited I-A expression. TEA, which did not block Ca^{++} mobilization, failed to inhibit I-A expression. Thus, the results indicate that the increased I-A expression is also dependent upon Ca^{++} influx and that the release of Ca^{++} from the endoplasmic reticulum is insufficient to generate this response.

Conclusion

The data presented indicate that anti-Ig–induced Ca^{++} influx in B lymphocytes is partially electrically balanced by, and dependent on, K^+ efflux from the cell. These findings are consistent with the recent report of Choquet et al. (1987) that B cells express voltage-dependent K channels after long-term culture with lipopolysacharride. Our results with intact, anti-Ig–stimulated B cells indicate that a previously unknown event may occur during B cell activation.

References

Bijsterbosch, M. K., J. C. Meade, G. A. Turner, and G. G. B. Klaus. 1985. B lymphocyte receptors and polyphosphoinositide degradation. *Cell.* 41:999–1006.

Bijsterbosch, M. K., K. P. Rigley, and G. G. B. Klaus. 1986. Crosslinking of surface immunoglobulin on B lymphocytes induces intracellular Ca^{++} release and Ca^{++} influx: analysis with indo-1. *Biochemical and Biophysical Research Communications.* 137:500–508.

Chandy, K. G., T. E. DeCoursey, M. D. Cahalan, C. McLaughlin, and S. Gupta. 1984. Voltage-gated potassium channels are required for human lymphocyte activation. *Journal of Experimental Medicine.* 160:369–380.

Chen, Z. Z., K. M. Coggeshall, and J. C. Cambier. 1986. Translocation of protein kinase C during membrane immunoglobulin-mediated transmembrane signalling in B lymphocytes. *Journal of Immunology.* 136:2300–2306.

Choquet, D., P. Sarthou, D. Primi, P. Cazenave, and H. Korn. 1987. Cyclic AMP-modulated potassium channels in murine B cells and their precursors. *Science.* 235:1211–1214.

Coggeshall, K. M., and J. C. Cambier. 1985. B cell activation. VI. Studies of modulators of phospholipid metabolism suggest an essential role for diacylglycerol in transmembrane signalling by mIg. *Journal of Immunology.* 134:101–107.

Coggeshall, K. M., J. G. Monroe, J. T. Ransom, and J. C. Cambier. 1985. Mechanisms of transmembrane signal transduction during B cell activation. *In* B Lymphocyte Differentiation. J. C. Cambier, editor. CRC Uniscience Series, Boca Raton, FL. 1–22.

Fukushima, Y., and S. Hagiwara. 1983. Voltage-gated Ca^{++} channel in mouse myeloma cells. *Proceedings of the National Academy of Sciences.* 80:2240–2243.

Grupp, S. H., E. C. Snow, and A. K. Harmony. 1988. Phosphatidylinositol response is an early event in physiologically relevant activation of antigen specific B lymphocytes. *Cellular Immunology.* In press.

Hannum, Y. A., C. R. Loomis, and R. M. Bell. 1985. Activation of protein kinase C by Triton X-100 mixed micelles containing diacylglycerol and phosphatidylserine. *Journal of Biological Chemistry.* 260:10039–10044.

Hornbeck, P., and W. E. Paul. 1986. Anti-immunoglobulin and phorbol ester induce phosphorylation of proteins associated with plasma membrane and cytoskeleton in murine B lymphocytes. *Journal of Biological Chemistry.* 261:14817–14821.

Klaus, G. G. B., A. O'Garra, M. K. Bijsterbosch, and M. Holman. 1986. Activation and proliferation signals in mouse B cells. VIII. Induction of DNA synthesis in B cells by a combination of calcium ionophores and phorbol myristate acetate. *European Journal of Immunology.* 16:92–97.

Monroe, J. G., and J. C. Cambier. 1983. B cell activation. I. Receptor crosslinking by anti-immunoglobulin antibodies induces a rapid decrease in B cell plasma membrane potential. *Journal of Experimental Medicine.* 157:2073–2086.

Monroe, J. G., J. E. Niedel, and J. C. Cambier. 1984. B cell activation. IV. Induction of cell membrane depolarization and hyper I-A expression by phorbol diesters suggests a role for protein kinase C in murine B lymphocyte activation. *Journal of Immunology.* 132:1472–1478.

Rabinovitch, P. S., C. H. June, A. Grossmann, and J. A. Ledbetter. 1986. Heterogeneity among T-cells in intracellular free calcium responses after mitogen stimulation with PHA or anti-CD3. Simultaneous use of Indo-1 and immunofluorescence with flow cytometry. *Journal of Immunology.* 137:952–960.

Ransom, J. T., and J. C. Cambier. 1986. B cell activation. VII. Independent and synergistic effects of mobilized calcium and diacylglycerol on membrane potential and I-A expression. *Journal of Immunology.* 136:66–72.

Ransom, J. T., L. K. Harris, and J. C. Cambier. 1986. Anti-Ig induces release of inositol 1,4,5-trisphosphate which mediates mobilization of intracellular Ca^{++} stores in B lymphocytes. *Journal of Immunology.* 137:708–714.

Ransom, J., M. Chen, V. M. Sandoval, J. A. Pasternak, D. DiGiusto, and J. C. Cambier. 1988. Increased plasma membrane permeability to Ca^{2+} in anti-Ig-stimulated B lymphocytes is dependent on activation of phosphoinositide hydrolysis. *Journal of Immunology.* 140:3150–3155.

Rink, T. J., C. Montecucco, T. R. Hesketh, and R. Y. Tsien. 1980. Lymphocyte membrane potential assessed with fluorescent probes. *Biochimica et Biophysica Acta.* 595:15–30.

Snow, E. C., J. D. Fetherston, and S. Zimmes. 1986. Induction of the c-*myc* after antigen binding to hapten-specific B cells. *Journal of Experimental Medicine.* 164:944–950.

Tsien, R. Y., T. Pozzan, and T. J. Rink. 1982. Calcium homeostasis in intact lymphocytes: cytoplasmic free calcium monitored with a new intracellularly trapped fluorescent indicator. *Journal of Cell Biology.* 94:325–331.

Wolf, M., H. LeBine, W. S. May III, P. Cuatrecasas, and N. Sahyoun. 1985. A model for intracellular translocation of protein kinase C involving synergism between Ca^{++} and phorbol ester. *Nature.* 317:546–549.

Chapter 22

Optical Measurement in Single Cells of Membrane Potential Changes Linked to T and B Lymphocyte Antigen Receptors

H. Alexander Wilson, David Greenblatt, Fred D. Finkelman, and Thomas M. Chused

Laboratory of Microbial Immunity and Laboratory of Immunology, National Institute of Allergy and Infectious Diseases, Bethesda, Maryland, and the Department of Medicine, Uniform Services University of the Health Sciences, Bethesda, Maryland

Introduction

The finding that unmanipulated (resting) lymphocytes maintain a large electrical potential difference (E_m) across the plasma membrane, and early indications that depolarization occurs after exposure to mitogens (reviewed in Freedman and Laris, 1988) have stimulated interest in the possibility that changes in E_m function as signals in lymphocyte activation and differentiation pathways. Recent approaches to this question have been influenced both by modifications of E_m measurement techniques and by advances in the understanding of antigen receptor function. Using fluorescent potentiometric dyes, E_m can be measured in unperturbed single lymphocytes, and it has become apparent that the use of anionic probes circumvents certain major difficulties associated with the intracellular accumulation of cationic indicators (Wilson and Chused, 1985; Wilson et al., 1985). In place of mitogenic lectins, specific reagents (antibodies) are now available for stimulation of T and B lymphocyte antigen receptors. Cross-linkage of these receptors initiates the degradation of phosphatidyl-inositol-4,5-bisphosphate (PIP$_2$), with the resultant formation of two intracellular messengers: diacylglycerol, an activator of protein kinase C (PKC) (Nishizuka, 1984), and calcium-releasing inositol polyphosphates (IP$_3$) (Berridge and Irvine, 1984). Because antigen is the driving force in antibody production, the sequences triggered by and interactions among PIP$_2$ hydrolysis products are of central interest to lymphocyte activation and differentiation. In this chapter, we will focus on experiments that characterize antigen-induced membrane potential changes measured by flow cytometry, and show that quite large changes in the membrane's electric field are effected by both PKC activation and a rise in $[Ca^{2+}]_i$.

Technical Considerations: Dyes, Signal Calibration, and Cell Preparations

A recent symposium in this series (DeWeer and Salzberg, 1986) considered the variety of techniques available for optical measurement of E_m in small cells. To date, only "slow-response" indicator dyes have been used in lymphocytes. These are fluorescent probes (Waggoner, 1979; Bashford and Smith, 1979; Freedman and Novak, 1987) with a single delocalized charge, which confers both membrane permeability and pH insensitivity, that redistribute across membranes in response to transmembrane voltage changes. Fluorescence efficiency increases in an apolar microenvironment and the dyes are lipophilic. At appropriate dye concentrations, changes in cell fluorescence register changes in E_m. The dyes divide into two major classes, cationic cyanine and anionic oxonol dyes. In lymphocytes, as would be expected from prior studies in other cell types and organelles, cyanine dyes preferentially accumulate in mitochondria, which have a high inside-negative potential. More troublesome, these dyes diminish the plasma membrane's K$^+$ selectivity and, even at low nanomolar concentrations, depolarize B cells within several minutes. Other perturbations of direct relevance to the study of cell activation include inhibition of Na/K ATPase and elevation of $[Ca^{2+}]_i$ to micromolar levels (Wilson et al., 1985; Wilson and Chused, 1985). Hence, cyanine dyes both block and mimic events in the activation cascade. The toxic effects attributable to mitochondrial accumulation would appear to be characteristic of all cationic probes examined (Azzone et al., 1984). In contrast, oxonol dyes at concentrations yielding adequate fluorescence signals exhibit none of these undesirable qualities, and have been preferred to cyanines by several groups of investigators who have compared the two dye classes in lymphocytes (Rink et al., 1980; Tatham and Delves, 1984).

The voltage-sensitive fluorescence of diBa-C_4(3), the preferred dye for flow cytometry measurements using an argon laser, has been compared directly and shown to agree with microelectrode measurements (Bräuner et al., 1984). Nonetheless, studies continue to appear that report putative E_m changes in resting lymphocytes measured by uncalibrated cyanine fluorescence signals (Cambier and Ransom, 1987; Aldo-Benson, 1986).

To be reasonably certain that E_m is actually being measured, it is necessary to calibrate the fluorescence signal (Freedman and Novak, 1987), preferably by two independent methods. We have analyzed oxonol and cyanine fluorescence signals in lymphocytes in detail elsewhere (Wilson et al., 1985; Wilson and Chused, 1985). Several approaches to calibration and some potential pitfalls are briefly noted. DiBa-C_4(3)-valinomycin interactions exclude standard fluorescence null point titrations accomplished by varying $[K^+]_o$ in the presence of the K^+ ionophore. Somewhat more cumbersome methods are therefore required. The fluorescence shifts of oxonol (and cationic indicator dyes) occurring with manipulation of medium ions and Na^+ pump inhibition indicate that the lymphocyte plasma membrane is K^+ selective, and at $[K^+]_o > 10$ mM, E_m is very close to E_K (Deutsch et al., 1979; Rink et al., 1980; Grinstein et al., 1982). Hence, depolarization from resting potentials can be estimated by comparison with a series of graded K^+ solutions. Below 10 mM $[K^+]_o$, the ratio P_K/P_{Na} decreases in unstimulated cells such that E_m remains constant. Substitution of choline for Na^+ in the lower $[K^+]_o$ range restores a near linear relation between the log $[K^+]_o$ and oxonol fluorescence. The calibration suggested by these manipulations can be verified with ionophores. In T cells, elevation of $[Ca^{2+}]_i$ by Ca^{2+} ionophores activates a Ca^{2+}-sensitive K^+ conductance (discussed below) such that a linear relation exists between the log $[K^+]_o$ (2.5–120 mM) and fluorescence, which is unaffected by ouabain and Na^+ or Cl replacement. Since this permeability is not dominant in B cells, we routinely calibrate by using a variation of the valinomycin null point method: the cation ionophore gramicidin D (0.2 μg/ml) is used to induce K^+ equilibrium potentials in cells suspended in Na^+-free media. This procedure yields stable oxonol histograms for at least several minutes. However, by defining E_m-dependent and E_m-insensitive mitochondrial cyanine dye accumulation (Wilson et al., 1985), it can be shown that higher (micromolar) gramicidin D permeates cells and discharges mitochondrial dye. The resulting decrease in cyanine fluorescence under these conditions has been misinterpreted to indicate a large voltage-dependent signal. An alternative calibration method, proven useful in other cell types, is the pH null point technique (Akerman and Wikstrom, 1976; Bashford and Pasternak, 1984).

Validation of fluorescence measurements also requires care to exclude possible voltage-independent explanations for observed fluorescence shifts. For example, changes in pH_i occurring with cell activation could alter fluorescence by changing proportionality constants for bound to free dye within the cellular compartment. This can be shown not to occur with diBa-C_4(3): cell fluorescence remains constant under zero voltage conditions (140 mM extracellular K^+ and gramicidin D) as pH_i is manipulated between 6.8 and 7.8 by varying pH_o in the presence of FCCP. The distinct advantages of oxonol over cyanine dyes for studies of human or murine lymphocytes should not be uncritically extended to other cell types. Finding the best available dye for a given cell preparation and experimental protocol may require the consideration of several options.

The variable effects of mitogens on E_m as reported in early studies would appear largely explicable by problems with the choice of voltage-sensitive probe, the use in some instances of nonphysiologic stimulus concentrations, and the necessity of inferring single cell responses from bulk suspension measurements. Some additional pitfalls in the design and interpretation of experiments have been identified. Preparative techniques, particularly involving exposure of cells to anisotonic solutions (Grinstein et al., 1984), or the use of simple salts solution for cell suspension (Laris et al., 1978), may alter resting E_m and membrane permeability characteristics. Even in highly enriched media and with careful control of temperature and CO_2 tension, a time-dependent depolarization occurs (before cell death determined by dye exclusion assays), which will bias or invalidate bulk suspension measurements. Finally, while clonal cell lines, tumors, and hybridomas have been immensely important in the study of lymphocytes, and, for electrophysiological studies, have the ostensible advantages of hardiness, size, and uniformity over normal cells, available data suggest that they generally do not exhibit the same complement of E_m, permeability, and ion channel characteristics as do normal resting cells. Whether any of these cells represents, electrophysiologically, a stage of normal differentiation is speculative.

T Cell Receptor Cross-Linkage Induces Hyperpolarization by Two Mechanisms

Using the highly fluorescent Ca^{2+} indicator dyes indo-1 and fura-2 (Grynkiewicz et al., 1985), it has been possible to show that antigen or antireceptor antibodies induce a rise in $[Ca^{2+}]_i$ from resting 100 nM to >1 μM in both T (Rabinovitch et al., 1986) and B lymphocytes (Wilson et al., 1987a, b). The initial phase of Ca^{2+} elevation is due to Ca^{2+} release from endoplasmic reticulum (ER). Whether inward Ca^{2+} transport (Ca^{2+} channels) has a role in sustained Ca^{2+} elevation, other than preventing intracellular Ca^{2+} depletion, is currently unclear (Wilson et al., 1987b). The elevation of $[Ca^{2+}]_i$ is one link to observations by a number of investigators that T cell mitogens or Ca^{2+} ionophores induce plasma membrane hyperpolarization in T cells. Using the oxonol diSBa-C_2(3), Tsien et al. (1982) reported graded membrane hyperpolarizations with concanavalin A and Ca^{2+} ionophore, which correlated with the degree of Ca^{2+} elevation. The hyperpolarization was attributed to Ca^{2+} activation of a K^+ conductance because Ca^{2+} elevation increased the K^+ selectivity of the membrane, and the hyperpolarization was blocked by quinine. We subsequently extended these observations in flow cytometry experiments done with diBa-C_3(4) (Wilson and Chused, 1985). After excluding depolarized or non-T cells by gating techniques, we observed the induction of K^+ equilibrium potentials (a shift from -70 to -90 mV in 5 mM K^+ medium) by Ca^{2+} ionophores. Cells depolarized by 4-aminopyridine (4-AP), at concentrations completely blocking the voltage-sensitive K^+ conductance identified by patch-clamp techniques (DeCoursey et al., 1984; Matteson and Deutsch, 1984), were repolarized by ionophore-induced Ca^{2+} elevation. Cells depolarized by quinine did not repolarize, which indicates that quinine blocks both conductances in T cells. The utility of these compounds in cell culture experiments is more problematic than their use in microelectrode recording solutions: 4-AP would be expected to alkalinize cells, and quinine is toxic to mitochondria (Wilson and Chused, 1985), but their observed

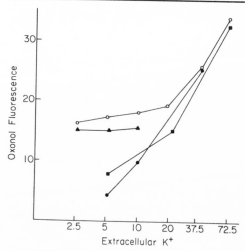

Figure 1. $[Ca^{2+}]_i$ elevation increases the K^+ selectivity of the T cell plasma membrane. Data points represent mean diBaC$_4$(3) fluorescence values (20,000 cells) for the T cell hybridoma 2B4 (provided by L. E. Samelson). A decrease in cell fluorescence indicates increased electrical potential. Curves were obtained under the following conditions: (○) control; (●), cross-linkage of antigen receptors (increasing $[Ca^{2+}]_i$ to >1 μM) with antireceptor antibody (the same curve is generated by 100 nM ionomycin); (▲), same as (●) plus 20 nM charybdotoxin (provided by C. Miller); (■), substitution of choline for Na^+ and 0.2 μg/ml gramacidin D.

inhibition of mitogenesis has been interpreted as demonstrating a requirement for K^+ channel activation in cell cycle transitions.

Ca^{2+}-mediated hyperpolarization to K^+ equilibrium potentials is characteristic of the majority (>90%) of T cells from various sources: murine spleen or thymus, human peripheral blood, and some T cell tumors. Recordings taken up to 5 min after Ca^{2+} elevation show no evidence of current inactivation. Fig. 1 shows that in even a relatively depolarized T cell hybridoma (resting E_m of -40 mV), the Ca^{2+}-activated conductance drives the cells very close to E_K. More recent studies with channel blocking agents suggest that neither Ca^{2+} nor voltage-sensitive K^+ channels primarily determine resting E_m. Charybdotoxin potently blocks the voltage-sensitive channels (Lewis and Cahalan, 1987) and the Ca^{2+}-induced hyperpolarization (Fig. 1), but does not significantly alter E_m in unstimulated cells (the largest effect we have observed is a 5-mV depolarization of mouse thymocytes). The mitogenic effect of Ca^{2+} ionophore and phorbol myristate acetate (PMA) on thymocytes is only modestly (~20%) reduced by charybdotoxin (H. A. Wilson, unpublished observations).

A second antigen receptor–linked mechanism for hyperpolarization is mediated by PKC, independently of Ca^{2+} elevation. Fig. 2 shows E_m changes induced in

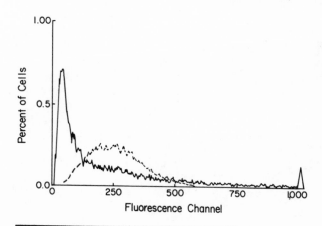

Figure 2. PKC activation induces T and B lymphocyte hyperpolarization. Dashed-line curve shows a control diBaC$_4$(3) fluorescence histogram of splenic lymphocytes. Solid curve shows the response 7 min after the addition of PMA (10 ng/ml).

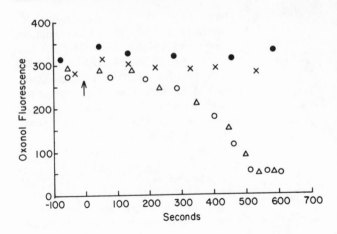

Figure 3. Electrogenic Na^+/K^+ exchange explains the hyperpolarization induced by PKC activation. Data points represent mean values for histograms in Fig. 2. PMA, added for each condition, is indicated by the arrow. (O), PMA (10 ng/ml), mean values calculated for responding cells only; (△), medium K^+ increased to 12 mM, mean values as in (O); (●) oubain (0.1 mM); (X) C-1 inhibitor (0.1 mM/ml).

unfractionated splenic lymphocytes (T and B cells respond similarly) by 10 ng/ml of PMA. The hyperpolarization (20 mV) is unaffected by raising $[K^+]_o$ to 12 mM, making E_K equivalent to the resting potential, but is abolished either by the PKC inhibitor C-1 (Gerard et al., 1986) or by ouabain (Fig. 3). Amiloride (not shown) also inhibits the hyperpolarization, which suggests that the step proximal to electrogenic Na/K ATPase activation is inward Na^+ transport by the Na^+/H^+ exchanger (cf. Besterman et al., 1985). Ca^{2+}-induced hyperpolarizations are unaffected by C-1 or ouabain. Hence, in T cells, both PKC and IP_3 mediate early plasma membrane hyperpolarization.

Antigen Receptor–induced Depolarization of B Cells Is Cation Independent

Within a few seconds of antigen receptor cross-linkage by high concentrations (10–30 μg/ml) of anti-Fab or anti-δ, murine splenic B cells undergo a depolarization from -70 mV resting potential to -15 ± 5 mV. Typical flow cytometric tracings are shown in Fig. 4. Ion replacement studies indicate that the depolarization cannot be attributed to a classic cationic current: replacement of Na^+ (choline) and/or reduction of $[Ca^{2+}]_o$ by EGTA to ≤100 nM just before antibody addition (such that E_{Ca} for activated cells

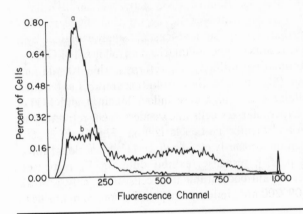

Figure 4. Antigen receptor cross-linkage depolarizes B lymphocytes. Small B lymphocytes were prepared as described (Wilson et al., 1987a). DiBa-C_4(3) histograms for 10,000 cells: (*a*) control; (*b*) 30 s after addition of 30 μg/ml of F(ab)$_2$ rabbit anti–mouse IgD.

initially approximates the resting potential) do not significantly affect the degree of depolarization. Evaluation of Cl^- by this technique is more equivocal (E_{Cl} cannot be manipulated over a wide range), but cells reach more positive potentials than E_{Cl} values calculated from estimates of $[Cl^-]_i$ in lymphocytes, and Cl^- replacement by gluconate does not appreciably alter the depolarization. In parallel patch-clamp studies with these cells (Sheehy, P., D. Greenblatt, and H. A. Wilson, submitted for publication), a noninactivating current of similar nonselectivity has been identified. This channel does not appear to be present in B cell tumors (Heikkilä et al., 1985; LaBaer et al., 1986; H. A. Wilson, unpublished observations).

Membrane depolarization can be specifically linked to antigen receptor function. The B cell mitogen lipopolysaccharide (LPS), which bypasses the antigen receptor, does not induce PIP_2 hydrolysis and does not depolarize cells (this result contrasts with experiments done with cationic probes [Kiefer et al., 1980; Monroe and Cambier, 1983]). Also, the negative-feedback receptor for antigen-driven responses is the receptor for the Fc fragment of IgG (FcγR). In vivo, when sufficient antibody levels are generated in response to a challenge, the antigen is then presented to B cells in the

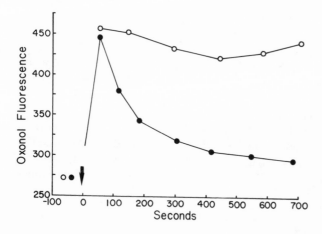

Figure 5. Fcγ receptor cross-linkage to the antigen receptor attenuates anti-Ig–induced B lymphocyte depolarization. Curves represent diBa-C$_4$(3) mean fluorescence for responding cells as in Fig. 4: (○) F(ab)$'_2$ rabbit anti–mouse IgD (30 μg/ml); (●) intact rabbit anti–mouse IgD (45 μg/ml).

form of antigen-antibody complexes, which bridge antigen and Fcγ receptors. Such cross-linkage inhibits IP_3 formation and attenuates both $[Ca^{2+}]_i$ elevation (Wilson et al., 1987a) and membrane depolarization (Fig. 5).

As shown above, the initial effect of isolated PKC activation on B cells is membrane hyperpolarization. Our recent experiments suggest that depolarization requires a Ca^{2+}-dependent (IP_3-mediated) step, but is at least not solely explained by a Ca^{2+}-activated conductance. These results can be summarized as follows. If EGTA is added simultaneously with the simulating antibody, E_m returns within 10 min and slightly overshoots the resting level. $[Ca^{2+}]_i$ under these conditions returns to baseline within 3 min. Buffering intracellular Ca^{2+} (high-dose indo-1 loading) and EGTA addition to the medium slightly hyperpolarizes cells and renders them refractory to anit-Ig–induced $[Ca^{2+}]_i$ elevation and membrane depolarization (Fig. 6). In experiments in which $[Ca^{2+}]_i$ and E_m are simultaneously measured, certain monoclonal anti-δ antibodies that induce identical $[Ca^{2+}]_i$ responses exhibit differing E_m responses. Similarly, $[Ca^{2+}]_i$ and E_m values in antibody-stimulated cells show only a weak correlation. Finally, ionomycin (100–200 nM) induces a depolarization, comparable to

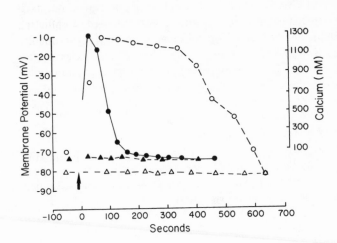

Figure 6. Effect of $[Ca^{2+}]_i$ buffering on B lymphocyte E_m and anti-Ig–induced depolarization. E_m (*open symbols*) and $[Ca^{2+}]_i$ (*closed symbols*) were measured simultaneously with DiSBa-$C_4(5)$ and indo-1. Arrow indicates $F(ab')_2$ rabbit anti–mouse Fab (30 μg/ml) addition. (○, ●) $[Ca^{2+}]_o$ reduced to ≤100 nM by EGTA simultaneously with antibody addition; (△, ▲) cells were loaded with high dose (10 μM) indo-1 (see Wilson et al., 1987a) and EDTA was added to the medium before antibody addition.

anti-Ig, which is also independent of Na^+ and Cl^-. In experiments with this ionophore, a nonspecific increase in ion permeabilities has not been excluded, but interestingly, the ionophore-induced depolarization is preferentially blocked by oligomycin C.

B cells remain depolarized for at least several hours. The mechanisms of protracted depolarization have not been characterized but would appear to differ from that described in initially stimulated resting cells. PMA-treated cells, after an initial hyperpolarization (Fig. 1), gradually depolarize and by 30–60 min reach values (-15 mV) similar to anti-Ig–stimulated cells. The ionic basis of this depolarization is under investigation.

Do Changes in E_m Matter?

This remains a somewhat open question, which until recently, has been framed by conflicting results and limited experimental approaches. It now appears clear that the initial changes in E_m in T and B cells induced by antigen receptor cross-linkage are in opposite directions and that the induced electrical field changes ($2-5 \times 10^4$ V/cm) are impressive. These findings encourage a careful evaluation of the role of voltage-sensitive transport processes in lymphocyte activation, and, given the variety of surface receptors and major histocompatibility complex gene products involved in activation, it is attractive to speculate that such large electrical forces might reorient membrane proteins to facilitate specific cell surface interactions. Values for the steady potentials provided by the experiments described in this chapter offer clues for the design of experimental approaches to these questions. However, the task of relating E_m changes to cell function will be simpler and, more importantly, differently perceived, when E_m perturbations are more precisely defined. For example, the observation, made by continuous recordings from stimulated B lymphocytes (Wilson et al., 1987b), that $[Ca^{2+}]_i$ elevation is resolvable into repetitive transients, provides a new set of potential regulatory sites for mediators of lymphocyte activation. Should E_m be found to oscillate in a similar way (as is predictable for initial T cell activation by antigen), then the utility of one conventional approach to determine the role of E_m in activation, fixed depolarization effected by manipulating $[K^+]_o$, ought to be reassessed. Conflicting inferences have been made from rather subtle differences in such experimental protocols (Deutsch and Price, 1982; Gelfand et al., 1984). These uncertainties suggest to us that elucidation of the role of membrane potential in lymphocyte activation may await better characterization in single cells of E_m changes and the specific underlying conductances.

Acknowledgements

This work was supported by an Investigator Award from the Arthritis Foundation (H. A. Wilson), by Uniform Services University of the Health Sciences Protocol No. R08308 (F. D. Finkelman), and by National Institutes of Health grant 1RO1-21328 (F. D. Finkelman).

References

Akerman, K. E. O., and M. K. F. Wikstrom. 1976. Safranine as a probe of the mitochondrial membrane potential. *FEBS Letters* 68:191–197.

Aldo-Benson, M. 1986. Membrane depolarization is induced in tolerant B lymphocytes by stimulation with antigen. *Cellular Immunology.* 103:417–425.

Azzone, G. F., D. Peitroban, and M. Zoratti. 1984. Determination of the proton electrochemical gradient across biological membranes. *Current Topics in Biophysics and Bioengineering.* 13:1–77.

Bashford, C. L., and C. A. Pasternak. 1984. Plasma membrane potential of Lettre cells does not depend on cation gradients but on pumps. *Journal of Membrane Biology.* 79:275–284.

Bashford, C. L., and J. C. Smith. 1979. The use of optical probes to monitor membrane potential. *Methods in Enzymology.* 60:569–586.

Berridge, M. J., and R. F. Irvine. 1984. Inositol trisphosphate, a novel second messenger in cellular signal transduction. *Nature.* 312:315–321.

Besterman, J. M., W. S. May, H. LeVine, E. J. Cragoe, and P. Cuatrecasas. 1985. Amiloride inhibits phorbol ester-stimulated Na^+/H^+ exchange and protein kinase C. An amiloride analogue selectively inhibits Na^+/H^+ exchange. *Journal of Biological Chemistry.* 260:1155–1159.

Bräuner, T., D. F. Hülser, and R. J. Strasser. 1984. Comparative measurements of membrane potentials with microelectrodes and voltage-sensitive dyes. *Biochimica et Biophysica Acta.* 771:208-216.

Cambier, J. C., and J. T. Ransom. 1987. Molecular mechanisms of transmembrane signaling in B lymphocytes. *Annual Review of Immunology.* 5:175–199.

DeCoursey, T. E., K. G. Chandy, S. Gupta, and M. D. Cahalan. 1984. Voltage-gated K^+ channels in human T lymphocytes: a role in mitogenesis? *Nature.* 307:465–468.

Deutsch, C. J., A. Holian, S. K. Holian, R. P. Daniele, and D. F. Wilson. 1979. Transmembrane electrical and pH gradients across human erythrocytes and human peripheral lymphocytes. *Journal of Cellular Physiology.* 99:79–94.

Deutsch, C., and M. Price. 1982. Role of extracellular Na and K in lymphocyte activation. *Journal of Cellular Physiology.* 113:73–79.

De Weer, P., and B. M. Salzberg, editors. 1986. Optical Methods in Cell Physiology. Wiley-Interscience, New York, NY.

Freedman, J. C., and P. C. Laris. 1988. Optical potentiometric indicators for nonexcitable cells. *In* Spectroscopic Membrane Probes. Vol. 3. L. Loew, editor. CRC Press, Boca Raton, FL.

Freedman, J. C., and T. S. Novak. 1987. Optical measurement of membrane potential in cells, organelles, and vesicles. *Methods in Enzymology.* 172:In press.

Gelfand, E. W., R. K. Cheung, and S. Grinstein. 1984. Role of membrane potential in the regulation of lectin-induced calcium uptake. *Journal of Cellular Physiology.* 121:533–539.

Gerard, C., L. C. McPhail, A. Marfat, N. P. Stimler-Gerard, D. A. Bass, and C. E. McCall. 1986. Role of protein kinases in stimulation of human polymorphonuclear leukocyte oxidative metabolism by various agonists. Differential effects of a novel protein kinase inhibitor. *Journal of Clinical Investigation.* 77:61–65.

Grinstein, S., C. A. Clarke, A. Dupre, and A. Rothstein. 1982. Volume induced increase in anion permeability in human lymphocytes. *Journal of General Physiology.* 80:801-823.

Grinstein, S. A., A. Rothstein, B. Sarkadi, and E. W. Gelfand. 1984. Responses of lymphocytes to aniostonic media: volume-regulating behavior. *American Journal of Physiology.* 246:C204-215.

Grynkiewicz, G. M., M. Poenie, and R. Y. Tsien. 1985. A new generation of Ca^{2+} indicators with greatly improved fluorescence properties. *Journal of Biological Chemistry.* 260:3440–3450.

Heikkilä, R., J. Iversen, and G. Godal. 1985. No correlation between membrane potential and increased cytosolic free Ca^{2+} concentration, ^{86}Rb influx or subsequent thymidine incorporation in neoplastic human B cells stimulated with antibodies to surface Ig. *Acta Physiologica Scandinavica.* 124:107–115.

Kiefer, H., A. J. Blume, and H. R. Kaback. 1980. Membrane potential changes during mitogenic stimulation of mouse spleen lymphocytes. *Proceedings of the National Academy of Sciences.* 77:2200–2204.

LaBaer, J., R. Y. Tsien, K. A. Fahey, and A. L. DeFranco. 1986. Stimulation of the antigen receptor on WEHI-231 B lymphoma cells results in a voltage-independent increase in cytoplasmic calcium. *Journal of Immunology.* 137:1836–1844.

Laris, P. C., M. Bootman, H. A. Pershadsingh, and R. M. Johnstone. 1978. The influence of cellular amino acids and the Na^+-K^+ pump on the membrane potential of the Ehrlich ascites tumor cell. *Biochimica et Biophysica Acta.* 512:397–414.

Lewis, R. S., and M. D. Cahalan. 1987. Diversity of K^+ channel expression in developing T lymphocytes. *Journal of General Physiology.* 90:27a–28a. (Abstr.)

Matteson, D. R., and D. Deutsch. 1984. K channels in T lymphocytes: a patch clamp study using monoclonal antibody adhesion. *Nature.* 307:468–471.

Monroe, J. G., and J. C. Cambier. 1983. B cell activation. I. Anti-immunoglobulin-induced receptor cross linking results in a decrease in plasma membrane potential of murine B lymphocytes. *Journal of Experimental Medicine.* 157:2073–2086.

Nishizuka, Y. 1984. The role of protein kinase C in cell surface signal transduction and tumor promotion. *Nature.* 308:693–698.

Philo, R. D., and A. A. Eddy. 1978. The membrane potential of mouse ascites tumor cells studied with the fluorescent probe 3,3′ dipropyloxadicarbocyanine. *Biochemical Journal.* 174:801–810.

Pozzan, T., P. Arslan, R. Y. Tsien, and T. J. Rink. 1982. Anti-immunoglobulin, cytoplasmic free calcium, and capping in B lymphocytes. *Journal of Cell Biology.* 94:335–340.

Rabinovitch, P. S., C. H. June, A. Grossman, and J. A. Ledbetter. 1986. Heterogeneity among T cells in free calcium responses after mitogen stimulation with PHA or anti-CD3. *Journal of Immunology.* 137:952–961.

Rink, T. J., C. Montecucco, T. R. Hesketh, and R. Y. Tsien. 1980. Lymphocyte membrane potential assessed with fluorescent probes. *Biochimica et Biophysica Acta.* 595:15–30.

Tatham, P. E. R., and P. J. Delves. 1984. Flow cytometric detection of membrane potential changes in murine lymphocytes induced by concanavalin A. *Biochemical Journal.* 221:137–146.

Tsien, R. Y., T. Pozzan, and T. J. Rink. 1982. T-cell mitogens cause early changes in cytoplasmic free Ca^{2+} and membrane potential in lymphocytes. *Nature.* 295:68–70.

Waggoner, A. S. 1979. Dye indicators of membrane potential. *Annual Review of Biophysics and Bioengineering.* 8:47–68.

Wilson, H. A., and T. M. Chused. 1985. Lymphocyte membrane potential and Ca^{2+}-sensitive potassium channels described by oxonol dye fluorescence measurements. *Journal of Cellular Physiology.* 125:72–81.

Wilson, H. A., D. Greenblatt, C. W. Taylor, J. W. Putney, R. Y. Tsien, F. D. Finkelman, and T. M. Chused. 1987a. The B lymphocyte calcium response to anti-Ig is diminished by membrane immunoglobulin cross-linkage to the Fc receptor. *Journal of Immunology.* 138:1712–1718.

Wilson, H. A., D. Greenblatt, M. Poenie, F. D. Finkelman, and R. Y. Tsien. 1987b. Cross linkage of B lymphocyte surface immunoglobulin by anti-Ig or antigen induces prolonged oscillation of intracellular ionized calcium. *Journal of Experimental Medicine.* 166:601–606.

Wilson, H. A., B. E. Seligmann, and T. M. Chused. 1985. Voltage sensitive cyanine dye fluorescence signals in lymphocytes: plasma membrane and mitochondrial components. *Journal of Cellular Physiology.* 125:61–71.

Chapter 23

Platelet Membrane Potentials and Their Significance in Monitoring Stimulus Response Coupling

Elizabeth R. Simons, Theresa A. Davies, Sheryl M. Greenberg, Judith M. Dunn, and William C. Horne

Department of Biochemistry, Boston University School of Medicine, Boston, Massachusetts

Introduction

Like most mammalian cells investigated to date, platelets exhibit a negative membrane potential. Other chapters in this book and elsewhere (Freedman and Laris, 1988; Freedman and Novak, 1988) have recently dealt with the overall concept of membrane potential measurement in small cells in suspension, as well as with the indirect probes used for such measurements and the controls required for data evaluation. We have recently reviewed the literature specific to platelets (Simons and Greenberg-Sepersky, 1987) and will hence summarize it only briefly here before discussing some of the more recent findings. The magnitude of the platelet resting potential at physiological pH and ion concentrations is in the vicinity of -50 to -60 mV. The measured value is independent of the method of determination used, provided the concentrations of probe are kept sufficiently low to prevent poisoning of the cells, yet sufficiently high so that the potential, and not the quantity of available probe, controls the dye distribution. For lipophilic cationic or anionic probes, the appropriate concentration range is usually from 10^{-8} to 2×10^{-6} M, but it must be checked for each cell and probe used (Freedman and Hoffman, 1979). When these probe concentration precautions are obeyed, the values for the human platelet resting potential in isotonic buffers at pH 7.4, whether obtained by the distribution of isotopically or fluorescently labeled lipophilic cations (Horne and Simons, 1978b; MacIntyre et al., 1978; Friedhoff and Sonenberg, 1981, 1983; Horne et al., 1981; MacIntyre and Rink, 1982), by that of isotopically labeled anions such as Cl^- (Horne et al., 1981), or by the null point of potential change upon addition of valinomycin (Horne and Simons, 1978b; MacIntyre et al., 1978; Friedhoff and Sonenberg, 1981, 1983) have been similar, although the latter technique has been reported to yield slightly more negative values (-72 mV) under different buffer conditions.

Since the initial report by Bramhall and co-workers (1976), stimulus-induced changes in platelet membrane potential have been amply documented (Horne and Simons, 1978b; Larsen et al., 1979; Friedhoff and Sonenberg, 1981, 1983; Horne et al., 1981; MacIntyre and Rink, 1982; Greenberg-Sepersky and Simons, 1984; Pipili, 1985); a typical curve is shown in Fig. 1. The change in potential has been observed with α-thrombin, ADP, and epinephrin, but appears to be absent when fibrillar collagen is the stimulus (Horne and Simons, 1978b; Larsen et al., 1979; Friedhoff and Sonenberg, 1981, 1983; Horne et al., 1981; MacIntyre and Rink, 1982; Pipili, 1985; Banga et al., 1986). It is dose dependent and, for α-thrombin, is maximal at concentrations ($\sim$0.02 U/ml, or 3.6 nM) approaching those in actively coagulating blood (Larsen et al., 1979; Horne et al., 1981). We have demonstrated a similar α-thrombin dose dependence for Na^+ influx, for intracellular H^+ and Ca^{++} concentration changes, as well as for serotonin and β-glucuronidase release (Fig. 2) (Horne and Simons, 1978a, b; Larsen et al., 1979; Horne et al., 1981; Greenberg-Sepersky and Simons, 1984; Davies et al., 1987a, b). The magnitude of the membrane potential change induced by saturating concentrations of thrombin ($>$0.02 U/ml), as monitored with isotopically labeled TPP^+, is a depolarization of 35 mV which indicates that fully thrombin-stimulated platelets retain a potential of -15 mV.

While changes in membrane potential as well as in intracellular cation concentrations are part of the platelets' stimulus response, their role in the activation process is not yet completely clear. Since cells can be artificially depolarized without spontaneous activation (Feinberg et al., 1977; MacIntyre and Rink, 1982; Greenberg-Sepersky and

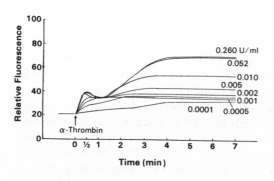

Figure 1. Representative tracings of the dose-dependent fluorescence changes in a platelet suspension induced by thrombin. DiS-C_3(5)-equilibrated gel-filtered human platelets were stimulated with various doses of thrombin and the suspension fluorescence changes were monitored continuously. As in all of the following figures, the conditions and concentrations of platelets and of probe used were those detailed under Materials and Methods. From Greenberg-Sepersky and Simons, 1984.

Simons, 1984) by immersion in isotonic buffers depleted in [Na$^+$] and repleted with high [K$^+$], the implication is that depolarization is not sufficient for initiation of platelet activation. Furthermore, since collagen does not initiate a change in membrane potential yet clearly activates platelets when in the fibrillar, albeit not in the monomeric form (Horne and Simons, 1978b; Simons, 1978), depolarization is also not a necessary component of platelet activation. Therefore, depolarization is neither necessary nor sufficient as an initiator—and thus is not, in itself, an indicator—of platelet activation. Depolarization does, however, accompany activation by most stimuli; its role in the overall process is still being explored.

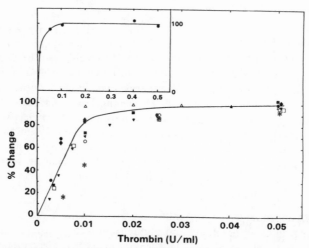

Figure 2. Thrombin-induced dose-response curve. Curve represents similarities among thrombin-induced dose responses of human platelets as measured by several parameters. Parameters evaluated are: slope/F_o measured with diO-C_5-3 (*), $\Delta F/F_o$ measured with diS-C_3(5) (●), slope measured with diS-C_3(5) (□), membrane potential measured with [^{3}H]tetraphenylphosphonium bromide (○), pH change measured with 9-aminoacridine (△), pH change measured with 6-carboxyfluorescein (▲), lysosomal granule secretion as β-glucuronidase (◊), and dense granule secretion as serotonin (■). The figure is presented as percent maximal change vs. thrombin concentration (0.0025–0.05 U/ml thrombin). Inset shows an expansion of the curve above 0.05 U/ml thrombin.

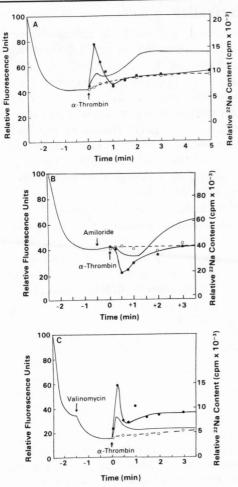

Figure 3. (*A*) Thrombin-induced platelet responses. DiS-C$_3$(5)-equilibrated human platelets were stimulated with a saturating dose of thrombin, and the ensuing changes in the suspension fluorescence were monitored for 5 min. The membrane potential change was complete within 30 s. (——) A tracing of the fluorescence changes. Twice-concentrated gel-filtered human platelets were mixed with an equal volume of HEPES buffer containing ^{22}Na with (●) or without (○) a saturating dose of thrombin. Aliquots were taken periodically and counted for the ^{22}Na$^+$ content of the platelets. From Greenberg-Sepersky and Simons, 1984. (*B*) Effects of amiloride L591–605 on the thrombin-induced platelet responses. DiS-C$_3$(5)-equilibrated human platelets were incubated with amiloride L591–605 for 1 min before stimulation by a saturating dose of thrombin. After stimulation, the thrombin-induced fluorescence changes were monitored for several minutes, and a typical tracing of the fluorescence is shown here (——). Gel-filtered platelets were preincubated with the fast acting amiloride analog L591–605 for 1 min in the presence of ^{22}Na$^+$. Aliquots were taken periodically for assay of the ^{22}Na$^+$ content of the thrombin-stimulated platelets (●) and compared with control platelets which had not been stimulated with thrombin (○). From Greenberg-Sepersky and Simons, 1984. (*C*) Effect of valinomycin on the thrombin-induced platelet responses. DiS-C$_3$(5)-equilibrated platelets were treated with 5 μM valinomycin until the probe had reequilibrated. A saturating dose of thrombin was then used to stimulate the platelets, and the subsequent membrane potential change was monitored; a typical tracing is shown (——). Twice-concentrated gel-filtered platelets were treated with 10 μM valinomycin (also twice the usual concentration) for 3 min. They were then mixed with an equal volume of buffer containing a saturating dose of thrombin at time 0. Aliquots were taken periodically and assayed for the ^{22}Na$^+$ content. The ^{22}Na$^+$ content of the thrombin-stimulated valinomycin-pretreated platelets (●) was compared with that of the control valinomycin-pretreated platelets (○). (——) Used to indicate that the apparent increase at 1 min appeared only in this one of six similar experiments and may be attributable to experimental error. From Greenberg-Sepersky and Simons, 1984.

    With stimuli such as thrombin, depolarization always accompanies and precedes aggregation; it has therefore proven to be not only one of the earliest but also one of the best indicators of functional integrity of the platelets, and it has become a convenient initial verifying control. Incidentally, such controls have allowed us to monitor and detect such problems as endotoxin or trace organic solvent-contaminated buffers in a timely fashion.

Origin of Stimulus-induced Depolarization

We, as well as others, have attempted to delineate the origin of the stimulus-induced membrane depolarizations that, when induced by α-thrombin, are detectable in stirred thermostated samples within <3 s, exhibit maximal rates by that time, and are greatest <30 s after thrombin exposure (Horne et al., 1981; Friedhoff and Sonenberg, 1983; Greenberg-Sepersky and Simons, 1984). It is important to indicate that this rapid response is detectable for bovine thrombin (purified from commercial Parke-Davis bovine thrombin by the method of Lundblad et al., 1977) (Horne et al., 1981; Greenberg-Sepersky and Simons, 1984) or for human thrombin (prepared by Fenton et al., 1977) (Friedhoff and Sonenberg, 1981, 1983; Gear and Burke, 1982) freed of β- and γ-thrombin. A rapid response is not exhibited by γ-thrombin, which induces a much slower and smaller one (McGowan and Detwiler, 1986).

The origin of the membrane potential change is still the subject of some controversy. The maximal rate (as well as the extent) of depolarization can be fully correlated not only with the dose of thrombin used, but also with the influx of $^{22}Na^+$ (Feinberg et al., 1977; Greenberg-Sepersky and Simons, 1984). Furthermore, blockage of Na^+ influx with brief exposure (1 min) to 10^{-4} M dimethylamiloride abrogates not only the thrombin-induced rapid influx of Na^+ but also the depolarization (Fig. 3

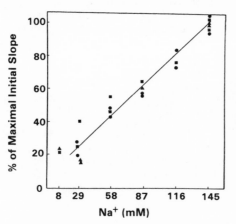

Figure 4. Sodium gradient dependence of the thrombin-induced membrane potential change. Gel-filtered human platelets were diluted in isotonic choline buffers containing various concentrations of sodium and then equilibrated with diS-C$_3$(5). The initial slope of the membrane potential change in response to a saturating dose of thrombin was measured as the indicator of the platelet's ability to depolarize in response to stimulation. The symbols (●, ■, and ▲) represent the measured responses of three separate experiments. From Greenberg-Sepersky and Simons, 1984.

A, B) (Horne and Simons, 1978a; Greenberg-Sepersky and Simons, 1984). At this high concentration, dimethylamiloride, which has a greater affinity for Na^+/H^+ antiports than for passive Na^+ channels (Cragoe et al., 1967) would nevertheless be expected to block both. Both depletion of Na^+ in the extracellular milieu and the resultant inhibition of Na^+ influx therefore reduce the initial rapid membrane depolarization induced by exposure to α-thrombin (Fig. 4). Surprisingly, although serotonin secretion is somewhat reduced, β-glucuronidase release is unaffected under these conditions, implying that the platelet release reaction, at least as far as the lysosomal granules are concerned, is not dependent upon the existence of a Na^+ gradient across the plasma membrane. Conversely, however, we reported at the same time (Greenberg-Sepersky and Simons, 1984) that valinomycin, which causes reduction of the transmembrane K^+ gradient by permeabilization of plasma as well as organellar membranes, abrogates the release of β-glucuronidase without inhibiting the depolarization or the Na^+ influx, which remains linked (Fig. 4, Table I). Indeed the apparent depolarization is slightly enhanced, presumably because in the absence of a K^+ gradient, the Na^+/K^+ ATPase makes no attempt to transport Na^+ back to the external medium once it has passively entered the cytoplasm (Greenberg-Sepersky and Simons, 1984). Thus we have reported that thrombin-induced platelet depolarization seems to depend exclusively on the Na^+ gradient (Fig. 3 C) while degranulation is controlled, at least for the lysosomal granules, by the K^+ gradient (Feinstein et al., 1977; Greenberg-Sepersky and Simons, 1984). Thus, as shown in Table I, β-glucuronidase secretion is unaffected when extracellular Na^+ is replaced by choline while the K^+ concentration remains equal to the control. Conversely, lowering the K^+ gradient by increasing extracellular K^+ and decreasing extracellular Na^+ buffer concentrations while maintaining isotonicity does inhibit β-glucuronidase secretion. The reasons for these separate cation controls are, as yet, unclear, particularly in light of the fact that other blood cells (granulocytes) that also exhibit stimulus-induced membrane depolarization and degranulation, exhibit control by both of these cations of both of these disparate functions (Simchowitz et al., 1982; Luscinskas et al., 1988).

TABLE I
Summary of Effects of Isotonic Choline and Potassium Buffers on Resting and Thrombin-stimulated Platelets

Buffer	Na^+	diS-C$_3$-(5) uptake	Initial slope $\pm SD$	[^{14}C]Serotonin secretion $\pm$ SD	β-Glucuronidas secretion $\pm$ SD
	mM	%	$\Delta F/min$	%	%
Control	145	57.7	31.8 ± 1.1	90.7 ± 2.1	25.6 ± 2.9
68 mM choline	77	56.7	17.7 ± 0.8	95.6 ± 1.4	25.3 ± 6.9
106 mM choline	39	55.5	13.7 ± 1.0	98.3 ± 1.7	27.7 ± 4.5
68 mM K^+	77	49.9	13.0 ± 1.0	97.7 ± 4.8	19.7 ± 2.9
106 mM K^+	39	48.7	13.0 ± 0.5	87.0 ± 9.6	20.7 ± 2.5

Gel-filtered ^{14}C-labeled serotonin-loaded platelets were equilibrated with diS-C$_3$(5) in buffers containing various concentrations of choline chloride or potassium chloride substituting for part of the extracellular sodium. The platelets were then stimulated with a saturating dose of thrombin and then assayed for their ability to respond to stimulation under these conditions. (From Greenberg-Sepersky and Simons, 1984.)

Na/H Exchange

For platelets, as for other cells, the existence of a Na^+/H^+ antiport has been postulated (Horne et al., 1981; Grinstein and Rothstein, 1986; Siffert and Akkerman, 1987). We showed several years ago that the thrombin-initiated membrane depolarization and Na^+ influx in platelets (Horne and Simons, 1978b; Horne et al., 1981; Greenberg-Sepersky and Simons, 1984) are temporally and mechanistically linked to the concomitant alkalinization of the platelet cytoplasm (Fig. 5) (Horne et al., 1981). Dimethylamiloride blocks both events at 10^{-4} M, as does amiloride itself at the same concentration. These results are compatible with the existence of a Na^+/H^+ antiport (Horne et al., 1981; Siffert and Akkerman, 1987), but give no indication of the stoichiometry involved. We have recently reexamined this question by a novel approach based upon the large difference in transmembrane diffusion rates between H^+ and $^2H^+$ (Davies et al., 1987a). By comparing rates of depolarization, alkalinization, and degranulation of control cells with those equilibrated with ^2H_2O for 3 min, we

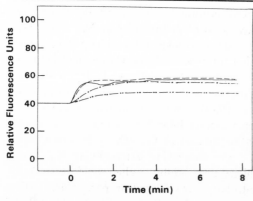

Figure 5. Typical tracing of change in fluorescence of 6-carboxyfluorescein-containing platelets upon addition of α-thrombin. λ_{em} = 518 nm, λ_{exc} = 492 nm. Line for λ_{em} = 518, λ_{exc} = 464 nm is omitted for simplicity as fluorescence change is small upon thrombin addition. (—··—) 0.005 U/ml, (—·—) 0.01 U/ml, (– – –) 0.025 U/ml, (——) 0.05 U/ml. From Horne et al., 1981.

have been able to demonstrate that, while all are altered in the presence of heavy water, the extent and the direction of ^2H_2O effects on these parameters are not comparable and therefore allow some distinctions to be made between these parameters (Fig. 6) (Davies et al., 1987*a*). Specifically, we have now been able to demonstrate that the substitution of ^2H_2O for H_2O-based external buffers leads to decreased rates and extents of alkalinization and depolarization in response to a saturating concentration of α-thrombin (and increased lysosomal degranulation).

The observed inhibition of the intracellular pH change caused by $^2H_2$ substitution for H_2O is larger (Fig. 7) and the rate of alkalinization is more rapid (Fig. 6) than the corresponding rate of depolarization. Both depolarization and alkalinization, induced by thrombin, remain subject to amiloride or dimethylamiloride inhibition at 10^{-4} M concentrations in ^2H_2O buffers. We have found that the effect cannot be mimicked by artificial cytoplasmic alkalinization (to the same extent) achieved via NH_4^+ treatment. We interpret these results to implicate passive transport of H^+, followed by Na^+

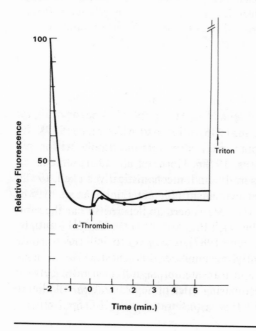

Figure 6. Fluorimetric analysis of platelet depolarization by thrombin. Superimposed tracings of the response of diS-C_3(5)-equilibrated platelets in H_2O (——) or D_2O (—●—) HEPES to 0.05 U/ml thrombin. From Davies et al., 1987*a*.

(because H^+ is faster and the D_2O effect greater) through interdependent but neither stoichiometric nor electrically neutral channels opened as a consequence of thrombin stimulation. We therefore conclude that a classical electrically silent antiport is not involved in the initial facets of platelet thrombin response, and that nonequivalent transport of Na^+ into—and H^+ out of—the platelets could contribute to the membrane potential change observed upon thrombin stimulation of human platelets (Davies et al., 1987*a*).

Flow Cytometry

There is thus agreement that the three cations discussed above do, as described by the Goldman equation (Hoffman and Laris, 1974; Freedman and Hoffman, 1979; Freedman and Laris, 1988; Freedman and Novak, 1988), play a role in the membrane potential of most mammalian cells, including the platelet. Rapid measurement of these

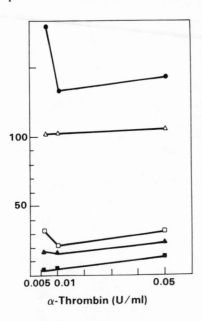

Figure 7. Relative effect of D_2O on the platelet-thrombin response. Relative effect of D_2O on the initial rate of depolarization (□), the β-glucuronidase secretion (●), and the serotonin secretion (△) by diS-C_3(5)-equilibrated platelets upon thrombin stimulation. Relative effect of D_2O on 6-carboxyfluorescein-loaded (■) and BCECF-loaded (▲) platelets after thrombin stimulation. Thrombin doses were 0.005, 0.01, and 0.05 U/ml. From Davies et al., 1987*a*.

changes has generally been accomplished continuously in cell suspensions with fluorescent probes (Hoffman and Laris, 1974; Sims et al., 1974; Waggoner, 1976; Waggoner and Grinvald, 1977; Waggoner et al., 1977; Freedman and Hoffman, 1979; Freedman and Laris, 1988; Freedman and Novak, 1988) or in aliquots with rapid quenching systems (Gear and Burke, 1982). Since the initial measurements of individual neutrophil membrane potentials by flow cytometry (Seligmann et al., 1981), examination of the kinetics of stimulus-induced membrane potential changes by flow cytometry in a number of cell systems has been reported (Seligmann et al., 1981; Lazzari et al., 1986; Sullivan et al., 1987). The advantage of such measurements by flow cytometry rests primarily in the ability to evaluate simultaneously the response and the population responding (Fig. 8). This is particularly useful when subsaturating concentrations of stimuli demonstrate the existence of a response threshold (in terms of receptor occupancy) (Fig. 9), or of a positive cooperativity among receptors leading to a separation into responding and nonresponding (or not yet responding) populations.

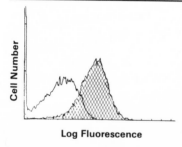

Log Fluorescence

Figure 8. Flow microfluorimetric analysis of RITC-TLCK-thrombin noncovalently bound to platelets. Histogram representing log fluorescence vs. cell number of control platelets (*nonshaded*) and platelets noncovalently bound to 0.05 U/ml TLCK-RITC-thrombin (*shaded*). The histogram shift to the right indicates a fluorescence increase due to binding in a homogeneous population of platelets.

These types of studies can be performed either with one of the positively charged cyanine probes, diO-C$_5$(3) (Table I), which does not tend to aggregate and undergo fluorescence quenching in the intracellular milieu as does diS-C$_3$(5) (Sims et al., 1974; Waggoner et al., 1977), or with one of the oxonol series of negatively charged fluorescent anions (Bashford et al., 1985) (Fig. 10). Quenching is highly desirable when measurements are being performed on cell suspensions since the external probe alone and not a combination of the external and internal probe concentrations gives rise to the observed fluorescence. Quenching is clearly not satisfactory when one looks at individual cells during their passage in front of a laser beam, since the fluorescence in the cell, and not that of the medium, is measured in the flow cytometer. It is imperative, of course, that the control experiments already discussed above and elsewhere in this volume for optimal probe concentration and for linearity of response with log[K$^+$] be performed anew with the specific probe in the system of choice. Such experiments were successful in our hands for neutrophils (Lazzari et al., 1986), in which comparable results were obtained in suspensions with the cyanines diS-C$_3$(5) and diO-C$_5$(3) (Waggoner, 1977) and with bis-oxonol (Bashford et al., 1985) at concentrations from 0.2 to 20×10^{-7} M depending on the probe/cell ratio, and in flow cytometry with diO-C$_5$(3) and bis-oxonol at 0.2×10^{-7} M (Lazzari et al., 1986; Sullivan et al., 1987). In all cases, stimulation with subsaturating concentrations of

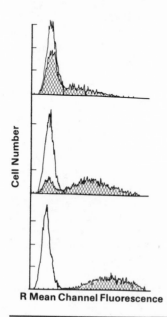

R Mean Channel Fluorescence

Figure 9. Histograms of platelets before and after thrombin-induced stimulation of human platelets. Histograms illustrating baseline indo-1 fluorescence on the FACS (*nonshaded*) and the change in fluorescence after stimulation with 0.0025 U/ml (*top*), 0.005 U/ml (*middle*), and 0.05 U/ml α-thrombin (*shaded*). Plotted is the R (405 nm/485 nm) ratio of mean channel fluorescence on the FACS vs. platelet number. A shift in relative fluorescence is indicated as a shift in the histogram to the right. From Davies et al., 1987*b*.

Cationic carbocyanines: 3,3'-dipropylthiadicarbo- cyanine iodide DiSC$_3$-5	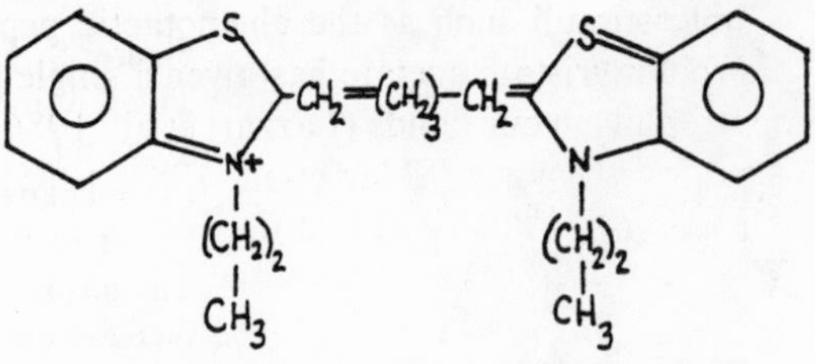
3,3'-dipentyloxacarbo- cyanine iodide DiOC$_5$-3	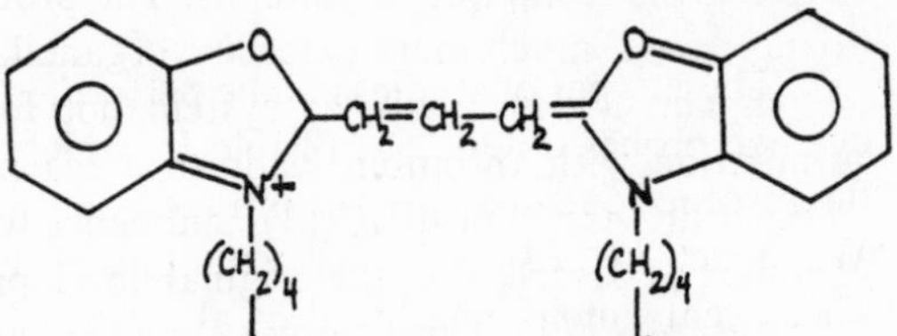
1,1'-dipropyl-3,3,3',3'- tetramethylindocarbocyanine iodide DiIC$_3$-3	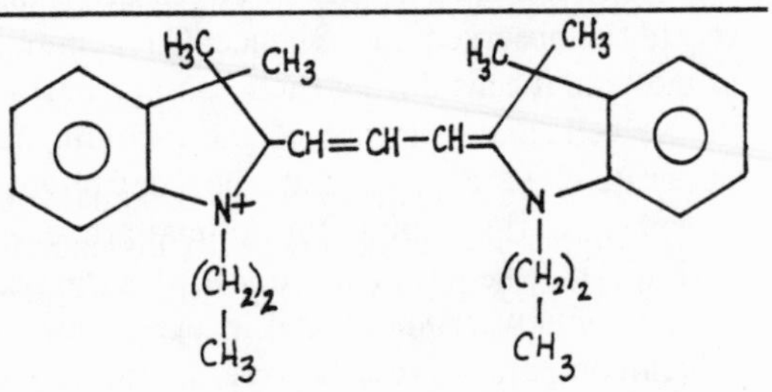
Anionic bis-oxonols: bis-(1,3 diethylthio- barbiturate) trimethine oxonol DiSBaC$_2$-3	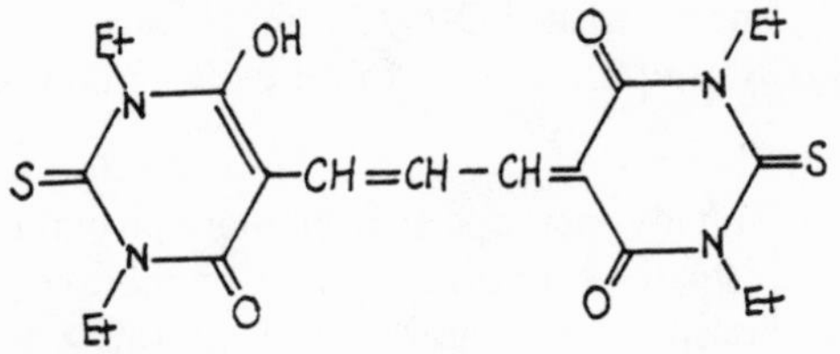
bis-(1,3-dibutylbarbituric acid) trimethine oxonol DiBaC$_4$-3	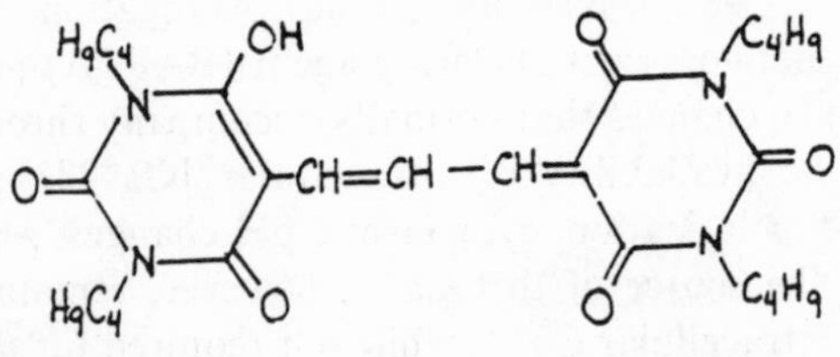

Figure 10. Fluorescent membrane potential probes. Structures of cationic and anionic fluorescent membrane potential probes. Cationic carbocyanine include 3,3'-dipropylthiadicarbocyanine iodide [diS-C$_3$(5)], 3,3'-dipentyloxacarbocyanine iodide [diO-C$_5$(3)], and 1,1'-dipropyl-3,3,3'3'-tetramethylindocarbocyanine iodide [diI-C$_3$(3)]. Anionic bis-oxonols include bis-(1,3-diethylthiobarbuturate) trimethine oxonol [DiBa-C$_2$(3)] and bis-(1,3-dibutyl-barbituric acid) trimethine oxonol [DiBa-C$_4$(3)].

soluble stimuli such as the chemotactic peptide f-met-leu-phe or the phorbol ester phorbol myristate acetate has given a single population of responding normal human neutrophils in our hands (Lazzari et al., 1986; Sullivan et al., 1987) as well as those of some investigators (Sklar et al., 1984, 1985) but not of all others (Seligmann et al., 1981; Fletcher, 1986). The differences may lie in the cell preparations, but are unlikely to be due to the probes themselves.

In contrast to this easy transferability of the techniques of suspension measurements to flow cytometry in neutrophils, we have had a much more difficult time adapting the technique to platelets. The problem appears to be intrinsic, lying in the existence of a much more extensive organellar system, including an open canalicular system and a dense tubular system not found in phagocytes. As a result, while stimulation with thrombin leads to a rearrangement of the probe within the cell, whether diO-C_5(3) or diI-C_3(3), and hence to a change in fluorescence, that change is complex, as described in the original development of the diO and diI series of cyanine (Sims et al., 1974). This complex change is composed of opposite effects that are probe concentration dependent: the fluorescence at low probe concentrations is reduced as the cells' interior becomes depolarized; conversely, at high probe concentrations, the fluorescence increases as the probe leaves the cells and there is less quenching. Thus, while measurements of thrombin-induced platelet depolarization can be made on suspensions of platelets with several of the cyanines as well as oxonols (Fig. 9), and can be shown to be dose dependent, the signals obtainable from individual platelets on the flow cytometer are too low with bis-oxonol, and marginal, albeit detectable, with diO-C_5(3). It is also likely that stimulus-induced volume changes, which occur in both systems (Akkerman and Holmsen, 1981; Grinstein et al., 1986), may alter the distribution in neutrophils less than in platelets whose open canalicular system allows a much greater expansion of cytoplasmic volume.

Role of Ca^{++}

Another cation that has been receiving increasing attention, Ca^{++} (Feinstein et al., 1977), plays perhaps an even more pivotal role in the platelet response but probably is not directly correlatable with membrane potential changes. While its role will be discussed in another chapter in this volume, it should be mentioned here that thrombin-elicited platelet cytoplasmic Ca^{++} concentration changes are detectable even more rapidly than those of their membrane potential or intracellular pH (Rink et al., 1982; Davies et al., 1988). Abrogation of these Ca^{++} changes with an intracellular nonfluorescent chelating agent (BAPTA) also abolishes the membrane potential and pH_i changes that normally accompany thrombin treatment of human platelets. Thus, the availability of intracellular [Ca^{++}] appears to be a requirement for platelet depolarization, cytoplasmic pH changes, and serotonin release (Davies et al., 1988). The source of this Ca^{++}, however, remains unknown. In contrast, the presence of extracellular Ca^{++}, while not required for thrombin-initiated platelet activation, does enhance the rate and extent of thrombin-induced intracellular [Ca^{++}] rise but does not appear to affect the resting concentration of Ca^{++} in the unstimulated human platelets (Davies et al., 1988). It should be emphasized that these studies were performed under conditions of very brief (15–30 s) Ca^{++} deprivation (EGTA treated). If, on the other hand, platelets are depleted of extracellular Ca^{++} for a long period of time, e.g., by exposure to EGTA or EDTA for 5 min or more, their resting state is affected as is their response to stimulation. We believe this is due to the removal of loosely bound Ca^{++} that, in nominally Ca^{++}-free media (in our hands, such media have concentrations at or below 12 μM), nevertheless plays a role in the stabilization of the intact resting cell.

Conclusions

The interrelationships between membrane potential, cytoplasmic calcium and cytoplasmic pH changes, and eventual degranulation and aggregation in response to thrombin mobilization remain unclear. Temporally, attainment of maximal intracellular $[Ca^{++}]$ (Rink et al., 1982) precedes that of membrane depolarization and alkalinization, but may follow a recently reported very rapid cytoplasmic acidification (Zavoico et al., 1986). Abrogation of the Ca^{++} change prevents depolarization and alkalinization, but not the acidification. Changes in cation concentration gradients and therefore in membrane potential are thus involved in the normal overall platelet activation process, and can be used as measures of that process. However, while cations clearly play a very major role in the process by which platelets undergo activation, the details of that role still await definition.

Acknowledgments

The authors would like to thank Virginia J. Phillips for her assistance in preparing the manuscript. This work was supported by National Institutes of Health grants HL-15335 and HL-07501.

References

Akkerman, J. W. N., and H. Holmsen. 1981. Interrelationships among platelet responses: studies on the burst in proton liberation, lactate production, and oxygen uptake during platelet aggregation and Ca^{++} secretion. *Blood.* 57:956–966.

Banga, H. S., E. R. Simons, L. F. Brass, and S. E. Rittenhouse. 1986. Activation of phospholipases A & C in human platelets exposed to epinephrine. *Proceedings of the National Academy of Sciences.* 83:9197–9201.

Bashford, C. L., G. M. Alder, M. A. Gray, K. J. Micklem, C. C. Taylor, P. J. Turek, and C. A. Pasternak. 1985. Oxonol dyes as monitors of membrane potential: the effect of viruses and toxins on the plasma membrane potential of animal cells in monolayer culture and in suspension. *J of Cellular Physiology.* 123:326–336.

Bramhall, J. S., J. L. Morgan, A. D. Perris, and A. Z. Britten. 1976. The use of a fluorescent probe to monitor alterations in transmembrane potentials in single cell suspensions. *Biochemical and Biophysical Research Communications.* 72:654–662.

Cragoe, Jr., E. J., O. W. Woltersdorf, Jr., J. B. Bicking, S. F. Kwong, and J. H. Jones. 1967. Pyrazine diuretics II. N-amido-3-amino-5-substituted-6-halopyrazinecarboxamides. *Journal of Medicinal Chemistry.* 10:66–75.

Davies, T. A., E. Katona, V. Vasilescu, E. J. Cragoe, Jr., and E. R. Simons. 1987a. Sequential sodium-proton exchange in thrombin-induced human platelets. *Biochimica et Biophysica Acta.* 903:381–397.

Davies, T. A., Dunn, J. M. and E. R. Simons. 1987b. Evaluation of changes in cytoplasmic pH in thrombin-stimulated human platelets. *Analytical Biochemistry.* 167:118–123.

Davies, T. A., D. Drotts, G. J. Weil, and E. R. Simons. 1988. Flow cytometric measurements of cytoplasmic calcium changes in human platelets. *Cytometry.* 9:138–142.

Feinberg, H., W. C. Sandler, M. Scorer, G. C. LeBreton, B. Grossman, and G. V. R. Born. 1977. Movement of sodium into human platelets induced by ADP. *Biochimica et Biophysica Acta.* 470:317–324.

Feinstein, M. B., E. G. Henderson, and R. I. Sha'afi. 1977. The effects of alterations of transmembrane Na^+ and K^+ gradients by ionophores (nigericin, monensin) on serotonin transport in human blood platelets. *Biochimica et Biophysica Acta.* 468:284–295.

Fenton, II, J. W., M. J. Fasco, A. B. Stackrow, D. L. Aronson, A. M. Young, and J. S. Finlayson. 1977. Human thrombins: production, evaluation, and properties of α-thrombin. *Journal of Biological Chemistry.* 252:3587–3598.

Fletcher, M. P. 1986. Modulation of heterogeneous membrane potential responses of neutrophils to N-formyl-methionyl-leucyl-phenylalanine by leukotriene LTB4. *Journal of Immunology.* 136:4213–4219.

Freedman, J. C., and J. F. Hoffman. 1979. Ionic and osmotic equilibrium of human red blood cells treated with nystatin. *Journal of General Physiology.* 74:187–246.

Freedman, J. C., and P. C. Laris. 1988. Optical potentiometric indicators for nonexcitable cells. *In Spectroscopic Membrane Probes.* L. Loew, editor. CRC Press, Inc., Boca Raton, FL. In press.

Freedman, J. C., and T. S. Novak. 1988. Optical measurement of membrane potential in cells, organelles and vesicles. *Methods in Enzymology.* 3: In press.

Friedhoff, L. T., and M. Sonenberg. 1981. The effect of altered transmembrane ion gradients on membrane potential and aggregation of human platelets in blood plasma. *Biochemical and Biophysical Research Communications.* 102:832–837.

Friedhoff, L. T., and M. Sonenberg. 1983. The membrane potential of human platelets. *Blood.* 61:180–185.

Gear, A. R. L., and D. Burke. 1982. Thrombin-induced secretion of serotonin from platelets can occur in seconds. *Blood.* 60:1231–1234.

Greenberg-Sepersky, S. M., and E. R. Simons. 1984. Cation gradient dependence of the steps in thrombin stimulation of human platelets. *Journal of Biological Chemistry.* 259:1502–1508.

Grinstein, S., W. Furuya, and E. J. Cargoe, Jr. 1986. Volume changes in activated human neutrophils: the role of Na^+/H^+ exchange. *Journal of Cellular Physiology.* 128:33–40.

Grinstein, S., and A. Rothstein. 1986. Mechanisms of regulation of the Na^+/H^+ exchanger. *Journal of Membrane Biology.* 90:1–12.

Hoffman, J. F., and P. C. Laris. 1974. Determination of membrane potentials in human and amphiuma red blood cells by means of a fluorescent probe. *Journal of Physiology.* 239:519–552.

Horne, W. C., N. E. Norman, D. B. Schwartz, and E. R. Simons. 1981. Changes in cytoplasmic pH and in membrane potential in thrombin-stimulated platelets. *European Journal of Biochemistry.* 120:295–302.

Horne, W. C., and E. R. Simons. 1978a. Effects of amiloride on the response of human platelets to bovine alpha-thrombin. *Thrombosis Research.* 13:599–607.

Horne, W. C., and E. R. Simons. 1978b. Probes of transmembrane potentials in platelets: changes in cyanine dye fluorescence in response to aggregation stimuli. *Blood.* 51:741–749.

Larsen, N. E., W. C. Horne, and E. R. Simons. 1979. Platelet interaction with active and TLCK-inactivated alpha-thrombin. *Biochemical and Biophysical Research Communications.* 87:403–409.

Lazzari, K. G., P. J. Proto, and E. R. Simons. 1986. Simultaneous measurement of stimulus-induced changes in cytoplasmic Ca^{2+} and in membrane potential of human neutrophils. *Journal of Biological Chemistry.* 261:9710–9713.

Lundblad, R. L., L. C. Uhteg, H. S. Kingdon, and K. G. Mann. 1977. Preparation and partial characterization of two forms of bovine thrombin. *Biochemical and Biophysical Research Communications.* 66:482–489.

Luscinskas, F. W., D. E. Mark, B. Brunkhorst, F. J. Lionetti, E. J. Cragoe, Jr. and E. R. Simons. 1988. The role of transmembrane cationic gradients in immune complex stimulation of human polymorphonuclear leukocytes. *Journal of Cellular Physiology.* 134:211–219.

MacIntyre, D. E., C. Montecucco, and T. J. Rink. 1978. Platelet membrane potential assisted with a fluorescent probe ($diSC_3$-5): has this potential a role in triggering aggregation? *Journal of Physiology.* 289:49P.

MacIntyre, D. E., and T. J. Rink. 1982. The role of platelet membrane potential in the initiation of platelet aggregation. *Thrombosis and Haemostasis.* 47:22–26.

McGowan, E. B., and T. C. Detwiler. 1986. Modified platelet responses to thrombin. *Journal of Biological Chemistry.* 261:739–746.

Pipili, E. 1985. Platelet membrane potential: simultaneous measurement of $diSC_3(5)$ fluorescence and optical density. *Thrombosis Haemostasis.* 54:645–649.

Rink, T. J., S. W. Smith, and R. Tsrin. 1982. Cytoplasmic free Ca^{2+} in human platelets: Ca^{2+} thresholds and CA-independent activation for shape change and secretion. *FEBS Letters.* 148:21–26.

Seligmann, B., T. M. Chused, and J. I. Gallin. 1981. Human neutrophil heterogeneity identified using flow microfluorometry to monitor membrane potential. *Journal of Clinical Investigation.* 68:1125–1131.

Siffert, W., and J. W. N. Akkerman. 1987. Activation of sodium-proton exchange is a prerequisite for Ca^{+2} mobilization in human platelets. *Nature.* 325:456–458.

Simchowitz, L., I. Spilberg, and P. De Week. 1982. Sodium and potassium fluxes and membrane potential of human neutrophils. *Journal of General Physiology.* 79:453–479.

Simons, E. R. 1978. Collagen requirements for initiation of platelet aggregation. *Collagen-Platelet Interaction.* H. Gastpar, editor. F. K. Schattauer Verlag, New York. 255–267.

Simons, E. R., S. M. Greenberg-Sepersky. 1987. Transmembrane monovalent cation gradients. *In* Platelet Responses and Metabolism. Vol. III. H. Holmsen, editor. CRC Press, Inc., Boca Raton, FL. 31–50.

Sims, P. J., A. S. Waggoner, C. H. Wang, and J. F. Hoffman. 1974. Studies of the mechanism by which cyanine dyes measure membrane potential in red blood cells and phosphatidylcholine vesicles. *Biochemistry.* 13:3315–3330.

Sklar, L. A., D. A. Finney, Z. G. Oades, A. J. Jesaitis, R. G. Painter, and C. G. Cochrane. 1984. The dynamics of ligand-receptor interactions. *Journal of Biological Chemistry.* 259:5661–5669.

Sklar, L. A., P. A. Hyslop, Z. G. Oades, G. M. Omann, A. J. Jesaitis, R. G. Painter, and C. G. Cochrane. 1985. Signal transduction and ligand-receptor dynamics in the human neutrophil. *Journal of Biological Chemistry.* 260:11461–11467.

Sullivan, R., D. A. Melnick, H. L. Malech, T. Meshulam, E. R. Simons, K. G. Lazzari, P. J. Proto, A.-S. Gadenne, J. L. Leavitt, and J. D. Griffin. 1987. The effects of phorbol myristate acetate and chemotactic peptide on transmembrane potentials and cytosolic free calcium in mature granulocytes evolve sequentially as the cells differentiate. *Journal of Biological Chemistry.* 262:1274–1281.

Waggoner, A. S. 1976. Optical probes of membrane potential. *Journal of Membrane Biology.* 27:317–334.

Waggoner, A. S., and A. Grinvald. 1977. Mechanisms of optical changes of potential sensitive dyes. *Annals of the New York Academy of Sciences.* 303:217–241.

Waggoner, A. S., C. H. Wang, and R. L. Tolles. 1977. Mechanism of potential-dependent light absorption changes in lipid bilayer membranes in the presence of cyanine or oxonol dyes. *Journal of Membrane Biology.* 33:109–140.

Zavoico, G. B., E. J. Cragoe, Jr., and M. B. Feinstein. 1986. Regulation of intracellular pH in human platelets. *Journal of Biological Chemistry.* 261:13160–13167.

Ion Channels and
Activation of Blood Cells

Chapter 24

Role of Potassium and Chloride Channels in Volume Regulation by T Lymphocytes

Michael D. Cahalan and Richard S. Lewis

Department of Physiology and Biophysics, University of California, Irvine, California

Introduction

Over the past five years, patch-clamp recording has been used to characterize ion channels expressed in several types of immune system cells, including T lymphocytes, B lymphocytes, and macrophages (see Cahalan et al., 1987; Gallin and Sheehy, 1987 for recent reviews). Several of the channels appear closely related to channels in the nervous system, based upon properties of gating, ion selectivity, and pharmacological sensitivity, but some channels not described elsewhere have been "discovered" in cells of the immune system. In the nervous system the functional significance of particular ion channels can usually be understood in terms of the concept of threshold for generating an action potential. Channels with reversal potentials above threshold are generally excitatory, while those with reversal potentials below threshold are generally inhibitory for action potential generation. Cells in the immune system carry out functions that protect us from a variety of pathogens, and these functions at least do not appear to involve rapid electrical signaling. "Activation" in the nervous system takes milliseconds, but in the immune system, cell activation events can take minutes, hours, or even days.

The cellular functions and mechanisms involving ion channels in the immune system remain largely unresolved. Lymphocytes in vivo mediate cellular and humoral immune responses in a complex network involving polypeptide mediators or antigens in solution, as well as contact between lymphocytes and foreign, virus-infected, or antigen-processing cells. Effector functions carried out by subsets of T lymphocytes include killing of virus-infected and foreign cells, secretion of lymphokines, chemotaxis, and cell proliferation. In each case, quiescent cells can be activated by appropriate stimuli, and in vitro assays of activation have been devised using purified or clonal cell populations. Several approaches to the problem of identifying a role for ion channels in these activation processes are possible. A type of channel might be identified through patch-clamp recording and characterized according to a property that can be tested in parallel on a functional assay. The channel property might be binding affinity for a series of pharmacological blockers, or a sequence of permeability ratios for various ions, for example.

In T lymphocytes the best-characterized ion channel is a potassium (K^+) channel activated by depolarization of the membrane (DeCoursey et al., 1984; Fukushima et al., 1984; Matteson and Deutsch, 1984; Cahalan et al., 1985). Voltage-activated K^+ channels appear to be required for a variety of effector functions, including mitogenesis, secretion of lymphokines, and cytolysis of target cells by natural killer and cytotoxic T cells (Chandy et al., 1984; Schlichter et al., 1986a; Chandy, K.G., T.E. DeCoursey, B. Sharma, M.D. Cahalan, and S. Gupta, unpublished observations). A series of compounds that block K^+ channels inhibits these functions in a parallel potency sequence. A mitogen-regulated calcium-permeable channel has also been identified in T cell lines (Kuno et al., 1986) and shown to be sensitive to intracellular inositol 1,4,5-trisphosphate (IP_3) (Kuno and Gardner, 1987). This channel may permit Ca^{++} entry after mitogen stimulation, but this conclusion must be regarded as tentative until pharmacological experiments are completed to facilitate comparison of dose-response relations for channel block and Ca^{++} entry. Pharmacological evidence has previously been presented to invoke the existence of voltage-activated calcium channels and calcium-activated potassium channels, and their functional requirement in a variety of effector mechanisms in the immune system, but patch-clamp studies have failed to detect these channels in T lymphocytes (Cahalan et al., 1985; Schlichter et al., 1986*b*).

Ion channels have also been implicated in the regulation of cell volume (Grinstein et al., 1982*a*, *b*). T cells placed in a hypotonic solution initially swell as water enters, but then lose K^+, Cl^-, and water as the cell reassumes its original volume. In this chapter we consider the role of ion channels in the regulation of cell volume during osmotic challenge, specifically in the regulatory volume decrease (RVD) observed after cells are placed in hypotonic solutions. At least two independent ionic transport pathways, one for K^+ and one for Cl^-, become activated during RVD (Grinstein et al., 1982*a*, 1984; Sarkadi et al., 1984*a*, *b*). Here, we review properties of several distinct types of K^+ channels in T lymphocytes, and suggest a mechanism by which one of these may mediate K^+ efflux during RVD. In addition, we describe a novel type of chloride channel that is induced by a transmembrane osmotic gradient. These osmotically activated chloride channels have escaped detection until recently, but under optimal conditions for induction, can be demonstrated to be widespread and abundant in lymphocytes. We present a hypothesis for RVD in which activation of chloride channels is the triggering event, resulting in depolarization, activation of K^+ channels, and loss of K^+, Cl^-, and water as the cell returns to its original volume following osmotic swelling.

Methods

Cells

Human T lymphocytes were separated from peripheral blood, as described in detail elsewhere (Chandy et al., 1984). The human T cell line Jurkat E6-1 was maintained in culture. Murine T lymphocytes and thymocytes were isolated from BALB/c mice following standard procedures described elsewhere (DeCoursey et al., 1987*a*; Lewis and Cahalan, 1988). In some experiments, subsets of murine T cells or thymocytes were identified just before recording by illumination of prestained cells through a fluorescein filter set on a Zeiss IM-35 microscope. Cells were prestained with the monoclonal antibodies (Becton Dickinson Immunocytometry Systems, Mountain View, CA) anti-Lyt-2 (conjugated with fluorescein) and anti-L3T4 (conjugated with phycoerythrin). These antibodies bind to surface glycoproteins CD4 and CD8, which are characteristic of helper or cytotoxic/suppressor T cells, respectively. With this double staining, $CD4^+CD8^-$ (helper) cells appear orange, $CD4^-CD8^+$ (cytotoxic/ suppressor) cells appear green, $CD4^+CD8^+$ (cortical thymocytes) appear yellow, and $CD4^-CD8^-$ (most primitive thymocytes) appear invisible under epi-illumination with a fluorescein filter set.

Patch Clamp

The whole-cell variant of the patch-clamp technique was used to control the intracellular milieu and to record ionic currents. Details of the methods have been described previously (Hamill et al., 1981; Cahalan et al., 1985). Briefly, cells in suspension were added to a glass chamber (250 μl volume), which in some cases had been pretreated with poly-D-lysine (0.2–2 mg/ml) to increase the fraction of cells that adhere to the glass. Poly-D-lysine pretreatment appeared to have no effect on ion channel properties. Pipettes filled with K^+ aspartate internal solution had resistances of 2–6 MΩ. No series resistance compensation was used. Junction potentials between pipette and bath solutions were corrected by appropriate adjustment of all applied potentials.

Solutions

The normal Ringer's solution bathing the cells during whole-cell recording had the following composition (concentrations in mM): 160 NaCl, 4.5 KCl, 2 $CaCl_2$, 1 $MgCl_2$, and 5 N-hydroxy-ethylpiperazine-N'-2-ethanesulfonic acid (HEPES), adjusted to pH 7.4 with NaOH. The osmolarity of Ringer's solution was 300 mosmol/kg. Na^+-free K^+ and tetramethylammonium (TMA) Ringer's solutions were titrated with KOH or TMAOH, respectively. 67% Ringer's was prepared by dilution of the Ringer's with distilled water. Hypertonic Ringer's solution (400 mosmol/kg) was made by adding 114 mM sucrose to the Ringer's solution. For anion substitution experiments, acid solutions of bromide, nitrate, aspartate, methanesulfonate, and ascorbate were mixed with hydroxides of Na^+, K^+, Ca^{++}, and Mg^{++} to achieve the same cation concentrations as in Ringer's solution, but with complete substitution of Cl^- by the test anion. The KF pipette solution consisted of (in mM): 140 KF, 1 $CaCl_2$, 2 $MgCl_2$, 11 EGTA, and 5 HEPES, titrated to pH 7.2 with KOH. The free $[Ca^{++}]$ of this solution is <2 nM, from the stability constant of CaF_2 (see Cahalan et al., 1985). The solution was prepared by first dissolving divalent salts, then adding EGTA and correcting the pH to buffer the Ca^{++} before adding KF. The K^+ aspartate pipette solution consisted of (in mM): 160 K^+ aspartate, 0.1 $CaCl_2$, 2 $MgCl_2$, 1.1 EGTA, and 10 HEPES, adjusted to pH 7.2 with KOH, with free $[Ca^{++}]$ of 10^{-8} M (315 mosmol/kg). K^+ aspartate + ATP was prepared by adding 4 mM Na_2ATP (grade I; Sigma Chemical Co., St. Louis, MO) to the K^+ aspartate solution, increasing the osmolarity to 330 mosmol/kg; frozen aliquots were thawed for each day's experiments. The KCl + ATP pipette solution was identical to K^+ aspartate + ATP, except with chloride replacing aspartate. Osmolarities were measured with a vapor pressure osmometer (Wescor Inc., Logan, UT).

Results and Discussion

Voltage-activated K^+ Channels in Human T Cells

Human T lymphocytes isolated from peripheral blood typically express 300–500 voltage-activated K^+ channels per cell (DeCoursey et al., 1984). In the experiment illustrated in Fig. 1, the membrane potential was held at -80 mV, and depolarizing voltage steps elicited outward K^+ currents. The K^+ channels open after a short delay and then slowly inactivate if the depolarization is maintained. A quantitative description of the channel's gating kinetics, along with the measured single-channel conductance, yielded an estimate of K^+ efflux at the normal resting potential in reasonable agreement with tracer flux measurements on quiescent human T cells (Cahalan et al., 1985). One interesting result of the calculations is that only a few of the cell's ensemble of channels are expected to be open at any given time in a cell not under voltage clamp. At normal resting potentials of -50 to -70 mV, estimated from a variety of dye measurements and from patch-clamp recordings (Deutsch et al., 1979; Rink et al., 1980; Felber and Brand, 1983; Tatham and Delves, 1984), the number of open K^+ channels will fluctuate from zero to a few channels. Current-clamp recordings not shown here reveal membrane potential fluctuations of ~10 mV, as individual K^+ channels open and close. If the membrane potential drifts in a positive direction, additional K^+ channels will open, resulting in repolarization. Agents such as quinine

A

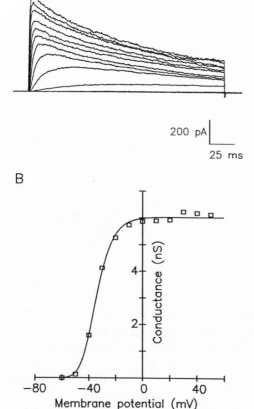

200 pA

25 ms

B

Figure 1. Voltage-gated K^+ currents in a human peripheral blood T lymphocyte. (*A*) A family of voltage-clamp currents elicited by steps from -50 to $+50$ mV in 10-mV increments, applied from a holding potential of -80 mV. External solution: Ringer's. Pipette solution: KF. (*B*) Conductance–voltage relation for the cell of *A*. For each current record the conductance was computed as the peak current divided by the difference between the membrane potential and the K^+ equilibrium potential. A Boltzmann function raised to the third power (see Cahalan et al., 1985) was fitted by a least-squares method. The potential, $V_{1/2}$, at which the conductance is half-maximal is -34 mV. Records were taken after equilibration of the pipette contents with the cell, 7 min after establishing the whole-cell recording configuration. The maximum conductance of 6.0 nS is somewhat higher than the average value for human T cells (Cahalan et al., 1985).

(see below), which block the K^+ channels, result in depolarization (Rink et al., 1980; Felber and Brand, 1983; Tatham and Delves, 1984). Thus, in these small 6-μm-diameter cells, a few open K^+ channels appear to control the membrane potential.

The kinetics and voltage dependence of the lymphocyte K^+ channel most closely resemble those of delayed rectifier K^+ channels found in skeletal muscle. The channels are opened by depolarization, not by intracellular Ca^{++}, even though agents believed to block Ca^{++}-activated K^+ channels also block the lymphocyte K^+ channel. In the records shown in Fig. 1 the pipette solution contained <2 nM free $[Ca^{++}]$, and if the intracellular $[Ca^{++}]$ is deliberately elevated to 10 μM, the K^+ conductance is decreased and the channel's inactivation rate is increased (Bregestovsky et al., 1986; see also Choquet et al., 1987 for similar results on K^+ channels in B lymphocyte cell lines).

The voltage-activated K^+ channel can be blocked by several classes of channel blockers, as summarized in Table I. The "classical" K^+ channel blockers, tetraethylammonium (TEA) and 4-aminopyridine, exhibit 50% block of K^+ current at 10 mM and 150 μM, respectively. Agents such as quinine and cetiedil, which block

TABLE I
Substances that Block T Cell K$^+$ Channels (and RVD)

Substance
Classical K$^+$ channel blockers
Tetraethylammonium (TEA)*
4-Aminopyridine*
Ca^{++}-activated K$^+$ channel blockers
Quinine*
Cetiedil*
Charybdotoxin (CTX)
Ca^{++} channel blockers
Diltiazem
Nifedipine
Verapamil*
Polyvalent cations: La^{+++}, Zn^{++}, Ni^{++}, Co^{++}, Mn^{++}
Calmodulin antagonists
Trifluoperazine*
Chlorpromazine*

*Agents that have also been shown to inhibit RVD (Cheung et al., 1982a; Grinstein et al., 1982b; Sarkadi et al., 1984a; Deutsch et al., 1986b).

Ca^{++}-activated K$^+$ channels in other cells (Schwarz and Passow, 1983; Berkowitz and Orringer, 1981), also block lymphocyte K$^+$ channels. These relatively low affinity agents might be expected to lack absolute specificity for a single type of channel; for example, quinidine, a stereoisomer of quinine, blocks both Na$^+$ and K$^+$ channels in squid axon (Yeh and Narahashi, 1976) and in frog node of Ranvier (Revenko et al., 1982). More surprisingly, charybdotoxin (CTX), a peptide component of *Leiurus quinquestriatus* scorpion venom, which has been shown to block two types of Ca^{++}-activated K$^+$ channels (Miller et al., 1985; Smith et al., 1986; Hermann and Erxleben, 1987), also potently blocks the lymphocyte K$^+$ channel (discussed below). Purified CTX, graciously provided by Dr. Chris Miller (Graduate Dept. of Biochemistry, Brandeis University, Waltham, MA), blocks 50% of the K$^+$ current at a concentration of 550 pM.

Agents that block voltage-activated calcium channels in cardiac cells also inhibit lymphocyte K$^+$ channels, as shown in Table I. Included are organic calcium channel antagonists, such as verapamil, nifedipine, and diltiazem, as well as inorganic blockers such as Ni^{++}, Co^{++}, and La^{+++}. This lack of specificity has resulted in confusion regarding the possible existence of voltage-activated Ca^{++} channels in T cells. For example, the ability of "Ca^{++} channel antagonists" to inhibit T cell activation has been taken as evidence in favor of a requirement for functional Ca^{++} channels in mitogen-stimulated T cell proliferation (Birx et al., 1984). Based upon the ability of a series of compounds that block K$^+$ channels to inhibit mitogenesis, an alternative interpretation is that the inhibition of mitogenesis by Ca^{++} channel blockers is due instead to block of K$^+$ channels.

Diversity of K$^+$ Channels in Murine T Cells

As in human T cells, K$^+$ channels are the most obvious channel type present in whole-cell recordings from murine thymocytes and mature T cells. However, in T lymphocytes from mice there are at least three distinct types of voltage-activated K⁺ channels (DeCoursey et al., 1987*a*; Lewis and Cahalan, 1988). Fig. 2 illustrates K⁺ currents evoked by membrane depolarization in three different thymocytes. Superficially, these records may appear similar, but a close examination of the voltage dependence, kinetics, and pharmacology of K⁺ channels from mouse thymocytes and T cells reveals that whole-cell current records can be grouped into three major classes. Fig. 2 *A* illustrates a cell that expressed primarily the same type of K⁺ channel found in human T cells—a channel that we have come to call type *n* for *normal*, since it is the most commonly found type of channel. Its voltage dependence, gating kinetics, single-channel conductance, and pharmacology are similar to human T cell K⁺ channels. The currents in Fig. 2 *B* are due to a second type of channel, termed *n'*, that is similar in several respects to type *n* channels, but differs in TEA sensitivity and

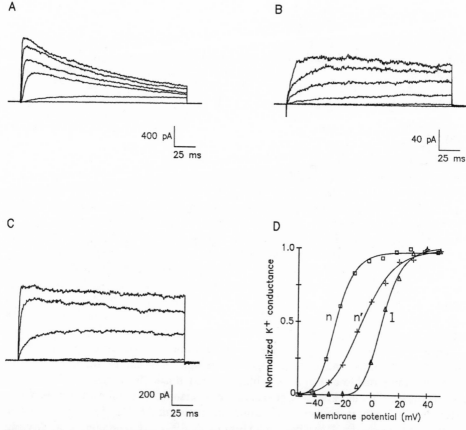

Figure 2. Three varieties of K⁺ currents in thymocytes. (*A, B*, and *C*) Currents were elicited by depolarizing pulses from −50 to +50 mV in 20-mV increments. (*A*) Type *n* current in a CD4⁺CD8⁺ thymocyte. (*B*) Type *n'* current in a CD4⁻CD8⁺ thymocyte. These data were recorded with 1 mM TEA externally to block current through type *l* channels. (*C*) Type *l* current in a CD4⁻CD8⁺ thymocyte. (*D*) Normalized conductance–voltage curves for the three thymocytes in *A–C*, showing differences in the voltage dependence of activation for the three types of K⁺ channels. The maximum conductances were 9.8, 0.72, and 5.8 nS for *A, B*, and *C*, respectively. Solutions as in Fig. 1.

288   *Cell Physiology of Blood*

TABLE II
Three Types of K⁺ Channels in Mouse T Cells

	Type n (normal)	Type l (large)	Type n' (n-ish)
Conductance	12-18 pS	21-27 pS	18 pS
Gating			
$V_{1/2}$	-30 mV	0 mV	-10 mV
Inactivation	Fast, use-dependent	Slow	Slow
τ at 40 mV	120 ms	300 ms	>300 ms
Tail current	Slow	Fast	Slow
τ at -80 mV	10–20 ms	1 ms	10–20 ms
Pharmacology			
TEA K_i	~10 mM	~100 μM	~100 mM
CTX K_i	550 pM	Not blocked	≪5 nM
Cobalt	Blocked by 1 mM	Not blocked	Not tested

Channel conductances were determined from measurements on isolated outside-out patches; all other properties were determined from whole-cell currents in which one channel type predominated. $V_{1/2}$ refers to the membrane potential at which half of the channels are activated. Inactivation time constants represent single-exponential fits to current decay during depolarizing pulses to $+40$ mV. Use dependence refers to a progressive decline in the peak K⁺ current amplitude during repetitive pulses to 30 mV applied once per second. Tail-current time constants were determined by fitting single exponential curves to currents measured upon repolarization to -80 mV, following a brief depolarization to activate the channels. See DeCoursey et al., 1987a, and Lewis and Cahalan, 1988, for experimental details.

inactivation kinetics. The currents illustrated in Fig. 2 *C* represent the activity of yet a third variety of K⁺ channel, which we have named type *l* for its *larger* single-channel conductance. As seen in Fig. 2 *D*, the voltage dependence of type *l* currents differs from that of *n* and *n'*; stronger depolarizing pulses are required to activate the channels.

Table II summarizes the properties of these K⁺ channels. Compared with type *n* channels, *l* channels become activated at more depolarized potentials, have a larger single-channel conductance (Fig. 3), close more rapidly during "tail currents" upon repolarization, inactivate more slowly, and do not accumulate inactivation with repetitive pulsing. Type *n'* channels are similar to type *n* in the voltage at which channels begin to open (Fig. 2), in single-channel conductance (Fig. 3), and in closing kinetics, but do not inactivate with repetitive pulsing. In addition, the three types of channels can be distinguished on the basis of their sensitivity to block by TEA and charybdotoxin. Compared with type *n* channels, *l* channels are much more sensitive to block by TEA, whereas *n'* channels are more resistant (Table II). Type *n* and *n'* channels are both completely blocked by 8.5 nM CTX, which suggests underlying structural similarities, while type *l* channels are unaffected by much higher toxin concentrations (Fig. 4). In practice, the rates of channel closing during tail currents, the degree of inactivation by repetitive pulsing, and the sensitivity to block by TEA or CTX provide a "fingerprint" for the types of K⁺ channel expressed by any particular cell. Usually one of the three channel types predominates in any particular cell, but good evidence for mixtures of channels has also been obtained (Chandy et al., 1986; DeCoursey et al., 1987a).

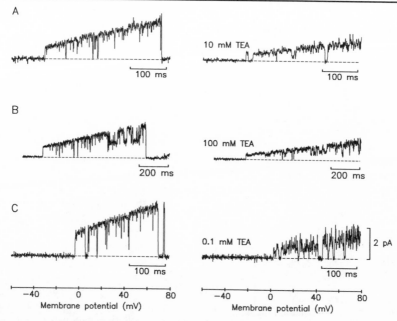

Figure 3. Unitary conductances of three types of K$^+$ channels. Channel type was determined from the voltage dependence of activation, cumulative inactivation, and sensitivity to block by TEA. Each of *A–C* shows multiple openings of a single K$^+$ channel in an excised, outside-out patch, stimulated by a voltage ramp from -60 to $+80$ mV in the absence or presence of bath-applied TEA. Pipette solution as in Fig. 1. (*A*) A type *n* K$^+$ channel with a slope conductance of 18 pS; 10 mM TEA blocks the channel by 56%. (*B*) Type *n'* K$^+$ channel with a conductance of 17 pS. A subconductance state of 10 pS is also evident. 100 mM TEA blocks the channel by 58% (CD4$^-$CD8$^+$ cell). (*C*) Type *l* K$^+$ channel with a conductance of 27 pS. The *l* channel is half-blocked by 0.1 mM TEA. The increased noise in the presence of TEA represents rapid blocking and unblocking events (CD4$^-$CD8$^+$ cell). From Lewis and Cahalan, 1988.

The particular ensemble of K$^+$ channels expressed by T lymphocytes and their precursors correlates consistently with developmental state, the subset of T cell, and the state of T cell activation. In general, the most rapidly proliferating T cells, including primitive precursors in the thymus and mitogen-stimulated T cells, express hundreds of type *n* K$^+$ channels per cell (McKinnon and Ceredig, 1986; DeCoursey et al., 1987*b*), similar to human T cells from peripheral blood. Mature quiescent cells have lower numbers of channels, with helper T cells expressing primarily type *n* K$^+$ channels. Type *l* and *n'* channels are normally restricted to the suppressor/cytotoxic T cell lineage (Lewis and Cahalan, 1988). Type *l* K$^+$ channels are found abundantly in functionally aberrant T cells from a strain of mouse, MRL-*lpr/lpr*, with both lymphoproliferative and autoimmune disorders (Chandy et al., 1986). Interestingly, after treatment with mitogens, actively dividing cells of either helper or cytotoxic/suppressor variety have large numbers of type *n* K$^+$ channels.

Of the three types of K$^+$ channels, it is most likely that *n* channels mediate the loss of K$^+$ during RVD. T cells depolarize in hypotonic medium to potentials estimated to be between -53 and -33 mV (Grinstein et al., 1982a); this depolarization would not be sufficient to increase appreciably the activation of type *l* K$^+$ channels, but could open either type *n* or *n'* channels (Fig. 2 *D*). Moreover, as summarized in Table I, the

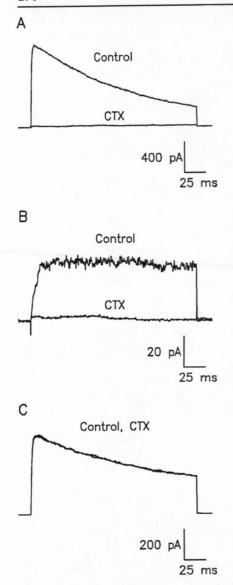

Figure 4. Charybdotoxin (CTX) blocks type n and n' channels, but not type l. K^+ currents are shown for depolarizing pulses to $+30$ mV. (*A*) Current in a CD4$^-$CD8$^-$ thymocyte, showing nearly complete block of K^+ channels by 8.5 nM charybdotoxin. The channels were identified as type n by kinetics of channel closing and inactivation (see text). The effect of the toxin was partially reversible upon prolonged washing with Ringer's solution. (*B*) In a CD4$^-$CD8$^+$ cell with predominantly type n' currents, identified on the basis of slow tail currents and lack of cumulative inactivation, 8.5 nM charybdotoxin blocks most of the K^+ current. (*C*) In another CD4$^-$CD8$^+$ cell, type l channels predominated, and 34 nM charybdotoxin had no effect. Type l channels were confirmed in this cell by rapid tail-current kinetics, lack of cumulative inactivation during repetitive pulsing, and sensitivity to 1 mM TEA. Solutions as in Fig. 1.

pharmacology of RVD most closely matches that of type n K^+ channels. In human T cells, in which most of the research on RVD in lymphocytes has been done, type n K^+ channels predominate; type l and n' channels have not been seen in patch recordings from human T cells, although small numbers of these channels per cell or a highly restricted subset distribution cannot at present be ruled out. Finally, the ability of lymphocytes to volume regulate in hypotonic media correlates with the surface density of type n K^+ channels. Volume responses by human B lymphocytes are limited by a lower K^+ efflux compared with human T cells (Cheung et al., 1982*b*; Grinstein et al., 1983), and human B cells also express fewer type n K^+ channels than T cells (Chandy et al., 1985; Deutsch et al., 1986*a*). Similarly, quiescent murine T cells of the L2 clone display low numbers of type n K^+ channels and are deficient in volume regulation; treatment with interleukin 2 increases n channel expression and the ability to volume regulate in parallel (Lee et al., 1987). Thus several lines of correlative evidence argue for a role of type *n* K$^+$ channels in mediating the K$^+$ efflux during RVD.

Maxi- and Mini-Chloride Channels

Although K$^+$ channels are the most frequently seen channels in T cells, chloride channels can also be induced to open under appropriate conditions. Fig. 5 illustrates a large-conductance, voltage-dependent chloride channel (the "maxi-chloride" channel) that has been described previously in macrophages (Schwarze and Kolb, 1984), as well as in B cells (Bosma, 1986), myotubes (Blatz and Magleby, 1983; Schwarze and Kolb, 1984), and epithelia (Krouse et al., 1986). The channel's conductance is nearly 400 pS, possibly the largest value for single-channel conductance of any channel under physiological ionic conditions. The channel is weakly anion selective. In T cells and thymocytes we found that the maxi-Cl$^-$ channel is normally shut, but can be induced to open by a prolonged period of depolarization, e.g., to +20 or 40 mV for 2–3 min. With this type of stimulus in whole-cell records, the maxi-Cl$^-$ channel suddenly begins to flicker open to a variety of different conductance levels, and then opens fully. The

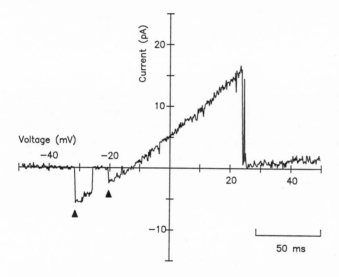

Figure 5. Maxi-Cl$^-$ channel in a CD4$^+$CD8$^-$ splenic T cell. In this whole-cell recording, the voltage was ramped from +50 to −50 mV (right to left on the displayed axis), following a prolonged period of depolarization (~2 min at +60 mV) to open the channel. The triangles indicate two of the channel's closing events. When open, the channel has a conductance of ~400 pS and a reversal potential of −13 mV. Solutions as in Fig. 1.

channel can then be closed by hyperpolarizing the membrane, as illustrated in Fig. 5. The normal physiological role of the maxi-Cl$^-$ channel is unknown. Calcium ionophores can increase the opening of these channels in macrophages, possibly through a rise in intracellular [Ca^{++}] initiating intracellular biochemical events, since maxi-Cl$^-$ channels in excised inside-out patches are not directly opened by elevated free [Ca^{++}] (Schwarze and Kolb, 1984).

T cells also have a very different type of chloride channel with low single-channel conductance—a "mini-Cl$^-$" channel. The conductance of these channels is too small to resolve clearly in whole-cell recording. However, in an osmotic gradient many of these channels became activated, resulting in a macroscopic current described more fully below. We believe that the properties of these mini-Cl$^-$ channels make them ideally suited to be the initial trigger for RVD in hypotonic medium.

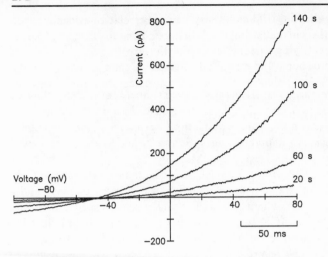

Figure 6. Activation of mini-chloride channels during whole-cell recording from a murine splenic T cell. Currents elicited by voltage-ramp stimuli from −100 to +80 mV are shown, beginning 20 s after "break-in" to obtain the whole-cell recording configuration, and at 40-s intervals thereafter. Bath solution: Ringer's. Pipette solution: K^+ aspartate + ATP.

Slow Activation of Mini-Chloride Channels during Whole-Cell Recording

Chloride conductance in T cells becomes spontaneously activated during whole-cell recordings made with a K^+ aspartate/ATP pipette solution that is hypertonic relative to Ringer's solution. Absent just after establishing the whole-cell recording configuration (hereafter referred to as "break-in"), the conductance slowly develops as the pipette solution dialyzes the cytoplasm. Fig. 6 illustrates an example of chloride-channel activation in a splenic T cell during whole-cell recording with a 30 mosmol/kg osmotic gradient. (Osmotic gradients are expressed as intra relative to extracellular.) The traces show whole-cell currents during four successive voltage-ramp stimuli at 40-s intervals, beginning 20 s after break-in. The first record illustrates that few voltage-dependent channels are present; the cell has an input resistance of ~10 GΩ and

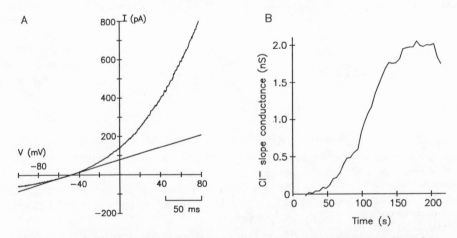

Figure 7. Quantifying the increase in chloride conductance. (*A*) After subtraction of current during the initial ramp after break-in, when no chloride conductance had yet activated, a linear regression to the points from −60 to −40 mV was fitted to determine the slope conductance. The illustrated trace with 1.62 nS conductance was collected 130 s after whole-cell recording began. (*B*) The slope conductances at various times after break-in (time = 0) illustrate the time course for induction of the chloride conductance. Same cell as in Fig. 6.

a capacitance of 0.9 pF, reflecting the small size of resting splenic T cells. Upon subsequent ramp stimuli, the current-voltage relation can be seen to "pivot" about − 50 mV, as an outwardly rectifying current develops with time.

To characterize the time course and magnitude of chloride channel activation, the slope conductance at the reversal potential of − 50 mV was calculated from a series of ramp currents recorded at various times after break-in (Fig. 7). In this cell, the conductance began to develop after an initial delay of ∼40 s, reaching a maximum of nearly 2 nS within several minutes of whole-cell recording. Although the time course of Cl⁻ channel activation varied somewhat from cell to cell, the time course pictured here is representative of small cells, such as quiescent splenic T lymphocytes.

The mini-chloride channels are ubiquitous among various types of T lymphocytes

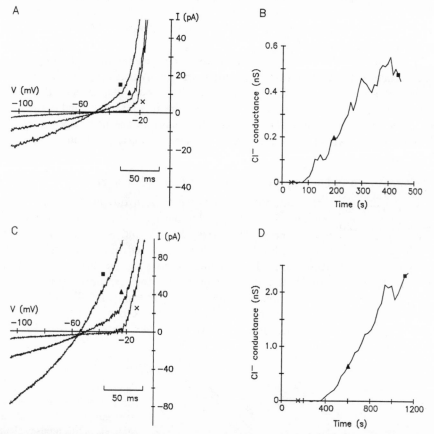

Figure 8. Activation of chloride channels in a human T cell (*A* and *B*) and in a Jurkat cell (*C* and *D*). (*A*) Ramp currents recorded 45 s (x), 205 s (▲), and 445 s (■) after break-in. Initially, the outward rectification seen at − 20 mV is due to K⁺ channels opening. At later times chloride channels also contribute, and the currents are seen to cross at about − 50 mV, the reversal potential for the developing chloride conductance. Cell membrane capacitance is 1.5 pF. (*B*) Time course of chloride channel activation for the cell in *A*. The symbols indicate measurements from the ramp records shown in *A*. (*C*) Ramp currents in a Jurkat cell, recorded at 150 s (x), 600 s (▲), and 1,140 s (■) after beginning whole-cell recording. Cell membrane capacitance is 6.0 pF. (*D*) The time course for chloride channel activation in the Jurkat cell. Solutions as in Fig. 6.

and related cells. As in the example shown above, most splenic T cells are quiescent and express very few voltage-dependent K^+ channels; in these cells, the induced chloride conductance predominates and can be studied in relative isolation. In murine thymocytes or activated T cells, which often express hundreds of K^+ channels per cell, the induced Cl^- conductance is comparable in magnitude to the maximal K^+ conductance, and current records usually represent the sum of currents through K^+ and Cl^- channels. Similar results are obtained with human T lymphocytes and related cell lines. Fig. 8 *A* illustrates induction of chloride current in a human T lymphocyte. Shortly after break-in only voltage-dependent K^+ current is apparent, but within minutes chloride current also contributes. The time course of chloride channel activation for this cell is shown in Fig. 8 *B*. The human T lymphoma cell line Jurkat E6-1 grows continuously in culture, and the cells are larger than resting T cells or thymocytes. Fig. 8, *C* and *D*, illustrates the induction of Cl^- channels in a Jurkat cell. Perhaps as a consequence of their larger cell volume, activation of Cl^- channels begins after a longer initial delay after break-in.

The opening of chloride channels in mouse splenic T cells appears to depend upon two main factors. First, an osmotic gradient across the membrane tending to swell the cells promotes the activation of Cl^- channels; the osmotic dependence is described further below. Second, Cl^- channels are induced only if ATP is included in the pipette solution. Without ATP, only voltage-gated K^+ channels are normally present, and chloride channels fail to open even in an imposed osmotic gradient of up to 100 mosmol/kg.

Ion Permeation and Unitary Conductance of Chloride Channels

The ion selectivity of the induced conductance was investigated by varying the concentrations of external ions. In cells bathed in normal Ringer's solution containing 160 mM Cl^-, the current reverses sign near -50 mV and is outwardly rectifying. Replacing external Na^+ ions with K^+ or TEA ions had no effect on the current–voltage relation, suggesting that the induced conductance is not selective for cations. However, when aspartate is substituted for external chloride ions, the current reverses near 0 mV and the current–voltage relation becomes nearly linear (Fig. 9). The simplest explanation of this result is that the induced conductance is primarily permeable to chloride ions, which contribute outward current as they enter the cell. Reversal-

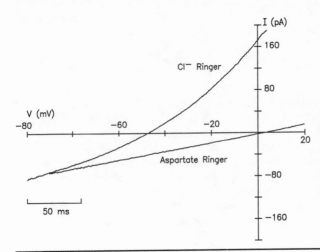

Figure 9. Effect of anion substitution on the whole-cell chloride current. After chloride channels in this mouse splenic T cell had reached maximal activation, chloride in the bathing solution was replaced by aspartate. The ramp currents shown have been corrected for the difference in junction potentials between the two solutions. Solutions as in Fig. 6.

potential measurements indicate that aspartate is ~0.12 times as permeant as chloride ions, based on the Goldman-Hodgkin-Katz equation (Hodgkin and Katz, 1949). The current–voltage curve rectifies outwardly due to the open channels' permeation properties, primarily because aspartate ions from the pipette dialyzing the cytoplasm are relatively impermeant. Voltage-dependent gating apparently does not contribute to this rectification, as depolarizing steps of membrane potential elicit no significant time-dependent current relaxations. Even a reasonably large anion—methanesulfonate, often used as an impermeant anion substitute—was significantly permeant. Judging from measured reversal potentials, the sequence of permeability among anions tested thus far is $NO_3^- > Br^-$, Cl^-, $F^- >$ methanesulfonate > ascorbate > aspartate.

Even though the chloride channel is measurably permeable to fairly large anions, the size of single-channel events is quite small, making detection of unitary currents during whole-cell recording difficult. To obtain a size estimate of the single-channel current we have measured whole-cell current fluctuations in a cell dialyzed with a KCl pipette solution. Dividing the current variance by the mean current level as the chloride conductance develops with time results in an estimated unitary current at -80 mV of -0.21 pA. This estimate for the single-channel current corresponds to a chord conductance of 2.6 pS. For a typical whole-cell conductance of 2 nS induced by a 30 mosmol/kg osmotic gradient, as in Figs. 6–8, the single-channel conductance estimate implies that on the order of 1,000 chloride channels are activated. The surprising conclusion from this analysis is that chloride channels appear to be the most abundant channel type in T lymphocytes.

A

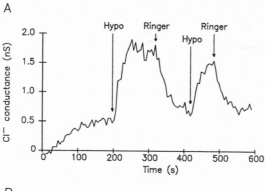

B

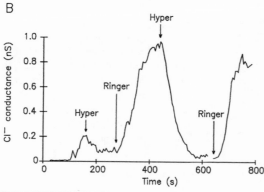

Figure 10. Osmotic regulation of chloride channels. (*A*) Enhancement of Cl^- channel activation by hypotonic (67%) Ringer's solution. After a 3-min period of spontaneous chloride channel induction, dilution of the bath solution, as indicated, reversibly increased the chloride conductance. $CD8^+$ mouse splenic T cell. Pipette solution: K^+ aspartate + ATP, pH 6.2. (*B*) Suppression of chloride channels by hypertonic Ringer's solution in a BALB/c splenocyte. Hypertonic Ringer's (Ringer's + 114 mM sucrose), applied at the indicated times, caused a reversible decrease in Cl^- conductance.

Dependence on Osmotic Gradients and Suction

Figs. 6–8 above demonstrate spontaneous induction of chloride channels after break-in, with a steady osmotic gradient of 30 mosmol/kg, the internal solution being hypertonic relative to the external solution. If the osmotic gradient is the stimulus that drives Cl⁻ channel activation, then dilution of the extracellular Ringer's solution should augment Cl⁻ conductance. Fig. 10 *A* illustrates a test of this idea in a splenic T cell. In this cell, the chloride conductance increased about threefold after the initial dilution to 67% Ringer's solution. The change in chloride conductance was reversible and repeatable upon further changes in tonicity. Fig. 10 *B* illustrates that the chloride conductance can be shut off by reversing the osmotic gradient. In this experiment, activation of chloride channels in normal Ringer's was arrested by addition of hypertonic Ringer's (made 100 mosmol/kg hypertonic by addition of 114 mM sucrose). On returning to normal Ringer's solution, chloride channel induction progressed, but could be reversed again by hypertonic Ringer's. The process of turning the chloride channels on or off is thus reversible, and depends upon the transmembrane osmotic gradient, rather than the ionic strength or osmolarity of one compartment alone. It should also be noted that the conductance changes induced by varying the tonicity are not instantaneous; solution changes are completed within 20 s, while conductance changes sometimes take >100 s to reach a final level.

What is the mechanism that causes Cl⁻ channels to open? If the signal for activating chloride channels were the swelling produced by osmotically driven water entry, as suggested previously by Sarkadi and colleagues (1984b), perhaps stretch on the membrane is part of the mechanism. Preliminary evidence consistent with this idea is illustrated in Fig. 11. In this experiment, suction applied to the pipette interior opposed the induction of chloride channels. Induction proceeded further after reequilibration to atmospheric pressure, but was reversed when suction was reapplied. These results imply that the Cl⁻ conductance can be modulated by changing the mechanical stress on the membrane.

A Working Hypothesis for RVD

Several properties of the mini-Cl⁻ channel, described in Figs. 6–11 above, are consistent with a role in mediating Cl⁻ efflux during the RVD under hypotonic conditions. First, the Cl⁻ channels are not gated by membrane potential, but can alter it by mediating charge transfer across the membrane, properties that are shared by the RVD Cl⁻ flux pathway (Grinstein et al., 1982a). In addition, the ability of NO_3^-, Br^-,

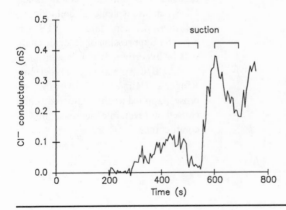

Figure 11. Effect on Cl⁻ channels of suction applied through the patch pipette. Suction of −20 inches H_2O was applied at the indicated times, evoking a reversible decrease in Cl⁻ channel activity. BALB/c splenocyte; solutions as in Fig. 6.

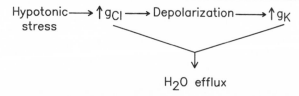

Figure 12. A diagram for the events of regulatory volume decrease (RVD).

and Cl^- to pass through mini-Cl^- channels agrees with their ability to support cell volume changes in hypotonic solutions (Grinstein et al., 1982*a*). Most importantly, the reversible control of Cl^- channel activation by the transmembrane osmotic gradient, possibly mediated through membrane stretch, is consistent with the anion permeability changes induced by osmotic swelling of intact cells (Sarkadi et al., 1984*b*).

The characteristics of Cl^- and K^+ channels in T cells described above prompt a hypothesis for the mechanism of RVD, in which activation of chloride channels is the triggering event (Fig. 12). RVD is associated with membrane depolarization, hypothesized to be the result of increased chloride conductance (Grinstein et al., 1982*a*; Sarkadi et al., 1984*b*). Intracellular $[Cl^-]$ determinations range from 31 to 38 mM for human lymphocytes, corresponding to a chloride equilibrium potential of -30 to -40 mV. Thus, an increase in Cl^- conductance induced by cell swelling would be expected to depolarize the cell from its normal resting (albeit fluctuating) membrane potential of -50 to -70 mV. The depolarization in itself would increase the driving force for K^+ efflux through open K^+ channels, but, more importantly, would increase activation of type *n* voltage-gated K^+ channels, further enhancing K^+ efflux. The membrane potential would thereby attain a value between the K^+ and Cl^- equilibrium potentials, driving efflux of both K^+ and Cl^- ions through their respective channels. As the cell loses KCl, water would follow, returning the cell toward its normal volume. Cell shrinkage would eventually turn off the Cl^- channels, causing the membrane to hyperpolarize and K^+ channels to close, and the cell would reach its steady-state volume. Thus, according to this hypothesis, Cl^- channels act as the sensor for cell swelling, and their influence on the membrane potential controls the activation of K^+ channels and the volume regulation process.

Although we favor the idea that mini-Cl^- channels represent the Cl^- efflux pathway in RVD, we cannot definitely exclude a role for the maxi-Cl^- channels. In fact, to account for the "all-or-none" volume regulation of individual T cells in a population, Sarkadi et al. (1984*b*) suggested that a small number of highly conductive anion channels might be switched on during RVD. The required number of open maxi-Cl^- channels can be estimated by the following calculation. The loss of Cl^- during RVD has been estimated as 60 mmol/liter of peripheral blood lymphocytes (5×10^{12} cells) in a 2-min period (Sarkadi et al., 1984*b*). Thus, averaged over this period, each cell loses 60×10^6 Cl^- (and K^+) ions per second. Considering the 400 pS single-channel conductance of the maxi-Cl^- channel, and allowing a 10–20 mV driving force (membrane potential minus chloride equilibrium potential) during RVD, one to three open maxi-Cl^- channels would be sufficient to account for the flux. However, none of the known stimuli for activating maxi-Cl^- channels are likely to be encountered during RVD. For example, the activation of maxi-Cl^- channels in macrophages is enhanced by an increase in intracellular $[Ca^{++}]$ (Schwarze and Kolb, 1984); yet, intracellular $[Ca^{++}]$ appears to remain constant during exposure of lymphocytes to hypotonic solutions (Rink et al., 1983). In addition, we have found that maxi-Cl^- channels can be opened in whole-cell recording by prolonged depolarization above about +20 mV. However, it is doubtful that these membrane potentials would ever be achieved in intact T lymphocytes. Alternatively, the Cl^- flux calculated above could be generated by the opening of 200–600 mini-Cl^- channels, a number not inconsistent with estimates discussed above that T lymphocytes possess >1,000 of these mini-Cl^- channels. Although large-conductance channels might be one way for a cell to generate all-or-none RVD behavior, biochemical switching mechanisms are also possible.

It has been proposed by others that Ca^{++}-activated K^+ channels mediate the cation component of RVD (Grinstein et al., 1982*b*; Sarkadi et al., 1984*a*). This conclusion is based largely on observations that quinine, cetiedil, and trifluoperazine, among other agents, inhibit RVD and Ca^{++}-activated K^+ transport, as typified by the "Gardos phenomenon" (Gardos, 1958). However, as these authors also point out, an unresolved difficulty with the hypothesis is the apparent constancy of intracellular free $[Ca^{++}]$ during RVD (Rink et al., 1983). The inhibitors mentioned above also block voltage-gated K^+ channels in T lymphocytes (Table I); hence, the pharmacological results are also consistent with the notion that voltage-activated type *n* K^+ channels mediate the K^+ efflux during RVD (DeCoursey et al., 1985; Deutsch et al., 1986 *a,b*). We have suggested above that depolarization may link activation of Cl^- and K^+ channels, but this cannot explain observations of osmotically activated cation fluxes under conditions where Cl^- flux is blocked (Grinstein et al., 1982*a*; Sarkadi et al., 1984*a*). Several possibilities might explain this discrepancy; for example, the type *n* K^+ channel may be subject to biochemical regulation during RVD, or a Ca^{++}-dependent K^+ channel may exist in T cells, but escape detection given our whole-cell recording procedures. Overall, the correlation between *n* channel expression in lymphocytes and the ability to volume regulate, as well as the parallel pharmacological sensitivities of type *n* K^+ channels and RVD, make them likely candidates for contributing to cation efflux during RVD.

Conclusions

The patch-clamp method has provided a means to study ion channels in lymphocytes with far better time resolution and sensitivity than tracer flux or microelectrode methods used previously. However, it should be kept in mind that in whole-cell or isolated-patch recording, the integrity of the cell is disrupted; essential cofactors for channels or enzymes might well be lost. An example described above is the osmotically activated mini-Cl^- channel, which eluded detection until recently, despite being the most abundant channel in the cell, in part because it requires the presence of ATP in the pipette solution for activation. Despite these caveats, the ability to control solution composition on both sides of the lymphocyte membrane with patch-clamp techniques will facilitate future studies of the molecular mechanism of Cl^- channel activation and the role of K^+ and Cl^- channels in controlling cell volume.

Acknowledgments

The authors would like to thank Dr. Thomas DeCoursey for valuable discussions, Ruth M. Davis for expert technical assistance, and Dr. Chris Miller for providing samples of purified charybdotoxin.

This work was supported by a National Institutes of Health postdoctoral fellowship (NS-08021) to R. S. Lewis, and by research grants from the Arthritis Foundation and NIH grants NS-14609 and AI-21808.

References

Berkowitz, L. H., and E. P. Orringer. 1981. Effect of cetiedil, an in vitro antisickling agent, on erythrocyte membrane cation permeability. *Journal of Clinical Investigation.* 68:1215–1220.

Birx, D. L., M. Berger, and T. A. Fleisher. 1984. The interference of T cell activation by calcium channel blocking agents. *Journal of Immunology.* 133:2904–2909.

Blatz, A. L., and K. L. Magleby. 1983. Single voltage-dependent chloride-selective channels of large conductance in cultured rat muscle. *Biophysical Journal.* 43:237–241.

Bosma, M. M. 1986. Chloride channels in neoplastic B lymphocytes. *Biophysical Journal.* 49:413*a*. (Abstr.)

Bregestovski P., A. Redkozubov, and A. Alexeev. 1986. Elevation of intracellular calcium reduces voltage-dependent potassium conductance in human T cells. *Nature.* 319:776–778.

Cahalan, M. D., K. G. Chandy, T. E. DeCoursey, and S. Gupta. 1985. A voltage-gated potassium channel in human T lymphocytes. *Journal of Physiology.* 358:197–237.

Cahalan, M. D., K. G. Chandy, T. E. DeCoursey, S. Gupta, R. S. Lewis, and J. B. Sutro. 1987. Ion channels in T lymphocytes. *In* Mechanisms of Lymphocyte Activation and Immune Regulation. S. Gupta, W. E. Paul, and A. S. Fauci, editors. Plenum Press, New York. 85–101.

Chandy, K. G., T. E. DeCoursey, M. D. Cahalan, C. McLaughlin, and S. Gupta. 1984. Voltage-gated K channels are required for T lymphocyte activation. *Journal of Experimental Medicine.* 160:369–385.

Chandy, K. G., T. E. DeCoursey, M. D. Cahalan, and S. Gupta. 1985. Electroimmunology: the physiologic role of ion channels in the immune system. *Journal of Immunology.* 135 (Suppl.):787*s*–791*s*.

Chandy, K. G., T. E. DeCoursey, M. Fischbach, N. Talal, M. D. Cahalan, and S. Gupta. 1986. Altered K[+] channel expression in abnormal T lymphocytes from mice with the *lpr* gene mutation. *Science.* 233:1197–1200.

Cheung, R. K., S. Grinstein, H.-M. Dosch, and E. W. Gelfand. 1982*a*. Volume regulation by human lymphocytes: characterization of the ionic basis for regulatory volume decrease. *Journal of Cellular Physiology.* 112:189–196.

Cheung, R. K., S. Grinstein, and E. W. Gelfand. 1982*b*. Volume regulation by human lymphocytes. Identification of differences between the two major lymphocyte subpopulations. *Journal of Clinical Investigation.* 70:632–638.

Choquet, D., P. Sarthou, D. Primi, P. A. Cazenave, and H. Korn. 1987. Cyclic AMP-modulated potassium channels in murine B cells and their precursors. *Science.* 235:1211–1214.

DeCoursey, T. E., K. G. Chandy, S. Gupta, and M. D. Cahalan. 1984. Voltage-gated K[+] channels in human T lymphocytes: a role in mitogenesis? *Nature.* 307:465–468.

DeCoursey, T. E., K. G. Chandy, S. Gupta, and M. D. Cahalan. 1985. Voltage-dependent ion channels in T-lymphocytes. *Journal of Neuroimmunology.* 10:71–95.

DeCoursey, T. E., K. G. Chandy, S. Gupta, and M. D. Cahalan. 1987*a*. Two types of potassium channels in murine T lymphocytes. *Journal of General Physiology.* 89:379–404.

DeCoursey, T. E., K. G. Chandy, S. Gupta, and M. D. Cahalan. 1987*b*. Mitogen induction of ion channels in murine T lymphocytes. *Journal of General Physiology.* 89:405–420.

Deutsch, C., A. Holian, S. K. Holian, R. P. Daniele, and D. F. Wilson. 1979. Transmembrane electrical and pH gradients across human erythrocytes and human peripheral lymphocytes. *Journal of Cellular Physiology.* 99:79–94.

Deutsch, C., D. Krause, and S. C. Lee. 1986a. Voltage-gated potassium conductance in human T lymphocytes stimulated with phorbol ester. *Journal of Physiology.* 372:405–423.

Deutsch, C., J. Patterson, M. Price, S. Lee, and M. Prystowsky. 1986b. Volume-regulation in cloned T-lymphocytes. *Biophysical Journal.* 49:162a. (Abstr.)

Felber, S. M., and M. D. Brand. 1983. Early plasma-membrane-potential changes during stimulation of lymphocytes by concanavalin A. *Biochemical Journal.* 210:885–891.

Fukushima, Y., S. Hagiwara, and M. Henkart. 1984. Potassium current in clonal cytotoxic T lymphocytes from the mouse. *Journal of Physiology.* 351:645–656.

Gallin, E. K., and P. A. Sheehy. 1988. Leukocyte ion channels and their functional implications *In* Inflammation: Basic Principles and Clinical Correlates. J. I. Gallin, I. M. Goldstein, and R. Snyderman, editors. Raven Press, New York. 651–664.

Gardos, G. 1958. The function of calcium in the potassium permeability of human erythrocytes. *Biochemica et Biophysica Acta.* 30:653–654.

Grinstein, S., C. A. Clarke, A. Dupre, and A. Rothstein. 1982a. Volume-induced increase of anion permeability in human lymphocytes. *Journal of General Physiology.* 80:801–823.

Grinstein, S., A. Dupre, and A. Rothstein. 1982b. Volume regulation by human lymphocytes: role of calcium. *Journal of General Physiology.* 79:849–868.

Grinstein, S., C. A. Clarke, A. Dupre, and A. Rothstein. 1983. Volume-induced anion conductance in human B lymphocytes is cation independent. *American Journal of Physiology.* 245:C160–C163.

Grinstein, S., A. Rothstein, B. Sarkadi, and E. W. Gelfand. 1984. Responses of lymphocytes to anisotonic media: volume-regulating behavior. *American Journal of Physiology.* 246:C204–C215.

Hamill, O.P., A. Marty, E. Neher, B. Sakmann, and F. J. Sigworth. 1981. Improved patch-clamp techniques for high-resolution current recording from cells and cell-free membrane patches. *Pfluegers Archiv European Journal of Physiology.* 391:85–100.

Hermann, A., and C. Erxleben. 1987. Charybdotoxin selectively blocks small Ca-activated K channels in Aplysia neurons. *Journal of General Physiology.* 90:27–47.

Hodgkin, A. L., and B. Katz. 1949. The effect of sodium ions on the electrical activity of the giant axon of the squid. *Journal of Physiology.* 108:37–77.

Krouse, M. E., G. T. Schneider, and P. W. Gage. 1986. A large anion-selective channel has seven conductance levels. *Nature.* 319:58–60.

Kuno, M., and P. Gardner. 1987. Ion channels activated by inositol 1,4,5-trisphosphate in plasma membrane of human T-lymphocytes. *Nature.* 326:301–304.

Kuno, M., J. Goronzy, C. M. Weyand, and P. Gardner. 1986. Single-channel and whole-cell recordings of mitogen-regulated inward currents in human cloned helper T lymphocytes. *Nature.* 323:269–273.

Lee, S., M. Price, M. Prystowsky, and C. Deutsch. 1987. Volume response of quiescent and interleukin 2-stimulated mouse T lymphocytes to hypotonicity. *Journal of General Physiology.* 90:27a. (Abstr.)

Lewis, R. S., and M. D. Cahalan. 1988. Subset-specific expression of potassium channels in developing murine T lymphocytes. *Science.* 239:771–775.

Matteson, D. R., and C. Deutsch. 1984. K channels in T lymphocytes: a patch clamp study using monoclonal antibody adhesion. *Nature.* 307:468–471.

McKinnon, D., and R. Ceredig. 1986. Changes in the expression of potassium channels during mouse T cell development. *Journal of Experimental Medicine.* 164:1846–1861.

Miller, C., E. Moczydlowski, R. Latorre, and M. Phillips. 1985. Charybdotoxin, a protein inhibitor of single Ca^{2+}-activated K^+ channels from mammalian skeletal muscle. *Nature.* 313:316–318.

Revenko, S. V., B. I. Khodorov, and L. M. Shapovalova. 1982. Quinidine blockage of sodium and potassium channels in myelinated nerve fibre. *Neurophysiology* (English Translation of *Neirofiziologiya*). 14:244–250.

Rink T. J., C. Montecucco, T. R. Hesketh, and R. Y. Tsien. 1980. Lymphocyte membrane potential assessed with fluorescent probes. *Biochimica et Biophysica Acta.* 595:15–30.

Rink T. J., A. Sanchez, S. Grinstein, and A. Rothstein. 1983. Volume restoration in osmotically swollen lymphocytes does not involve changes in cytoplasmic pH or free Ca^{2+} concentration. *Biochimica et Biophysica Acta.* 762:593–596.

Sarkadi, B., E. Mack, and A. Rothstein. 1984*a*. Ionic events during the volume response of human peripheral blood lymphocytes to hypotonic media. I. Distinctions between volume-activated Cl^- and K^+ conductance pathways. *Journal of General Physiology.* 83:497–512.

Sarkadi, B., E. Mack, and A. Rothstein. 1984*b*. Ionic events during the volume response of human peripheral blood lymphocytes to hypotonic media. II. Volume- and time-dependent activation and inactivation of ion transport pathways. *Journal of General Physiology.* 83:513–527.

Schlichter, L., N. Sidell, and S. Hagiwara. 1986*a*. Potassium channels mediate killing by human natural killer cells. *Proceedings of the National Academy of Sciences.* 83:451–455.

Schlichter, L., N. Sidell, and S. Hagiwara. 1986*b*. K channels are expressed early in human T-cell development. *Proceedings of the National Academy of Sciences.* 83:5625–5629.

Schwarz, W., and H. Passow. 1983. Ca^{2+}-activated K^+ channels in erythrocytes and excitable cells. *Annual Review of Physiology.* 45:359–374.

Schwarze, W. and H.-A. Kolb. 1984. Voltage-dependent kinetics of an anionic channel of large unit conductance in macrophages and myotube membranes. *Pfluegers Archiv European Journal of Physiology.* 402:281–291.

Smith, C., M. Phillips, and C. Miller. 1986. Purification of charybdotoxin, a specific inhibitor of the high-conductance Ca^{2+}-activated K^+ channel. *Journal of Biological Chemistry.* 261:14607–14613.

Tatham, P. E. R., and P. J. Delves. 1984. Flow cytometric detection of membrane potential changes in murine lymphocytes induced by concanavalin A. *Biochemical Journal.* 221:137–146.

Yeh, J. Z., and T. Narahashi. 1976. Mechanism of action of quinidine on squid axon membranes. *Journal of Pharmacology and Experimental Therapeutics.* 196:62–70.

Chapter 25

Intracellular pH Regulation in Neutrophils:
Properties and Distribution of the Na^+/H^+
Antiporter

Sergio Grinstein, Wendy Furuya, and Ori D. Rotstein

*Division of Cell Biology, Hospital for Sick Children, Department
of Biochemistry and Department of Surgery, University of Toronto,
Toronto General Hospital, Toronto, Ontario, Canada*

Cell Physiology of Blood © 1988 by The Rockefeller University Press

Introduction

Neutrophils migrate to sites of infection in response to chemotactic stimuli derived from bacteria, complement, and other inflammatory mediators. Their primary function is to ingest, kill, and degrade invading microorganisms. In this capacity, neutrophils are exposed to media of widely ranging pH, yet must be able to regulate their intracellular pH (pH_i) within the physiological range. The problem is compounded by the intracellular generation of protons by a variety of metabolic pathways, including glycolysis and the hexose monophosphate shunt. Thus, powerful and versatile regulatory mechanisms must exist to control pH_i. The purpose of this chapter is to review the properties of one of these regulatory mechanisms, the Na^+/H^+ exchanger or antiporter. We will first review the evidence for the existence of the antiporter in neutrophils and will briefly summarize its properties. Next, its intracellular distribution will be considered. Finally, we will discuss the importance of Na^+/H^+ exchange in particular and of pH_i homeostasis in general to neutrophil function.

Na^+/H^+ Exchange in Neutrophils

The presence of an electroneutral Na^+/H^+ antiporter in the plasma membrane of neutrophils was first demonstrated by Molski et al. (1980). The activity of this antiporter can be readily demonstrated by monitoring pH_i in acid-loaded cells, as shown in Fig. 1 for human blood neutrophils. For these experiments, the cells were loaded with the pH-sensitive probe 2',7'-*bis*(carboxyethyl)-5,6-carboxyfluorescein (BCECF) and suspended in Na^+-free, *N*-methyl-D-glucammonium$^+$ solution. Acid loading was accomplished by treatment of the cells with nigericin, an ionophore that catalyzes the electroneutral exchange of K^+ for H^+. Under the conditions of the experiment, there is a large, outwardly directed K^+ gradient that drives the intracellular accumulation of H^+, resulting in acid loading. In cells maintained in Na^+-free and nominally bicarbonate-free medium, removal of the ionophore by addition of defatted serum albumin was followed by only marginal pH_i recovery. In contrast, a marked alkalinization is recorded upon addition of extracellular Na^+, returning pH_i towards the original level (Fig. 1 *A*). The rate of Na^+/H^+ exchange can be quantified by multiplying the rate of change of pH_i times the buffering power of the cells, estimated to average 28 mM/pH in bicarbonate-free media in the pH range of 6.4 to 7.4 (Grinstein and Furuya, 1986*a*). In cells acid loaded to pH_i 6.65–6.8, the maximal rate of Na^+/H^+ exchange reached 9.2 mmol/liter cells/min. Such a flux could double the intracellular Na^+ content within a couple of minutes. Indeed, the fluxes through the antiporter can be so large that they are detectable, indirectly, by the changes in cell volume that result from the associated water uptake. When cells are acid loaded by suspending them in media containing the Na^+ salt of weak organic acids such as propionate, the activation of the antiporter leads to accumulation of intracellular Na^+ which, together with the uptake of the anionic form of the weak acid, produces osmotic gain, followed by water uptake and cell swelling, which are detectable by electronic cell sizing (Grinstein and Furuya, 1984). An alkalinization similar to that induced by Na^+ in Fig. 1 *A* can also be obtained by addition of Li^+, but not by Rb^+ (Fig. 1 *C*), K^+ or Cs^+ (Grinstein and Furuya, 1986*a*). Divalent cations and organic monovalent cations are similarly not transported, which indicates that the antiporter is very selective. Under certain conditions, e.g., when Na^+-loaded cells are suspended in Na^+-free media, the antiporter can operate in reverse, mediating the uptake of external H^+ in exchange for internal Na^+.

The rate of Na^+-induced alkalinization of acid-loaded neutrophils is a saturable function of extracellular Na^+ (apparent $K_m = 73$ mM) and is inhibited by extracellular H^+. As shown in Fig. 1 *B*, the Na^+-induced alkalinization is also blocked when the K^+-sparing diuretic amiloride is present in the medium, but is restored upon addition of the exogenous Na^+/H^+ exchanger monensin, which demonstrates the persistence of an inward Na^+ gradient. Inhibition by amiloride is competitive with extracellular Na^+

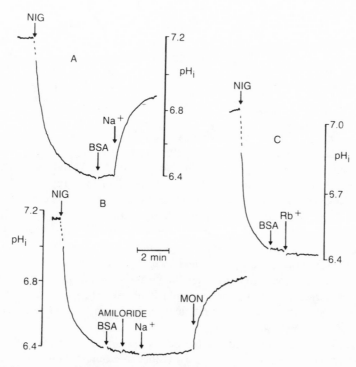

Figure 1. Evidence for Na^+/H^+ exchange in the plasma membrane of human neutrophils. (*A*) BCECF-loaded neutrophils were suspended in Na^+-free, *N*-methyl-D-glucammonium$^+$ medium. Where indicated, 1 μM nigericin (Nig) was added to initiate K_i^+/H_o^+ exchange. Acid loading was terminated by scavenging the ionophore with fatty acid–free bovine serum albumin (BSA; 5 mg/ml final). Finally, NaCl (70 mM) was added. The dotted line and discontinuities in the traces indicate intervals when the sample compartment was opened for additions. (*B*) BCECF-loaded cells were acid loaded with nigericin/albumin as in *A*. Where indicated, 200 μM amiloride was added, followed by 70 mM NaCl. Finally, 5 μM monensin (Mon) was added. (*C*) Cells were acid loaded as in *A*. Where noted, 70 mM RbCl was added. Traces are representative of at least three experiments. Temperature, 37°C. From Grinstein and Furuya (1986*a*).

and has an apparent K_i of 24 μM. Analogues of amiloride with alkyl or alkenyl chains attached to the nitrogen on position 5 of the pyrazine ring are up to one 100-fold more potent as inhibitors of the exchanger than the parent molecule.

Net forward exchange (external Na^+ for internal H^+) is also inhibited by elevating the intracellular Na^+ concentration, as would be expected for competition with intracellular H^+ for the inward-facing transport site of the antiporter. In contrast, elevating the intracellular H^+ concentration produces a marked activation of forward

Na$^+$/H$^+$ exchange (Grinstein and Furuya, 1986*a*). By analogy with the renal brush border system, this is probably a dual effect: more H$^+$ are available to occupy the transport site and, more importantly, an allosteric "modifier" site becomes protonated, stimulating the turnover of the transport site(s). The existence of a modifier site was initially postulated by Aronson et al. (1982) because: (*a*) the dependence of the rate of transport on the intracellular H$^+$ concentration ([H$^+$]$_i$) cannot be described by Michaelis-Menten type of kinetics, and (*b*) efflux of internal Na$^+$ can in fact be accelerated by elevating [H$^+$]$_i$. It is this putative modifier site that confers to the antiporter its peculiar pH$_i$ dependence: the rate of Na$^+$/H$^+$ exchange is highest when the cytoplasm is acidic, with transport declining gradually as pH$_i$ approaches the physiological level of 7.0–7.2 (the range reported for most mammalian cell types; for review see Roos and Boron, 1981). This behavior is consistent with a central role of Na$^+$/H$^+$ exchange in the regulation of pH$_i$, with the effector (the transport site[s]) becoming quiescent when the threshold for inactivation or "set point" of the allosteric site is attained.

Though regulated, pH$_i$ in neutrophils is not invariant. A number of conditions have been reported where transient or sustained pH$_i$ changes are reproducibly recorded. Some of these changes appear to be mediated by variations in the mode of operation of the Na$^+$/H$^+$ antiporter. Molski et al. (1980) first reported that stimulation of neutrophils with chemotactic peptides induces a biphasic pH change: a transient acidification followed by a more sustained alkalinization. A similar pattern was later reported for cells treated with phorbol esters, though a third phase of delayed acidification was frequently observed (Grinstein and Furuya, 1986*b*). Two lines of evidence suggested that the antiporter mediated the alkalinization. First, pH$_i$ failed to increase in response to the activators when Na$^+$ was omitted from the medium. Second, addition of amiloride or of its analogues similarly blocked the alkalinization. Indeed, in both instances, a more pronounced acidification was unmasked (Molski et al., 1980; Grinstein and Furuya, 1986*b*).

Stimulation of the antiporter by the activators seems to be twofold. It is partly indirect, in response to the initial rapid cytoplasmic acidification. However, other mechanisms must exist, since pH$_i$ "overshoots" the resting value, i.e., the antiporter continues to operate beyond the original pH$_i$ set point. Studies where the activity of the antiporter was measured with varying degrees of acid loading in untreated and stimulated cells revealed an apparent shift in the pH$_i$ dependence of transport, as if an alkaline shift of the set point had occurred (Grinstein and Furuya, 1984). We concluded that a change in the properties of the modifier site had occurred, inasmuch as this site largely determines the pH$_i$ sensitivity of the rate of transport through the exchanger. This change might be mediated by phosphorylation, since protein kinase activity is known to be greatly stimulated in activated neutrophils.

The source of the initial cytoplasmic acidification observed upon stimulation of the cells has been investigated in detail. It is not due to transmembrane uptake of external H$^+$ (or the equivalent efflux of OH$^-$), since the external medium does not alkalinize concomitantly. In fact, an extracellular acidification accompanies the internal acidification, which suggests that intracellular protons first accumulate in the cytoplasm and eventually are liberated into the medium. The protons that appear in the cytoplasm are not released from the acidic secretory granules. This was concluded from experiments where pH$_i$ was measured in enucleated and degranulated cell fragments called cytoplasts (Grinstein et al., 1985). Stimulation of cytoplasts in media containing amiloride or devoid of Na^+ produced an acidification indistinguishable from that obtained in intact cells. It was therefore concluded that the acid was metabolically generated by the cells during stimulation.

Three lines of evidence indicate that the protons causing the cytoplasmic acidification originate from the stimulation by the activators of the NADPH oxidase and/or the hexose monophosphate shunt, two pathways that are virtually quiescent in intact cells. First, an excellent correlation exists between the rates of acidification and the activity of the oxidase and the shunt. Inhibition of the latter by N-ethylmaleimide, 2-deoxy-D-glucose, pertussis toxin, or adenosine is accompanied by a commensurate reduction in the rate of cytoplasmic acidification (Grinstein and Furuya, 1986b, c). Second, the acidification can be mimicked by stimulating the hexose monophosphate shunt in cells that have not been activated by ligands. This can be accomplished by adding membrane permeable redox agents capable of oxidizing NADPH. The product of this oxidation, $NADP^+$, is the rate-limiting substrate for the shunt. Addition of one such redox agent, phenazine methosulfate, produced an instantaneous and quite pronounced fall in pH_i in otherwise unstimulated neutrophils (Grinstein and Furuya, 1986b).

The third and perhaps most convincing proof that acid generated during the respiratory burst can account for the change in pH_i was obtained from studies with neutrophils isolated from individuals suffering from chronic granulomatous disease (CGD). These patients are afflicted with recurrent infections, due to the inability of their phagocytic cells to kill invading microorganisms (Babior, 1978). This deficiency is largely due to the inability of the cells to generate bactericidal oxygen metabolites. Neutrophils from CGD patients lack one of the components of the NADPH oxidase system (X-linked CGD) or have defective stimulus-response coupling (autosomal recessive CGD), which results in failure to synthesize superoxide. This enabled us to test whether the phorbol ester–induced acidification is indeed associated with the burst in oxidative metabolism and stimulation of the hexose monophosphate shunt. We found that under conditions where Na^+/H^+ exchange is prevented by addition of amiloride or omission of Na^+, stimulated CGD cells displayed little change in pH_i (Grinstein et al., 1986), which indicates that oxidative metabolism is the main source of proton accumulation in the cytoplasm of activated neutrophils.

Are the Na^+/H^+ Exchangers Present in the Phagosomal Membrane?

The biogenesis and turnover of the Na^+/H^+ antiporters have not been investigated in detail. The fate of the antiporters after endocytosis is particularly interesting, since internalization of the plasma membrane is followed by marked acidification of the endosomal matrix. It is not clear if the exchangers are part of the endosomal membrane and, if so, whether they contribute to the observed pH changes. An extreme case is presented by cells such as neutrophils, which can rapidly internalize a large fraction of the surface membrane during phagocytosis of particles such as bacteria. For these reasons we undertook experiments designed to assess the presence of Na^+/H^+ exchangers in neutrophil phagosomes. The activity of the antiporter in the membranes of phagosomes was investigated measuring the intraphagosomal pH in intact cells. This was accomplished by inducing neutrophils to phagocytose bacteria (*Micrococcus lysodeikticus*) covalently labeled with a pH-sensitive fluorescent dye. Excess and extracellularly adhering bacteria were then removed by washing and treating the suspension with lysozyme, an enzyme that can digest the bacterial wall without affecting the neutrophil membrane. The washed cell suspension was then used for macrofluorimetric determinations of intraphagosomal pH (Grinstein and Furuya, 1988).

When the labeled neutrophils were suspended in media at 37°C, a spontaneous acidification of the phagosomes was recorded. This change has generally been attributed to activation of a proton-pumping ATPase (Al-Awqati, 1986). It was of interest to determine whether the Na^+/H^+ antiporters, if present in the phagosomal membranes, participated in the generation or maintenance of this pH gradient. Two

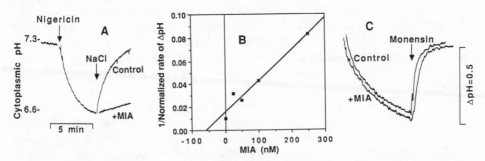

Figure 2. Na^+/H^+ exchange in the plasma membrane of neutrophils and its role in phagosomal acidification. (*A*) BCECF-loaded neutrophils were suspended in *N*-methyl-D-glucammonium$^+$–rich medium with or without 100 nM methyl-isobutyl-amiloride (MIA) and were used for spectroscopic measurement of pH_i. Where indicated, the cells were acid loaded by addition of 0.1 μM nigericin, a K^+/H^+ exchanger. Next, NaCl (final 70 mM) was added to initiate Na^+/H^+ exchange. pH_i was calibrated by the nigericin/K^+ method. The trace is a composite, representative of >10 experiments. (*B*) Concentration dependence of the inhibitory effect of methyl-isobutyl-amiloride. Experiments such as those in *A* were performed at varying concentrations of the inhibitor. The inverse of the rate of ΔpH_i, measured as the maximal rate of Na^+-induced alkalinization, is plotted against the concentration of the inhibitor (Dixon plot). The rate of alkalinization is expressed relative to the untreated control, taken as 100%. The plot is representative of three experiments and the line was fitted by least squares. (*C*) Effect of methyl-isobutyl-amiloride on the spontaneous acidification of the phagosomes. Neutrophils containing covalently fluoresceinated bacteria were suspended in Na^+ medium in the presence or absence of 1 μM methyl-isobutyl-amiloride. Where indicated, 5 μM monensin was added to both samples. Representative of three experiments. From Grinstein and Furuya (1988).

possibilities were envisaged: the coupled activity of the Na^+/H^+ antiporters and the Na^+/K^+ pumps could contribute to the acidification by exchanging intraphagosomal Na^+ for cytoplasmic H^+. Conversely, if the combined Na^+ and H^+ gradient is reversed by the large acidification of the phagosome, the exchanger could in fact dissipate (uncouple) the pH gradient created by the H^+ pump. Several lines of evidence indicate that the Na^+/H^+ antiporters and/or the Na^+/K^+ pumps are not required for phagosomal acidification. The rate and extent of the acidification were not affected by: (*a*) addition of methyl-isobutyl-amiloride, a permeable and very potent amiloride analogue, at concentrations severalfold higher than required to inhibit the antiporter of the surface membrane (Fig. 2; Grinstein and Furuya, 1988); (*b*) omission of Na^+ from the medium in which phagocytosis occurred; (*c*) trapping of micromolar amiloride inside the phagosome during internalization of the bacteria; (*d*) addition of concentrations of digitoxin, a permeable agent, known to fully inhibit the pump, and (*e*) pretreatment of the cell with subsequent trapping of ouabain in the phagosome.

We similarly failed to obtain evidence of dissipation of the pH gradient by reverse exchange of cytoplasmic Na^+ for intraphagosomal H^+ through the antiporter. The thermodynamic feasibility of this phenomenon is illustrated by the alkalinizing effects of monensin, an exogenous Na^+/H^+ exchanger (Fig. 2 *C*). Operation of the endogenous Na^+/H^+ antiporter in this mode would reduce the rate and/or the extent of

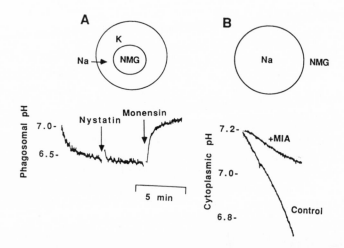

Figure 3. Assessment of reverse Na^+/H^+ exchange activity in the surface and phagosomal membranes. (*A*) Evidence against reverse exchange in the phagosomal membrane. The conditions of the experiment are schematized above the trace: cells were allowed to phagocytose fluoresceinated bacteria in *N*-methyl-D-glucammonium$^+$ solution. The cells were then suspended in Na^+ medium and phagosomal pH was determined fluorimetrically as in Fig. 2 *C*. After the spontaneous phagosomal acidification reached a steady state, the cytoplasmic Na^+ concentration was suddenly elevated by treatment of the cells with 50 μg/ml of nystatin. Where indicated, 5 μM monensin was added. The trace is representative of three experiments. (*B*) Evidence for reverse Na^+/H^+ exchange in the surface membrane. BCECF-stained neutrophils were loaded with Na^+ by treatment with nystatin and ouabain in Na^+ medium, followed by washout of the antibiotic. The traces start upon resuspension of the Na^+-loaded cells in *N*-methyl-D-glucammonium$^+$ medium in the presence or absence of 0.5 μM methyl-isobutyl-amiloride (MIA). The composite is representative of three experiments. From Grinstein and Furuya (1988).

acidification of the phagosome. However, as mentioned above, these parameters were unaffected by agents or conditions expected to impair Na^+/H^+ exchange. Therefore, we concluded that intrinsic Na^+/H^+ exchange does not seem to uncouple the pH gradient created by the proton pumps.

In light of our inability to demonstrate effects of Na^+/H^+ exchange on the spontaneous phagosomal acidification, we questioned whether functional antiporters were at all present in the internalized membranes. Three different types of experiments failed to demonstrate any detectable exchange activity sensitive to amiloride or methyl-isobutyl-amiloride across the phagosomal membrane. First, as illustrated in

Fig. 3, a cytoplasm-to-phagosomal matrix Na$^+$ gradient did not induce intraphago-somal alkalinization, which would be expected from the reverse operation of the antiporter. In these experiments, the cytoplasmic Na$^+$ concentration was suddenly increased by permeabilizing the plasma membrane with nystatin. At the concentrations used, this agent was found to increase the permeability of the surface membrane, without affecting internal membranes. The latter was confirmed measuring the partition of acridine orange into intracellular acidic compartments (Grinstein and Furuya, 1988). Nystatin, unlike nigericin or ammonium, failed to release the accumulated acridine. Fig. 3 *B* also shows that, under comparable conditions, the antiporter at the surface membrane can operate in the reverse mode. Two other findings argue against the presence of functional exchangers in the phagosomal membrane. In the presence of phagosomal Na$^+$, a cytoplasmic acidification did not induce a methyl-isobutyl-amiloride–sensitive acidification of the phagosomal matrix. As before, the surface membrane antiporter can be shown to be activated under similar conditions. Finally, induction of a phagosome-to-cytoplasm Na$^+$ gradient did not produce acidification of the phagosomal interior.

Taken together, the results summarized above indicate that the Na$^+$/H$^+$ antiporter is not functional in the phagosomal membrane. Two possible mechanisms could account for these observations. A segregation process may exclude the exchangers from the areas of the membrane that are being internalized. This could conceivably occur through direct or indirect interactions with cytoskeletal components. Alternatively, the antiporters may indeed be internalized, but become functionally inactivated. This might result from a chemical modification such as dephosphorylation. As mentioned above, the activity of the antiporter is known to be regulated by phosphorylation/dephosphorylation processes. However, other posttranslational modifications such as proteolysis or transglutamination could also account for the inactivation of transport.

Importance of pH$_i$ Homeostasis to Neutrophil Function

Regulation of pH$_i$ appears to be important to the normal bactericidal and locomotor activity of the neutrophil. The activity of the Na$^+$/H$^+$ antiporter, which is largely responsible for pH$_i$ homeostasis in neutrophils stimulated in the nominal absence of bicarbonate, is essential for the development of a normal bactericidal response. This can be inferred from early reports describing reduced superoxide production in cells suspended in media devoid of Na$^+$ (Korchak and Weissmann, 1980). These findings were initially interpreted as indicative of a role of Na$^+$ in signal transduction, perhaps through changes in transmembrane potential. However, more recent experiments suggest that inhibition of the NADPH oxidase is secondary to alterations in pH$_i$ (Nasmith and Grinstein, 1986). As described above, stimulation of cells in Na$^+$-free media induces a marked acidification of the cytoplasm that, in the presence of external Na$^+$, is small and transient, and is in fact, superseded by an alkalinization. By manipulating pH$_i$ with ionophores, we were able to demonstrate that the pH$_i$ changes that develop in the media with different ionic composition are responsible for the varying rates of superoxide production (or oxygen consumption). Cells were suspended in K$^+$-rich solutions containing nigericin. Under these conditions $[K^+]_o/[K^+]_i \approx [H^+]_o/[H^+]_i$ and, because the intracellular K$^+$ concentration is similar to that of the medium, pH$_i$ will be approximately equal to the extracellular pH (pH$_o$).

Therefore, by varying pH_o we were able to study the intracellular pH dependence of phorbol ester–stimulated oxygen consumption. A sharp inhibition of the function of the NADPH oxidase was found as pH_o (and therefore pH_i) was reduced below 6.8, with complete inhibition at pH 6.0. Interpolation in such a pH dependence curve indicated that the acidification observed in Na^+-free media could account for the observed depression in the rate of oxygen consumption (Nasmith and Grinstein, 1986).

The latter conclusion is also supported by experiments where restoration of physiological pH_i, even in the absence of external Na^+, produced recovery of the rate of oxygen consumption. This was achieved stimulating cells in K^+ medium, initially in the absence of nigericin. As expected for cells suspended in Na^+-free media, pH_i dropped and, concomitantly, the rate of oxidative metabolism was reduced. Upon addition of nigericin, the intracellular pH rapidly increased towards that of the medium, as the ionophore catalyzed the exchange of external K^+ for internal H^+. This alkalinization was accompanied by the restoration of a high rate of oxygen consumption, as expected for a pH-dependent, reversible inhibition. Clearly, normal rates of NADPH-oxidase activity can be obtained in media devoid of Na^+, ruling out a specific requirement for this cation in the activation process (Nasmith and Grinstein, 1986).

Is pH_i regulation essential for neutrophil function in vivo? Recent observations suggest that alterations in neutrophil pH_i may contribute to the development of intra-abdominal abscesses. Anaerobic bacteria, particularly *Bacteroides* species, are frequently encountered in abscesses and contribute to the pathogenicity of mixed (polymicrobial) infections (Rotstein et al., 1985). Among the principal by products of the metabolism of anaerobic bacteria are short-chain fatty acids, which accumulate in the extracellular milieu, lowering its pH. In fact, the pH of human pus has been reported to be as low as 5.7 (Bryant et al., 1980). We therefore hypothesized that, in the abscess, neutrophil function might be impaired as a result of derangement of pH_i homeostasis induced by the bacterial products. This was tested experimentally in vitro by measuring superoxide production by human neutrophils after incubation in the presence or absence of short-chain fatty acids at normal and acidic pH. As shown in Fig. 4, neither lowering the external pH to 5.5, nor adding millimolar concentrations of the fatty acids while keeping pH at 7.4 was, individually, sufficient to affect the rate of superoxide generated in response to opsonized zymosan. However, simultaneous acidification and addition of fatty acids drastically reduced the respiratory burst. Similar results were obtained with a variety of mono- and di-carboxylic fatty acids, which suggests that their weak acid nature, rather than a specific conformation or chain length, is essential to their inhibitory effect (Rotstein et al., 1987).

The above observations can be explained by assuming that only the protonated form of the weak acid permeates into the cells, dissociating in the cytoplasm and thereby lowering pH_i. According to this model, succinate added at normal extracellular pH would enter the cells only in negligible amounts, therefore not affecting pH_i significantly. These predictions were tested experimentally, measuring pH_i in BCECF-loaded cells (Fig. 5). As expected, cells suspended in normal media at physiological pH maintained a constant pH_i for extended periods. Similarly, pH_i was unaffected by incubation with 30 mM succinate at pH 7.4. Suspending cells in medium at pH 5.5 induced a gradual intracellular acidification, with pH_i equilibrating at ≥ 6.0 after 10–15 min. This suggests that the membrane has a finite permeability to H^+ (OH^-). A markedly different result was obtained when cells were suspended in medium at pH 5.5 containing succinate. The cytoplasmic pH dropped very quickly, equilibrating at ~ 5.5

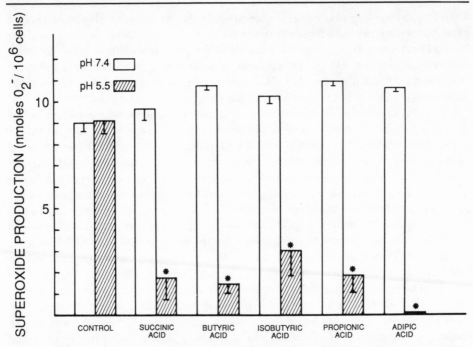

Figure 4. Effect of different short-chain fatty acids on superoxide production by neutrophils activated with opsonized zymosan in media of physiological or acidic pH. Cells were incubated in 30-mM solutions of various short-chain fatty acids at either pH 5.5 or 7.4 for 20 min at 37°C. The cells were then washed twice and suspended in Hank's balanced salt solution, pH 7.4, for testing of superoxide generation. Results are expressed as nanomoles of superoxide per 10^6 cells per 20 min. Data are means ± SE of at least four determinations. Asterisks indicate $p < 0.001$ versus sample with the same fatty acid at pH 7.4. From Rotstein et al. (1987).

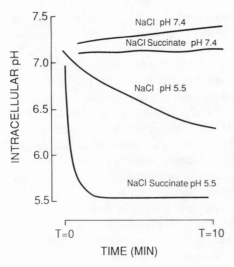

Figure 5. Effect of succinate on the cytoplasmic pH of human neutrophils suspended in media of varying pH. BCECF-loaded cells were suspended in Na^+ medium adjusted to pH 5.5 or 7.4. Where indicated, 30 mM succinate was added to the medium, which remained isoosmotic by omission of the appropriate amount of NaCl. The time course of pH_i was monitored fluorimetrically as described for Fig. 1. The traces are representative of at least four experiments. From Rotstein et al. (1987).

within ~ 3 min (Rotstein et al., 1987). The precise pH_i attained under these conditions cannot be determined accurately using BCECF, which has a pK_a of 6.95, but it was clearly much lower than 6.0, the pH required for full inhibition of the NADPH oxidase (see above). Under these conditions, Na^+/H^+ exchange cannot regulate the internal pH, since the high extracellular H^+ concentration inhibits the forward operation of the antiporter. Taken together, these observations indicate that, in this system, succinate is operating as a protonophore, shuttling H^+ across the plasma membrane. It is this protonophoric activity, together with the lowered extracellular pH, that may account for the inhibitory effect of short-chain fatty acids on neutrophil function in vitro, as well as in the abscess.

Several investigators have noted an inhibitory effect of reduced extracellular pH on neutrophil migration in vitro (Nahas et al., 1971; Simchowitz and Cragoe, 1986). Recent experiments have extended these observations by correlating reduced pH_i with inhibition of neutrophil chemotactic responses (Simchowitz and Cragoe, 1986). These authors postulated that pH_i may play an important role in the regulation of neutrophil locomotor responses via modulation of actin polymerization that occurs in response to chemotactic stimuli in neutrophils (White et al., 1983) and is known to be pH sensitive in other systems (Begg and Rebhun, 1979).

Conclusion

In summary, we have reviewed evidence demonstrating the presence, properties, and distribution of the Na^+/H^+ antiporter in neutrophils. The significance of antiporter activity to neutrophil functionality was also discussed. It was concluded that pH_i regulation is essential for the expression of the oxygen-dependent killing mechanisms of human neutrophils. Pending an assessment of the role of bicarbonate-dependent systems, Na^+/H^+ exchange appears to be the primary mechanism regulating pH_i in these cells. This is particularly evident after activation, when the excess metabolic acid generated mainly by the oxidative burst would otherwise accumulate in the cytoplasmic compartment. Finally, it is conceivable that derangements of pH_i homeostasis may underlie some pathological conditions of the immune system, such as the formation of abscesses.

Acknowledgments

The work summarized in this chapter was supported by the Medical Research Council and the National Cancer Institute of Canada. S. Grinstein is the recipient of a Medical Research Council Scientist Award.

References

Al-Awqati, Q. 1986. Proton-translocating ATPases. *Annual Review of Cell Biology.* 2:179–200.

Aronson, P. S., J. Nee, and M. A. Suhm. 1982. Modifier role of internal H in activating the Na/H exchanger in renal microvillus membrane vesicles. *Nature.* 299:161–163.

Babior, B. M. 1978. Oxygen-dependent microbial killing by phagocytes. *New England Journal of Medicine.* 12:659–668.

Begg, D. I. and L. I. Rebhun. 1979. pH regulates the polymerization of actin in the sea urchin egg cortex. *Journal of Cell Biology.* 83:241–248.

Bryant, R. E., A. L. Rashad, J. A. Mazza, and D. Hammond. 1980. Beta-lactamase activity in human pus. *Journal of Infectious Diseases.* 142:594–561.

Grinstein, S., B. Elder, and W. Furuya. 1985. Phorbol-ester-induced changes of cytoplasmic pH in neutrophils. Role of exocytosis in activation of Na/H exchange. *American Journal of Physiology.* 248:C379–C386.

Grinstein, S, and W. Furuya. 1984. Amiloride-sensitive Na/H exchange in human neutrophils: mechanism of activation by chemotactic factors. *Biochemical and Biophysical Research Communications.* 122:755–762.

Grinstein, S., and W. Furuya. 1986a. Characterization of the amiloride-sensitive Na-H antiport of human neutrophils. *American Journal of Physiology.* 250:C283–C291.

Grinstein, S., and W. Furuya. 1986b. Cytoplasmic pH regulation in phorbol ester activated human neutrophils. *American Journal of Physiology.* 251:C55–C65.

Grinstein, S., and W. Furuya. 1986c. Cytoplasmic pH regulation in activated human neutrophils: effects of adenosine and pertussis toxin on Na^+/H^+ exchange and metabolic acidification. *Biochimica et Biophysica Acta.* 889:301–309.

Grinstein, S., and W. Furuya. 1988. Assessment of Na^+/H^+ exchange activity in the phagosomal membrane of human neutrophils. *American Journal of Physiology.* 254:C272–C285.

Grinstein S., W. Furuya, and W. D. Biggar. 1986. Cytoplasmic pH regulation of normal and abnormal neutrophils. Role of superoxide generation and Na/H exchange. *Journal of Biological Chemistry.* 251:512–514.

Korchak, H. M., and G. Weissmann. 1980. Stimulus-response coupling in the human neutrophil. Transmembrane potential and the role of extracellular Na^+. *Biochimica et Biophysica Acta.* 601:180–194.

Molski T. F. P., P. H. Naccache, M. Volpi, L. M. Wolpert, and R. I. Sha'afi. 1980. Specific modulation of the intracellular pH of rabbit neutrophils by chemotactic factors. *Biochemical and Biophysical Research Communications.* 94:508–514.

Nahas, G. G., M. L. Tannieres, and J. F. Lennon. 1971. Direct measurement of leucocyte motility: effects of pH and temperature. *Proceedings of the Society for Experimental Biology and Medicine.* 138:350–352.

Nasmith, P. E., and S. Grinstein. 1986. Impairment of Na/H exchange underlies inhibitory effects of Na-free media on leukocyte function. *FEBS Letters.* 202:79–85.

Roos, A., and W. F. Boron. 1981. Intracellular pH. *Physiological Reviews.* 61:296–434.

Rotstein, O. D., P. E. Nasmith, and S. Grinstein. 1987. The *Bacteroides* by-product succinic acid inhibits neutrophil respiratory burst by reducing intracellular pH. *Infection and Immunity.* 55:864–870.

Rotstein, O. D., T. L. Pruett, and R. L. Simmons. 1985. Lethal microbial synergism in intra-abdominal infections: *Escherichia coli* and *Bacteroides fragilis. Archives of Surgery.* 120:146–151.

Simchowitz, L., and E. J. Cragoe, Jr. 1986. Regulation of human neutrophil chemotaxis by intracellular pH. *Journal of Biological Chemistry.* 261:6492–6500.

White, J. R., P. H. Naccache, and R. I. Sha'afi. 1983. Stimulation by chemotactic factors of actin association with the cytoskeleton in rabbit neutrophils. Effects of calcium and cytochalasin B. *Journal of Biological Chemistry.* 258:14041–14047.

Chapter 26

Potassium Conductances in Macrophages

Elaine K. Gallin and Leslie C. McKinney

Department of Physiology, Armed Forces Radiobiology Research Institute, Bethesda, Maryland

Cell Physiology of Blood © 1988 by The Rockefeller University Press

Introduction

Macrophages are long-lived leukocytes, originating from bone marrow progenitor cells that migrate into the bloodstream as monocytes and then into tissues where they mature into macrophages (Van Furth, 1985). The macrophage, a phagocytic, motile, and secretory cell, plays a pivotal role in the defense against infective agents and neoplastic cells as well as in the removal of antigen–antibody complexes and damaged cells. They respond to numerous signaling agents that can modify or induce functional responses, including lymphokines, complement components, neuropeptides, and bacterial products (Van Furth, 1985). In general, the mechanisms involved in signal transduction in the macrophage are poorly defined. In many other secretory and motile cells, signal transduction is initiated by the opening of ionic channels in the plasma membrane (Eckert and Brehm, 1979; Peterson and Maruyama, 1984). There is evidence that, at least for one stimulus, this may be the case for macrophages as well (Young et al., 1983). With the recent development of the patch-clamp technique, the ionic conductances in leukocytes have begun to be characterized and their relationship to function has begun to be examined (for a review see Gallin and McKinney, in press).

Patch-clamp studies on macrophages from several sources have shown that different ionic conductances are present in cells, depending on the origin of the cell as well as the in vitro growth conditions (Ypey and Clapham, 1984; Gallin and Sheehy, 1985; Randriamampita and Trautmann, 1987; Gallin and McKinney, 1988). They include four different K conductances, a large conductance Cl channel (Schwarze and Kolb, 1984), and two cation channels, one of which is gated by the binding and cross-linking of the Fc receptor (Young et al., 1983), and one that may be gated by calcium (Lipton, 1986). No voltage-dependent calcium or sodium channels have been described in macrophages. This paper focuses on the four types of K conductances present in the macrophage. We will first briefly review early intracellular microelectrode studies and then describe more recent patch-clamp studies.

Intracellular Microelectrode Studies

Early studies using intracellular microelectrodes to record from macrophages demonstrated two types of K conductances: (*a*) one that could be activated by mechanical stimulation, or exposure to the calcium ionophore A23187 or chemotactic factors (Gallin et al., 1975; Gallin and Gallin, 1977), and (*b*) an inwardly rectifying K conductance (Gallin, 1981; Gallin and Livengood, 1981). Both these conductances play a role in regulating membrane potential, although it is not yet clear whether changes in potential are necessary for signal transduction in the macrophage (Pfefferkorn, 1984; Gallin and McKinney, in press).

Spontaneous membrane hyperpolarizations have been observed in cultured mouse and guinea pig peritoneal macrophages and in cultured human peripheral blood monocytes (Gallin et al., 1975; Dos Reis and Oliviero-Castro, 1977; Gallin and Gallin, 1977). In all instances, the membrane hyperpolarizations were associated with an increase in membrane conductance that was ascribed to the activation of a Ca-activated K conductance since (*a*) the reversal potential (E_{rev}) of the hyperpolarizations was close to the K equilibrium potential (E_K) (Gallin et al., 1975), (*b*) hyperpolarization could be blocked by addition of EGTA (Gallin et al., 1975), and (*c*) the intracellular injection of calcium induced a similar hyperpolarization (Persechini et al., 1981). Data from a mouse spleen macrophage exhibiting spontaneous membrane hyperpolarizations are shown in Fig. 1. Initially the cell was current clamped. Microelectrode penetration (indicated by the first arrow) produced an initial hyperpolarization to -62 mV, after which the potential slowly changed to -43 mV. Transient hyperpolarizations (indicated by ●) were superimposed on the resting membrane potential (V_m). When the recording mode was switched to voltage clamp (holding potential (V_h) = -40 mV), outward currents of 5 nA or less were present. As is evident in the bottom current tracing, the transient outward currents decreased in amplitude as V_h was changed from -16 mV to -54 mV.

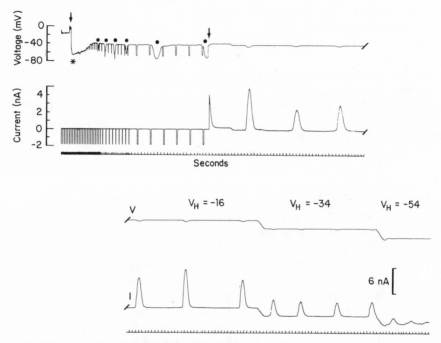

Figure 1. Chart recorder tracing of an intracellular microelectrode recording obtained using a single electrode voltage clamp from a mouse spleen macrophage cultured for 13 d. First arrow indicates impalement of the cell. Asterisk depicts initial hyperpolarization after electrode penetration. Initial segment of the recording was obtained under current clamp conditions; brief hyperpolarizing voltage steps were injected to monitor input resistance. Spontaneous hyperpolarizations are indicated (●). At the second arrow the recording mode was switched to voltage clamp and outward currents (upward deflections) are seen.

It is not clear whether spontaneous membrane hyperpolarizations occur physiologically in macrophages or whether they are due to the mechanical stimulation of microelectrode penetration and subsequent elevation of intracellular calcium ($[Ca]_i$). Ince et al. (1984) examined the rapid hyperpolarization induced immediately after electrode penetration, and concluded that it was an artifact of microelectrode damage. However, we have noted spontaneous current fluctuations that may correspond to membrane hyperpolarization in cell-attached patch recordings from cultured human macrophages where electrode-induced leak current was negligible (Gallin, 1988). In addition, Kruskal and Maxfeld (1988) have recently demonstrated oscillations in $[Ca]_i$

in macrophages immediately after adherence. These occur with a frequency similar to that of the spontaneous membrane hyperpolarizations. Taken together, these observations support the view that oscillations in membrane potential induced by the activation of a Ca-activated K conductance are not restricted to a microelectrode-induced damage and may occur under physiologically relevant conditions.

In addition to demonstrating a Ca-activated K conductance, intracellular studies in macrophages also revealed a prominent inward-rectifying K conductance in cultured mouse macrophages isolated form the peritoneal cavity or the spleen (Gallin, 1981; Gallin and Livengood, 1981). Cells exhibiting this conductance had current-voltage (I-V) relationships that showed inward rectification and often had a region of negative slope resistance between -60 and -30 mV. These cells usually had more negative resting membrane potentials (ranging from -65 to -95 mV) than did macrophages in which inward rectification was absent (-28 mV). The addition of either barium or rubidium to the bath inhibited inward rectification and depolarized cells by 20–30 mV (Gallin and Livengood, 1981). In some cells that showed prominent inward rectification and a region of negative slope resistance, two stable resting membrane potentials were present (Gallin, 1981). This behavior has been reported in cardiac ventricular cells and rat basophilic leukemia cells, which exhibit a similar K conductance (Gadsby and Cranefield, 1977; Lindau and Fernandez, 1986).

Patch-Clamp Studies
Calcium- and Voltage-activated K Conductance

Whole-cell current. Whole-cell, patch-clamp recordings of cultured human macrophages reveal an outward current that activates for voltage steps greater than $+20$ mV (Gallin and McKinney, 1988). Fig. 2 *A* shows current tracings from a whole-cell recording in which the patch electrode contained KCl Hanks' solution with 10^{-6} M [Ca]$_i$. Large outward currents characterized by a noisy baseline were present in response to voltage steps of $+40$ mV or greater, while hyperpolarizing steps produced small inward currents. The peak I-V relationship (Fig. 2 *B*) showed prominent outward rectification. The ionic selectivity of the outward current was determined from the instantaneous I-V relationship (data not shown). Cells were stepped to $+50$ mV for 200 ms to activate the outward current and then stepped down to various potentials. The resulting tail currents reversed at -75 mV, which was near E_K for these cells. Since no other ion present in the solution had a similar reversal potential, we concluded that the outward current was due predominantly to an increase in permeability to K.

Single-channel current. Cell-attached and excised patch recordings from human macrophages have demonstrated outward single-channel currents activating at depolarized potentials (Gallin, 1984). Fig. 3 *A* shows records from an inside-out excised patch illustrating the voltage and calcium sensitivity of these currents. At constant [Ca]$_i$, channel activity decreased as the voltage across the patch was made less positive; at constant V_m, channel activity decreased as [Ca]$_i$ was lowered from 3×10^{-6} M to 5×10^{-7} M. The calcium sensitivity of this channel is relatively low compared with similar channels in other secretory cells (Peterson and Maruyama, 1984). The open probability (p) of the channel at $+80$ mV in symmetrical KCl with 3×10^{-6} M [Ca]$_i$ was only 0.22 (mean of eight inside-out excised patches; Gallin and McKinney, 1988). Similarly, in cell-attached patches where resting [Ca]$_i$ is likely to be 10^{-7} M or lower, channel activity was present only at very depolarized pipette potentials of $+80$ mV or above (this represents a true membrane potential greater than or equal to $+30$ mV, assuming a resting V_m of -50 mV). Assuming that the calcium sensitivity of the channel in situ is similar to that of the excised patch, for channels to be open at negative membrane potentials [Ca]$_i$ must be $>3 \times 10^{-6}$ M. Thus, it is unlikely that these channels are open very often under normal physiological conditions.

In an excised patch with symmetrical KCl, the I-V relationship of the channel was linear (Fig. 3 *B*); channel conductance determined from the slope of the line was 250 pS. This value is similar to those reported for the large conductance Ca-activated K channel in other cells (Barrett et al., 1982; Guggino et al., 1987). The ionic selectivity

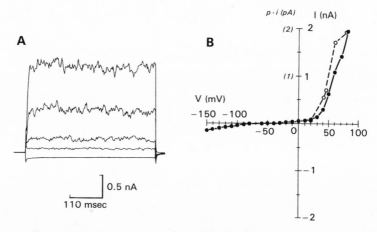

Figure 2. (*A*) Digitized whole-cell currents from a human macrophage grown in tissue culture for 10 d. Patch electrode contained 3×10^{-6} M Ca in KCl Hanks' solution. Cell capacitance was 83 pF. Holding potential (V_h) = the resting membrane potential (V_m) = -33 mV, defined as the zero current potential (ZCP). Voltage steps were to -140, 20, 40, 60, and 80 mV. (*B*) Peak I-V relation from cell shown in *A*. Dashed line represents mean single-channel current (p * I) obtained from excised inside-out patches with NaCl Hanks' solution in electrode and KCl Hanks' (3×10^{-6} M Ca) in bath. NaCl Hanks' solution consisted of 145 mM NaCl, 4.6 mM KCl, 1.1 mM MgCl$_2$, 1.6 mM CaCl$_2$, 10 mM HEPES, pH 7.3. KCl Hanks' solution contained 145 mM KCl, 10 mM NaCl, 1.1 mM MgCl$_2$, 1.1 mM EGTA, 1.07 mM CaCl$_2$, 10 mM HEPES, pH 7.3. (Gallin and McKinney, 1988).

of the channel was determined by changing the ionic constituents of the bath and/or electrode, and indicated that the channel was permeable to K and relatively impermeable to Cl or Na (Gallin, 1984).

To compare the voltage dependence of the single-channel current with that of the whole-cell outward current, the mean single-channel I-V relationship (refer back to Fig. 2 *B*, dashed line) of excised inside-out patches was measured under ionic conditions chosen to be similar to those of the whole-cell recording configuration, that is, [K]$_{pipette}$ = 4.6 mM and [K]$_{bath}$ = 148 mM. Mean single-channel current ($p * i$) was determined by measuring the single-channel open probability (p) and the single-channel current amplitude (i) at various voltages. If the number of channels (n) is constant, the mean single-channel current amplitude will be proportional to the

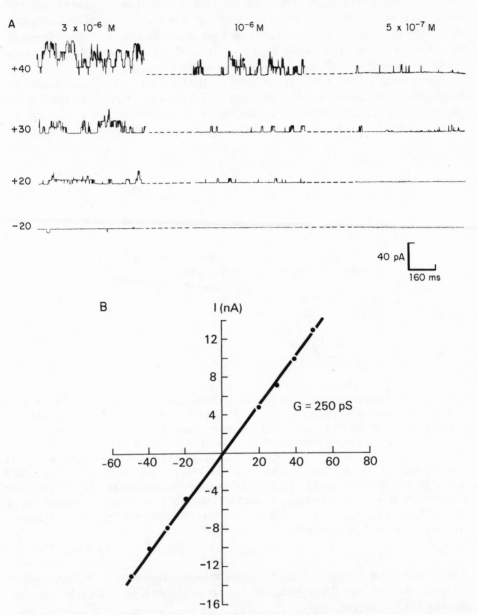

Figure 3. (*A*) Chart records of single-channel activity at four holding potentials from an excised inside-out patch from a human macrophage cultured for 14 d. Patch was exposed to symmetrical KCl Hanks' solution. Bath contained varying concentrations of $[Ca]_i$, as indicated. Patch contained a least six channels. In this and subsequent figures, upward deflections represent outward current. Dashed line represents closed current level. (*B*) I-V relationship of channels shown in *A*. (Gallin and McKinney, 1988).

whole-cell current amplitude (I) since $I = n * p * i$. It is evident from the two curves in Fig. 2 B that the single-channel and whole-cell currents have similar voltage dependencies, supporting the view that the single-channel Ca-activated K current fluctuations underlie the whole-cell outward current. From the relationship of the whole-cell current to the mean single-channel current we estimate the number of channels in the cell depicted in Fig. 1 to be 660.

Pharmacology. To show additional correspondence between single-channel and whole-cell Ca-activated K currents, we tested the action of two pharmacological agents, tetraethylammonium (TEA) and charybdotoxin, on the outward currents. Each is known to block Ca-activated K currents in other cells (Miller et al., 1985; Guggino et al., 1987). In the macrophage, addition of charybdotoxin ($\sim$25 nM) to the outside of the membrane blocked both whole-cell and single-channel currents (Gallin and McKinney, 1988). Addition of charybdotoxin to the inner surface of the membrane had no effect. Externally applied TEACl (10–15 mM) also blocked both the whole-cell and single-channel Ca-activated K current.

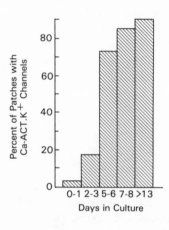

Figure 4. Relationship between time in culture and presence of large conductance Ca-activated K channels in human macrophages. Data are from excised inside-out patches in symmetrical KCl with 3×10^{-6} M Ca in bath. Number of patches in each group was 34, 16, 15, 22, and 49 (left to right). (Gallin and McKinney, 1988.)

Channel expression and functional relevance. Human peripheral blood monocytes differentiate into macrophages during the first week in tissue culture. This differentiation involves an increase in size, as well as changes in surface antigens (Zwadlo et al., 1985), enzyme content (Nichols et al., 1971), the extent of the oxidative burst, and phagocytosis (Nakagawara et al., 1981). So far, the data we have presented on the Ca-activated K conductance in the macrophage have been from cells that were cultured for 7 d or longer. To determine whether the expression of the Ca-activated K conductance changes during differentiation of the monocyte into a macrophage, monocytes were cultured for varying periods of time and the percent of patches containing Ca-activated K channels was measured. Fig. 4 shows a plot of the percentage of patches containing Ca-activated K channels versus time in culture. Only 1 of 34 patches from cells cultured for $\leq$24 h contained Ca-activated K channels, while 87% of the patches from cells cultured for >13 d did. Thus, the density of this channel in the membrane of monocytes changes in vitro as the cells are maturing.

Since the expression of other K conductances in the macrophage can be influenced by whether cells are grown under suspended or adherent conditions (Gallin and Sheehy, 1985), we also did experiments to determine whether the expression of the

Ca-activated K channel is affected by adherence. Cells grown in teflon jars (nonadherent conditions) for 2 wk and then plated for as little as several minutes had Ca-activated K channels, indicating that adherent growth conditions are not required for channel expression (Gallin and McKinney, 1988).

The functional role of the Ca-activated K channel is unknown. Freshly isolated monocytes have no Ca-activated K channels, yet are capable of both phagocytosis and chemotaxis. It is possible that activation of this conductance underlies the spontaneous membrane hyperpolarizations previously described. However, several observations make this seem unlikely. First, the voltage and calcium sensitivities of this channel (assuming that they are similar in situ and in the excised patch) are such that it will be closed at negative potentials unless $[Ca]_i$ is $> 3 \times 10^{-6}$ M. While large increases in $[Ca]_i$ may occur during penetration with intracellular microelectrodes, we have no evidence that such large increases occur in response to physiological stimuli. Second, we have noted oscillations in holding current in cell-attached patches where microelectrode damage is not present. Bursts of single-channel current were superimposed on these oscillations that were not due to the activity of the large conductance Ca-activated K channel, but instead were due to the activity of a distinctly different species of K channel (see the section on the bursting inward-rectifying K channel). Finally, exposure of cells to the calcium ionophore ionomycin during cell-attached patch recordings did not significantly increase the activity of the large conductance K channel ($V_m = -50$ mV), but it did increase the activity of a smaller conductance inward-rectifying K channel (see next section). Therefore, although the large conductance Ca-activated K channel will hyperpolarize the macrophage when $[Ca]_i$ is extremely high, it is not necessarily the primary channel involved in repetitive hyperpolarization.

Inwardly Rectifying K Conductances

Whole-cell current. An inwardly rectifying potassium current that activates at voltages negative to -60 mV was first described in mouse spleen and thioglycolate-induced macrophages cultured for several weeks (Gallin, 1981; Gallin and Livengood, 1981). This current is also present in cultured human peripheral blood monocytes (Gallin and McKinney, 1988), and in the mouse-derived macrophage-like cell line J774.1, where it has been characterized in some detail (Gallin and Sheehy, 1985; McKinney and Gallin, 1988). Fig. 5 shows whole-cell records obtained under three different conditions that illustrate the major properties of the inward current. Fig. 5 *A* shows currents recorded from a J774 cell bathed in normal Hanks' solution. Inward currents activated at potentials more negative than -70 mV and exhibited time-dependent inactivation for steps more negative than -120 mV. Fig. 5 B shows currents recorded from cells bathed in a Hanks' solution where all Na was replaced by K. Currents increased in size and showed less time-dependent inactivation. As is true for the inward-rectifying conductances of other tissues (Hagiwara and Takahashi, 1974; Leech and Stanfield, 1981), whole-cell conductance was proportional to the square root of external potassium concentration (in 4.5 and 145 mM $[K]_0$ the average conductance was 3.4 nS vs. 24 nS, respectively). Gating was also dependent on $[K]_0$. Activation of the current shifts to more positive potentials as $[K]_0$ increases (Gallin and Sheehy, 1985). Note that while time-dependent inactivation was reduced in Na-free media, it was not completely eliminated. The same amount of inactivation was observed when Na in the bath was replaced by *N*-methylglucamine (NMG) instead of K (Fig. 5 *C*), which

A

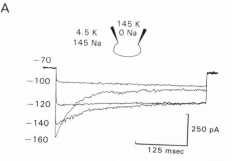

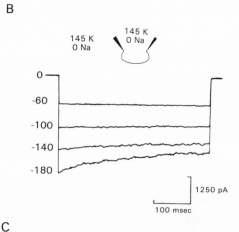

B

Figure 5. (*A*) Digitized whole-cell records from a J774 cell bathed in NaCl Hanks' solution. Inward currents were recorded in response to hyperpolarizing voltage steps to the potentials shown in the figure. $V_h = -70$ mV. (*B*) Whole-cell records from a cell bathed in Hanks' solution with all Na replaced by K. Average resting V_m of cells bathed in this solution was near 0 mV, so $V_h = 0$ mV. (*C*) Whole-cell records from a cell bathed in Hanks' solution with all Na replaced by NMG. Average resting V_m of cells in NMG Hanks' solution was -76 mV. $V_h = -80$ mV. Records in *A* and *B* have been corrected for leak and capacity current. Records in *C* were not corrected; leak conductance was $\sim 10\%$ of total.

C

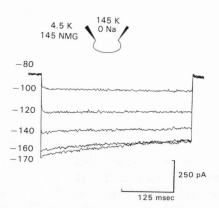

indicates that inactivation was not current dependent. Barium, a classic blocker of inward-rectifying K currents (Hagiwara et al., 1978; Standen and Stanfield, 1978) produced both concentration- and voltage-dependent inhibition of inward current when applied to the bath (data not shown).

Single-channel current. Inward-rectifying single-channel current fluctuations have been recorded in both human macrophages (Gallin and McKinney, 1988) and J774 cells (McKinney and Gallin, 1988). Fig. 6 shows single-channel currents recorded in the cell-attached patch mode from a J774 cell. Single-channel current was activated over the same voltage range as the macroscopic inward-rectifying current. Inward single-channel currents increased in amplitude as the patch was hyperpolarized, and did not reverse above E_K. Single-channel conductance in cell-attached patches with 145 mM K in the patch electrode was 30 pS and was proportional to the square root of $[K]_o$ (see Fig. 7, A, B, and C). E_{rev} was proportional to the log of $[K]_o$ with a slope of 63 mV per 10-fold change in $[K]_o$ (Fig. 7 D). Absolute values of the extrapolated E_{rev} are somewhat below the corresponding values of E_K across the patch membrane, probably due to the fact that the channel rectifies near E_K. More accurate E_{rev} values obtained using ramp stimuli were very close to E_K. Inward single-channel currents were blocked by the addition of barium to the external surface of the membrane (McKinney and Gallin, 1988). Neither macroscopic nor single-channel currents were affected by the replacement of external chloride by isethionate.

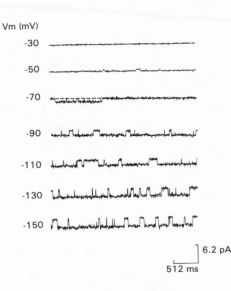

Vm (mV)

-30

-50

-70

-90

-110

-130

-150

6.2 pA

512 ms

Figure 6. Cell-attached patch recording of single-channel currents at various holding potentials from a J774 cell bathed in Na Hanks' solution. V_m across the patch is given as pipette potential (V_p) + resting V_m (approximately -70 mV). Dashed line represents closed current level. Records were filtered at 50 Hz (upper two traces) or 200 Hz and digitized at $>2\times$ the filtering frequency. Pipette contained K Hanks' solution consisting of (in mM) 145 KCl, 10 NaCl, 1 MgCl$_2$, 1.1 EGTA, 0.1 CaCl$_2$, and 10 HEPES, pH 7.3. Free Ca^{++} in this solution was $<10^{-8}$ M.

In Fig. 5, we show that inward-rectifying whole-cell currents inactivate at very negative potentials, even in the absence of Na. Fig. 8 shows that inactivation is a property of single-channel currents as well. In the cell-attached patch configuration, we repetitively applied hyperpolarizing pulses to the membrane, and observed single-channel activity that declined with time during the pulse. These traces were summed and averaged, and produced a current record that showed clear evidence of inactivation.

Consistent with the above finding is the observation that the open state probability of inward-rectifying channels is voltage dependent. Single-channel open probability was measured in eight patches containing only one channel and from two patches containing two channels. As the potential across the patch became more negative, channel closures were longer (up to tens of seconds). Open probability declined from an average of 0.87 at -40 mV to 0.43 at -120 mV.

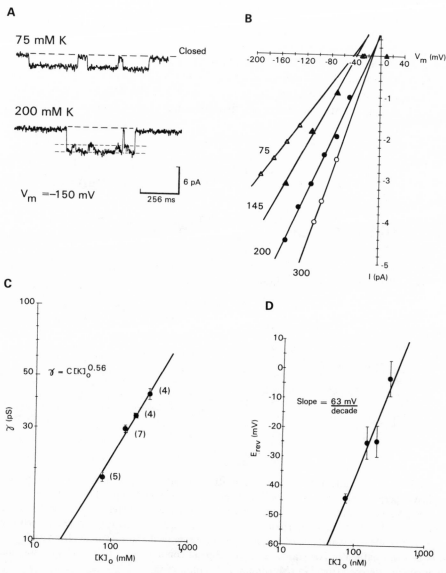

Figure 7. (*A*) Single-channel records from two cell-attached patches with different [K] in the pipette. Note subconductance level in the lower record. (*B*) Single-channel I-V relationships from four cells, each bathed in Na Hanks' solution, but with different [K] in the pipette; 75 ($\triangle$), 145 ($\blacktriangle$), 200 ($\bullet$), or 300 ($\circ$) mM. [NA] in the pipette was 75, 10, 10, and 10 mM, respectively. (*C*) Log-log plot of single-channel conductance vs. [K]$_o$. Number of experiments at each [K] is given adjacent to each point. Data were fit to the equation γ (single-channel conductance) = $C[K]_o^x$, where x was 0.56. Limiting conductance in 4.5 [K]$_o$ was 4.1 pS. (*D*) Extrapolated E_{rev} vs. log [K]$_o$, fitted by a straight line with a slope of 63 mV per 10-fold change in [K]$_o$ (McKinney and Gallin, 1988).

Thus far, we have shown that inward-rectifying whole-cell and single-channel currents correspond in the following ways: (*a*) voltage dependence of activation is the same, (*b*) each is K selective, (*c*) whole-cell and single-channel conductances are proportional to the square root of $[K]_o$, (*d*) each is blocked by similar concentrations of barium, (*e*) each shows voltage-dependent inactivation in the absence of sodium. We would like to conclude that the current through the 30-pS inward-rectifying channel can fully account for the behavior of the macroscopic inward-rectifying current, and in J774 cells, this may be the case. However, some recent observations made on human macrophages make this conclusion less firm. Another inward-rectifying K current, of

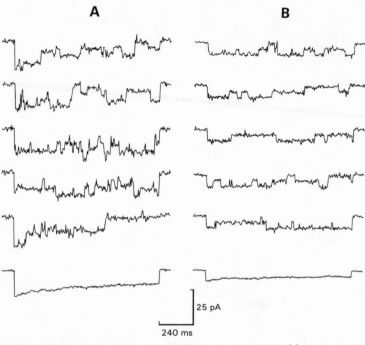

Figure 8. Single-channel records from a cell-attached patch elicited in response to repetitively applied hyperpolarizing steps. Patches were held at −30 mV and stepped to −190 (*A*) or −150 mV (*B*). At the bottom of each group of sweeps is the averaged current record for that voltage, comprising 40 or 48 sweeps applied at 5-s intervals. Leak and capacitative currents recorded in response to eight equal and opposite sized steps were averaged and subtracted from each record. J774 cell was bathed in Na Hanks' solution. Pipette contained Na-free K Hanks' solution to eliminate interference from Na (McKinney and Gallin, 1988).

slightly larger conductance, and possibly Ca activated, is present in human cells (see next section on bursting inward-rectifying K channel) and may account for some of the macroscopic inward current in these cells.

Expression and Functional Relevance. The only known function of the voltage-dependent inward-rectifying K conductance in macrophages is to help set the resting membrane potential of the cell near E_K. The resting membrane potential of adherent J774 cells in normal Hanks' solution is approximately −75 mV, a value in the same range as that obtained from measurements using the cationic lipophilic probe triphenylmethylphosphonium (−85 mV) (Sung et al., 1985). The dependence of channel gating on $[K]_o$ may be significant since the macrophage is present in sites of dead and dying tissue, and may well be exposed to elevated $[K]_o$. A rise in $[K]_o$ will increase the conductance, yet preserve a steep I-V relationship so that the cells remain sensitive to small current fluctuations.

The expression of inward-rectifying channels cells appears to be regulated; it is present in only one-third of human macrophages, which consequently have a slightly more negative resting membrane potential than cells that do not exhibit this conductance (-56 vs. -42 mV; Gallin and McKinney, 1988). In J774 cells, inward currents increase in size with time after adherence (Gallin and Sheehy, 1985), which may account for part of the difference between the resting potential in the suspended state vs. adherent state (-14 mV vs. -85 mV; Sung et al., 1985). However, the appearance of this conductance has not been correlated with the appearance of any specific function in J774.1 cells. Its presence is not necessary for J774 cells to be able to carry out chemotaxis or phagocytosis or to generate an oxidative burst; these functions still occur in the presence of 2 mM barium (our unpublished results). The involvement of the inward-rectifying channel in other macrophage functions, such as secretion or target cell lysis, has not been tested. The inward-rectifying conductance is absent in mouse peritoneal macrophages cultured for 4 d (Ypey and Clapham, 1984) but is present in peritoneal macrophages cultured for longer periods (Gallin and Livengood, 1981; Randriamampita and Trautmann, 1987).

Bursting Inward K Channel

Single-channel current. Exposure of human macrophages to the calcium ionophore ionomycin activates a second type of inwardly rectifying K channel (Gallin, 1988). Fig. 9 A shows current tracings from a cell-attached patch obtained from a human macrophage before and after exposing the cell to 10^{-6} M ionomycin, which was locally applied by pressure injection from a patch pipette. In the absence of ionomycin, only one type of channel was evident, the ~30-pS inwardly rectifying channel described in the preceding section. After exposure to ionomycin, a second type of inward current appeared that was slightly larger in amplitude and showed bursting behavior. The current showed inward rectification; outward currents of similar amplitude were not observed when the patch was depolarized. At very positive potentials ($+130$ mV), small outward current fluctuations (~0.5 nAmp) were present which may have been due to outward current through this channel. Ionomycin induced this channel in 9 of 10 cells tested.

Fig. 9 B shows the I-V curves of both the larger bursting and the smaller inward-rectifying channels obtained in the presence of ionomycin. Conductances were 28 pS and 36 pS, respectively. The reversal potentials of the two channels were similar ($V_{pipette} = 49$ and 54 mV). Assuming that the resting V_m was 50 mV, the potential across the patch was actually near zero, since $V_m = V_{pipette} +$ resting V_m. Since E_K was also near 0 mV, these data indicate that this channel is primarily K selective. Removal of Cl from that patch pipette had no effect on conductance or E_{rev}. We conclude that the 36-pS inward-rectifying K channel is distinct from the 30-pS inward-rectifying channel, since it has a different conductance, shows bursting kinetics and is found in patches with or without the 30-pS channel.

A similar conductance channel recently has been described in human macrophages (McCann et al., 1987) and mouse peritoneal macrophages (Somogyi et al., in press) in the absence of ionomycin. The channel in mouse peritoneal cells showed inward rectification. However, McCann et al. (1987) report that the 36-pS channel in human macrophages did not rectify. We occasionally saw activity of the 36-pS channel in patches from cells not exposed to ionomycin but in our studies the channel always rectified. In these instances, channel activity was present immediately after obtaining a tight seal, and then frequently subsided. Unlike the large conductance Ca-activated K channel, this channel appears to be present in human monocytes that have been cultured for as little as 1 d. We are currently characterizing its calcium and voltage sensitivities and its pharmacology. We speculate that this channel, and not the large conductance Ca-activated K channel, is responsible for the repetitive membrane hyperpolarizations previously described, since (*a*) the bursting pattern of this channel is similar to the pattern of the spontaneous hyperpolarizations, and (*b*) ionomycin induces both channel activity and membrane hyperpolarizations (Gallin et al., 1975). A similar Ca-activated inward-rectifying K channel of slightly larger conductance (50 pS in 200 mM [K]$_o$) has recently been described in HeLa cells (Sauve et al., 1986). In

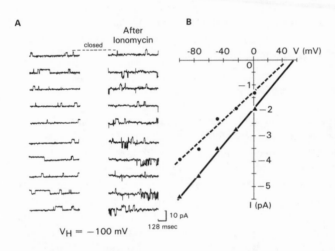

Figure 9. (*A*) Single-channel currents recorded from a cell-attached patch from a human macrophage before and after applying 10^{-6} M ionomycin from a second electrode. $V_h = -100$ mV. Data were filtered at 500 Hz and digitized at 1 kHz. Patch electrode contained KCl Hanks' solution; bath contained NaCl Hanks' solution. (*B*) I-V relationship of each of the channels present in the patch following exposure to ionomycin.

these cells the addition of histamine, which increases [Ca]$_i$, initiates oscillatory single-channel activity and membrane hyperpolarization (Sauve et al., 1987).

Outwardly Rectifying K Conductance

An outwardly rectifying K conductance has been described at the whole-cell current level in resident mouse peritoneal macrophages (Ypey and Claphan, 1984), cultured human monocytes (Nelson et al., 1986), and two macrophage cell lines, J774.1 (Gallin and Sheehy, 1985) and P388D1 (Sheridan and Bayer, 1986). This conductance activates at potentials positive to -50 mV, inactivates over a time course of seconds, and is blocked by 4-aminopyridine (Ypey and Clapham, 1984; Gallin and Sheehy, 1985). As is the case with the smaller inwardly rectifying K conductance, it appears to be variably expressed with time in culture. Ypey and Clapham (1984), using resident mouse peritoneal macrophages, reported that this conductance was absent during the first day after isolation but was present in 96% of the cells cultured for 1–4 d. It was not found in long-term (2–6 wk) cultured mouse peritoneal (Gallin and Livengood, 1981) and spleen macrophages (Gallin, 1981). These findings have recently been confirmed by Randriamampita and Trautmann (1987) who report that in mouse peritoneal macrophages this conductance decreases with time in culture. In J774 cells, outward current is present 1–8 h after adherence, and is not present in long-term adherent cultures where the inward-rectifying K current is predominant (Gallin and Sheehy, 1985). The current appears to be continuously present in the mouse-derived cell line P388D1 (Sheridan and Bayer, 1986). An example of outward currents recorded from J774 cells is shown in Fig. 10.

The functional relevance of the outwardly rectifying K channel in the macrophage has not yet been determined. Similar currents have been described in detail in T lymphocytes (Cahalan et al., 1985). In T lymphocytes and natural killer cells, agents that block this current inhibit mitogenesis and cytotoxicity, respectively, which implies that it plays at least a permissive role in these processes (Chandy et al., 1984; Deutsch et al., 1986). Nelson et al. (1986) have reported that this conductance is inhibited by phorbol myristate acetate (PMA) in cultured human monocytes.

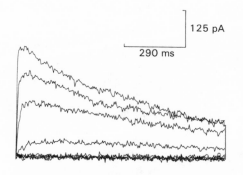

125 pA

290 ms

Figure 10. Digitized whole-cell records from a J774 cell plated for ~6 h and bathed in Na Hanks' solution. V_h = resting V_m = −66 mV. Depolarizing steps were applied in 10-mV increments from −60 to 0 mV.

Conclusion

To date, four different K conductances have been described in macrophages. They include a Ca-activated K conductance described at both the whole-cell and single-channel level in human macrophages that is not present in freshly isolated monocytes but is expressed during the first week in tissue culture. The absence of this conductance in monocytes implies that it is not required for phagocytosis, chemotaxis, or other functions of the immature macrophage. A second K conductance, described at both the whole-cell and single-channel level, is a voltage-dependent inward-rectifying K conductance that is present in a variety of macrophages, including cultured human peripheral blood monocytes, cultured mouse spleen and peritoneal macrophages, and J774 cells. It is important in setting the resting membrane potential of the cell, and in J774 cells is not necessary for chemotaxis or phagocytosis to occur. In addition to these two well-characterized K conductances, there are two other K conductances that have only been partially characterized. These are a second inward-rectifying K channel that may be activated by internal calcium, and an inactivating outward K current similar to that described in the T cell.

Acknowledgments

This work was supported by the Armed Forces Radiobiology Research Institute, Defense Nuclear Agency, under work unit 00020. Views presented in this paper are those of the authors; no endorsement by the Defense Nuclear Agency has been given or should be inferred.

References

Barrett, J. N., K. L. Magleby, and B. S. Pallotta. 1982. Properties of single calcium-activated potassium channels in cultured rat muscle. *Journal of Physiology.* 331:211–230.

Cahalan, M. D., K. G. Chandy, T. E. DeCoursey, and S. A. Gupta. 1985. A voltage-gated potassium channel in human T-lymphocytes. *Journal of Physiology.* 357:197–238.

Chandy, K. G., T. E. DeCoursey, and M. D. Cahalan. 1984. Voltage-gated K channels are required for human T lymphocyte activation. *Journal of Experimental Medicine.* 160:369–385.

Deutsch, C., D. Krause, and S. C. Lee. 1986. Voltage-gated potassium conductance in human T lymphocytes stimulated with phorbol ester. *Journal of Physiology.* 372:405–423.

Dos Reis, G. A., and G. M. Oliveiro-Castro. 1977. Potassium dependent slow membrane hyperpolarizations in mice macrophages. *Biochimica et Biophysica Acta.* 469:257–263.

Eckert, R., and P. Brehm. 1979. Ionic mechanisms of excitation in Paramecium. *Annual Review of Biophysics and Bioengineering.* 8:353–383.

Gadsby, D. C., and P. F. Cranefield. 1977. Two levels of resting potential in cardiac Purkinje fibers. *Journal of General Physiology.* 70:725–746.

Gallin, E. K. 1981. Voltage clamp studies in macrophages from mouse spleen cultures. *Science.* 214:458–460.

Gallin, E. K. 1984. Calcium and voltage-activated potassium channels in human macrophages. *Biophysical Journal.* 46:821–825.

Gallin, E. K. 1988. Inwardly rectifying K^+ channels induced by Ionomycin in cultured human macrophages. *Biophysical Journal.* 53:548a. (Abstr.)

Gallin, E. K., and J. I. Gallin. 1977. Interaction of chemotactic factors with human macrophages: Electrophysiology of the response. *Journal of Cell Biology.* 75:277–289.

Gallin, E. K., and D. R. Livengood. 1981. Inward rectification in mouse macrophages: evidence for a negative slope resistance region. *American Journal of Physiology.* 241:C9–C17.

Gallin, E. K., and L. C. McKinney. 1988. Patch clamp studies in human macrophages: Single-channel and whole-cell characterization of two K conductances. *Journal of Membrane Biology.* 103:55–66.

Gallin, E. K., and L. C. McKinney. 1989. Ionic conductances and transport mechanisms in phagocytic leukocytes. *In The Cellular Biochemistry and Physiology of the Neutrophil.* M. Hallet, editor. CRC Press, Inc., Boca Raton, FL. In press.

Gallin, E. K., and P. A. Sheehy. 1985. Differential expression of inward and outward potassium currents in the macrophage-like cell line J774.1. *Journal of Physiology.* 369:475–499.

Gallin, E. K., M. Wiederhold, P. Lipsky, and A. Rosenthal. 1975. Spontaneous and induced membrane hyperpolarizations in macrophages. *Journal of Cellular Physiology.* 86:653–662.

Guggino, S. E., W. B. Guggino, N. Green, and B. Sacktor. 1987. Blocking agents of Ca-activated K channels in cultured medullary thick ascending limb cells. *American Journal of Physiology.* 252:C128–C137.

Hagiwara, S., S. Miyaszki, W. Moody, and J. Patlak. 1978. Blocking effects of barium and hydrogen ions on the potassium current during anomalous rectification in the starfish egg. *Journal of Physiology.* 279:167–185.

Hagiwara, S., and K. Takahashi. 1974. The anomalous rectification and cation selectivity of the membrane of a starfish egg cell. *Journal of Membrane Biology*. 18:61–80.

Ince, C., P. C. Leijh, M. Meijer, E. VanBavel, and D. Ypey. 1984. Oscillatory hyperpolarizations and resting membrane potentials of mouse fibroblast and macrophage cell lines. *Journal of Physiology*. 352:625–635.

Kruskal, B. J., and F. R. Maxfeld. 1987. Cytosolic free calcium increases before and oscillates during frustrated phagocytosis in macrophages. *Journal of Cell Biology*. 105:2685–2693.

Leech, C. A., and P. R. Stanfield. 1981. Inward rectification in frog skeletal muscle and its dependence on membrane potential and external potassium. *Journal of Physiology*. 319:295–309.

Lindau, M., and J. Fernandez. 1986. A patch clamp study of histamine secreting cells. *Journal of Physiology*. 88:349–368.

Lipton, S. A. 1986. Antibody activates cationic channels via second messenger Ca^{2+}. *Biochimica et Biophysica Acta* 856:59–67.

McCann, F., T. Keller, and P. Guyre. 1987. Ion channels in human macrophages compared with the U-937 cell line. *Journal of Membrane Biology*. 96:57–64.

McKinney, L. C., and E. K. Gallin. 1988. Inwardly rectifying whole-cell and single channel K currents in the murine macrophage cell line J774. *Journal of Membrane Biology*. 103:41–53.

Miller, C., E. Moczydlowski, R. Latorre, and M. Phillips. 1985. Charybdotoxin, a protein inhibitor of single channel Ca-activated K channels from mammalian skeletal muscle. *Nature*. 313:316-318.

Nakagawara, A., C. Nathan, and Z. Cohn. 1981. Hydrogen peroxide metabolism in human monocytes during differentiation in vitro. *Journal of Clinical Investigation*. 68:1243–1252.

Nelson, D. J., L. Rufer, T. Nakayama, and J. M. Zeller. 1986. Phorbol ester block of voltage-dependent K current in monocyte-derived macrophages. *Biophysical Journal*. 49:164a. (Abstr.)

Nichols, B., D. F. Bainton, and M. G. Farqhuar. 1971. Differentiation of monocytes. Origin nature and fate of azurophilic granules. *Journal of Cell Biology*. 50:498–510.

Persechini, P. M., E. G. Araujo, and G. M. Oliveiro-Castro. 1981. Electrophysiology of phagocytic membranes: induction of slow membrane hyperpolarizations in macrophages and macrophage polykaryons by intracellular calcium injections. *Journal of Membrane Biology*. 61:81–90.

Peterson, O. H., and Y. Maruyama. 1984. Calcium-activated potassium channels and their role in secretion. *Nature*. 307:693–696.

Pfefferkorn, L. 1984. Transmembrane signaling: an ion flux-independent model for signal transduction by complexed Fc receptor. *Journal of Cell Biology*. 99:2231–2240.

Randriamampita, C., and A. Trautmann. 1987. Ionic channels in murine macrophages. *Journal of Cell Biology*. 105:761–769.

Sauve, R., C. Simoneu, R. Monette, and G. Roy. 1986. Single-channel analysis of the potassium permeability in HeLa cancer cells: evidence for a calcium-activated potassium channel of small unitary conductance. *Journal of Membrane Biology*. 92:269–282.

Sauve, R., C. Simoneau, L. Parent, R. Monette and G. Roy. 1987. Oscillatory activation of calcium-dependent potassium channels in HeLa cells induced by histamine H1 receptor stimulation. A single channel study. *Journal of Membrane Biology*. 96:199–208.

Schwarze, W., and H. A. Kolb. 1984. Voltage-dependent kinetics of an ionic channel of large unit conductance in macrophages and myotube membranes. *Pflügers Archiv European Journal of Physiology.* 402:281–291.

Sheridan, R. E. and B. M. Bayer. 1986. Ionic membrane currents induced in macrophages during cytolysis. *Federation Proceedings.* 45:1009a. (Abstr.)

Somogyi, R., J. Ubl, and H. A. Kolb. 1987. Zymosan and potassium induced activation of single voltage-dependent ion channels in membranes of peritoneal macrophages. *Advances in the Biosciences.* 66:275–284.

Standen, N. B., and P. R. Stanfield. 1978. A potential and time-dependent blockade of inward rectification in frog skeletal muscle fibres by barium and strontium ions. *Journal of Physiology.* 820:169–191.

Sung, S. S., J. D. Young, A. M. Origlio, J. M. Heiple, H. R. Kaback, and S. C. Silverstein. 1985. Extracellular ATP perturbs transmembrane ion fluxes, elevates cytosolic [Ca] and inhibits phagocytosis in mouse macrophages. *Journal of Biological Chemistry.* 260:13442–13449.

Van Furth, R. 1985. Mononuclear Phagocytes. Characteristics, Physiology and Function. Martinus Nihoff, Boston-Dordrecht-Lancaster.

Young, D. J., J. Unkeless, T. Young, A. Mauro, and Z. Cohn. 1983. Role of mouse macrophage IgG Fc receptor as ligand-dependent ion channel. *Nature.* 306:186–189.

Ypey, D. L., and D. E. Clapham. 1984. Development of a delayed outward rectifying K conductance in cultured mouse peritoneal macrophages. *Proceedings of the National Academy of Sciences.* 81:3083–3087.

Zwadlo, G., E. Brocker, D. von Bassewitz, U. Feige, and C. Sorg. 1985. Monoclonal antibody to a differentiation antigen present in mature human macrophages and absent from monocytes. *Journal of Immunology.* 134:1487–1492.

Chapter 27

The Events Leading to Secretory Granule Fusion

Guillermo Alvarez de Toledo and Julio M. Fernandez

*Department of Physiology, University of Pennsylvania,
Philadelphia, Pennsylvania*

Cell Physiology of Blood © 1988 by The Rockefeller University Press

Exocytosis, Endocytosis, and Membrane Capacitance

Exocytosis and endocytosis are crucial for the action of blood cells such as macrophages, neutrophils, basophils, and lymphocytes; however, these membrane recycling phenomena have resisted mechanistic explanations and their description has remained largely in the domain of electron microscopy.

Electron micrographs have shown that during exocytosis, secretory granules fuse with the plasma membrane, which causes an increase in the cell surface area. Since biological membranes have a fairly constant membrane capacitance of 1 $\mu F/cm^2$, measurements of the cell membrane capacitance have been used to study exocytosis (Cole, 1968; Jaffe et al., 1978; Gillespie, 1979; Neher and Marty, 1982; Fernandez et al. 1984; Peres and Bernardini, 1985; Breckenridge and Almers, 1987a, b; Zimmerberg et al., 1987). The most sensitive method uses a patch clamp amplifier (Hamill et al., 1981) in conjunction with a phase-sensitive detector (Neher and Marty, 1982; Joshi and Fernandez, 1988). This combination makes it possible to study the changes in the membrane area of a single cell, due to the fusion of individual secretory granules.

Fusion of individual secretory granules with the plasma membrane has been observed in chromaffin cells (Neher and Marty, 1982), rat peritoneal mast cells (Fernandez et al., 1984), and more recently in mast cells of mutant mice (Breckenridge and Almers, 1987a, b; Zimmerberg et al., 1987). These measurements have already provided a wealth of new information that questions widely accepted ideas such as the universal role of calcium in exocytosis (Fernandez et al., 1984) and the role of osmotic pressure and swelling (Breckenridge and Almers, 1987a; Zimmerberg et al., 1987). Capacitance measurements also revealed that secretory granules could fuse transiently with the plasma membrane ("flicker;" Fernandez et al., 1984; Breckenridge and Almers, 1987a, b) several times before undergoing complete exocytosis.

In this paper we use mouse peritoneal mast cells as a model secretory cell to explore in detail the mechanisms responsible for the "flicker" phenomenon, and the implications of these phenomena in defining the steps involved in the fusion of secretory granules with the plasma membrane.

Methods

Mast cells were obtained from the peritoneal cavity of adult beige mice or normal mice (Jackson Laboratories, Bar Harbor, ME) following a procedure similar to that used by Lindau and Fernandez (1986). In all the experiments shown here we used mast cells from beige mice, except in the experiments shown in Fig. 6, where we used normal mice. The cells were plated on glass coverslips and incubated at 37°C in a medium containing our standard extracellular saline plus 45 mM $NaHCO_3$ in a 5% CO_2 atmosphere for at least a half-hour before use.

All experiments were done using the whole cell mode of the patch clamp technique (Hamill et al., 1981). The extracellular saline contained 150 mM NaCl, 10 mM HEPES, 2.8 mM KOH, 1.5 mM NaOH, 1 mM $MgCl_2$, and 2 mM $CaCl_2$. The pipette solution contained: 140 mM K-glutamate, 10 mM KCl, 10 mM HEPES, 3.5 mM NaOH, 200 μM EGTA, 200 μM ATP, 7 mM $MgCl_2$, and typically, 20 μM GTPγS (Sigma Chemical Co., St. Louis, MO) to stimulate exocytosis (Fernandez et al., 1984). The patch clamp amplifier (EPC-7, List Electronics) was controlled by a PDP11/73 digital computer.

Detection of capacitive currents was done with a software-based phase detector operating at a frequency of $f = 833$ Hz. The phase detector has been described elsewhere (Joshi and Fernandez, 1988). After breaking into a cell (whole cell mode), the phase detector was adjusted such that one output reflected the real part of the changes in the input admittance of the cell, $Re(\Delta Y)$, and the second output reflected the imaginary part of the change in admittance, $Im(\Delta Y)$, only.

Frequently, the $Im(\Delta Y)$ trace will be called the capacitance trace, because this output was proportional to the cell capacitance and thus measured the cell membrane area. The calibration of the software based phase detector, as well as its theory of operation, has been described in detail by Joshi and Fernandez (1988).

Secretory Granule Fusion and the "Flicker" Phenomenon

Initial capacitance experiments produced an unexpected result. While stepwise increases in capacitance were observed as secretory granules fused with the plasma membrane, granule fusion measured this way did not appear to be irreversible, as sometimes we observed flicker: the transient formation of a water-filled channel between the plasma membrane and the secretory granule. A representation of a fusion event that accounts for the flicker phenomenon is shown in Fig. 1 *a*. We identify at

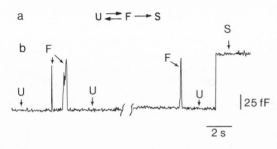

Figure 1. Fusion of secretory granules with the plasma membrane is described in three steps as shown in *a*: the unfused granule (U), the granule in the flicker state (F), and the irreversible secretory mode (S). (*b*) Recording of $Im(\Delta Y)$ showing the various stages during the fusion of a single secretory granule.

least three stages: the unfused granule (U), the granule in the flicker mode (F), and irreversibly fused granule associated with release of the granule content which we call call the secretory mode (S).

Normally the transition between U and S states is fast; however, in some cases the secretory granules remain for a relatively long time in the F state. The residence in the F state is evidenced by transient capacitance increases of an amplitude similar to that measured when the granule reaches the S state. A typical experiment displaying all three states is shown in Fig. 1 *b* where several transitions between the U and F states are observed before the final S state is reached.

These experiments define exocytosis as a two step process: the reversible formation of a water-filled pore between the plasma membrane and the secretory granule and a final, irreversible expansion of the pore causing complete fusion and secretion. Several questions arise: what are the molecular structures responsible for the formation of this pore (U to F transition)? What type of mechanism allows for a reversible pore (F to U and back)? And finally, what is the driving force for the final expansion of this pore (F to S state)? The molecular steps leading to the formation of this pore are likely to be electrically silent and we can only hypothesize their nature. Once the pore is formed, however, a measurable current will cross the secretory granule membrane and this can be followed with the phase-detector technique (Fernandez et al., 1984; Breckenridge and Almers, 1987*a*, *b*).

Phase-detector measurements typically monitor current at two phase angles ϕ and ϕ-90°. When the detector has been properly adjusted, current measured at ϕ-90° reflects the imaginary component of the change in the cell's input admittance, $Im(\Delta Y)$. The current measured at ϕ corresponds to the real part of the change in admittance, $Re(\Delta Y)$, only. Mast cells can be approximated by a simple RC equivalent circuit (Joshi and Fernandez, 1988); in this case, $Im(\Delta Y)$ measures changes in membrane capacitance only, and $Re(\Delta Y)$ measures resistive changes.

A striking feature of flicker is that the transient capacitance increases are always accompanied by transient changes in the resistive component of the current (see Fig. 2). Initially we thought that this indicated the transient opening of an ionic conductance; however, simultaneous measurements of steady-state current showed that during flicker no change in the DC current could be detected. Therefore, the transient

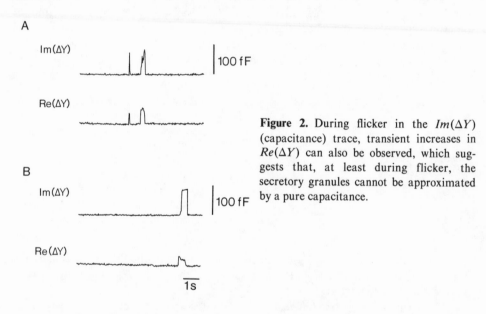

Figure 2. During flicker in the $Im(\Delta Y)$ (capacitance) trace, transient increases in $Re(\Delta Y)$ can also be observed, which suggests that, at least during flicker, the secretory granules cannot be approximated by a pure capacitance.

changes in conductance could not be due to the opening of ionic channels. Fig. 2 shows two typical examples of flicker in the $Im(\Delta Y)$ trace. Transient changes in $Re(\Delta Y)$ can be simultaneously observed.

The Equivalent Circuit for Granule Fusion

If the membrane capacitance was the only electrical parameter that changed during granule fusion (Fig. 3 *a*), then fusion of the granules with the plasma membrane would produce a change in the $Im(\Delta Y)$ component only. In contrast to this expectation, the results of Fig. 2 show that at least during flicker, secretory granules do not behave as pure capacitors, which indicates the need for a more complete equivalent circuit of granule fusion. Cole (1968) anticipated a more realistic equivalent circuit by assuming that secretory granules fuse by forming a connecting bridge that can be expected to have a resistance R_p. Accordingly, a modified equivalent circuit for secretory granule fusion is shown in Fig. 3 *b*. The boxed elements (R_p and C_g) correspond to the parameters associated with granule fusion.

As suggested by the equivalent circuit enclosed by the box in Fig 3 *b*, the voltage drop across the resistor and across the capacitor will depend on their relative values. If the impedance of the resistor is much higher than that of the capacitor ($R_p \gg 1/\omega C_g$), most of the voltage drop will occur across the pore resistance, and then the granule will appear to behave as a pure resistive element. Eventually, when R_p drops to zero due to the expansion of the pore (complete fusion), $R_p \ll 1/\omega C_p$, and then most of the voltage drop will occur across the capacitor; only then will the fusing granule appear as a pure capacitive element. The phase detector separates both elements by measuring the purely resistive [in $Re(\Delta Y)$ axis] or the purely capacitive [in the $Im(\Delta Y)$ axis] contributions. Since the value of C_g is fixed and is determined only by the granule membrane area, any changes in both the $Re(\Delta Y)$ and $Im(\Delta Y)$ projections are due

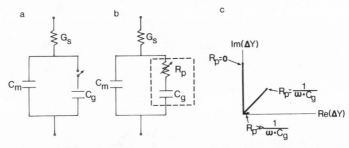

Figure 3. Simplified electrical equivalent circuit diagrams of the fusion of a secretory granule with the plasma membrane. G_s represents the access conductance into the cell through the patch pipette, C_m is the cell membrane capacitance and C_g is the granule membrane capacitance. The membrane conductance is assumed to be negligible. (*a*) Fusion is represented by the action of connecting the switch, which generates an instantaneous step increase in membrane capacitance; (*b*) shows a more realistic description where R_p represents the resistance of the pore connecting the secretory granule with the plasma membrane. Fusion is represented by a change in R_p, from infinity to zero. (*c*) Vector diagram showing that the projection of the granule admittance on the $Im(\Delta Y)$ and $Re(\Delta Y)$ axes depends on R_p. Only when $R_p = 0$ is the projection on the $Im(\Delta Y)$ axis only, and the granule behaves as a pure capacitance.

entirely to changes in R_p. Phase detector measurements like these are therefore able to follow the time course of the pore access resistance to the secretory granule during the various stages of fusion.

Quantitatively, the contributions of R_p and C_g can be determined by separating the real [$Re(\Delta Y)$] and imaginary [$Im(\Delta Y)$] parts of the admittance contributed by the secretory granule. During fusion the change in the cell's input admittance due to the equivalent circuit elements (R_p and C_g) representing granule fusion (Fig. 3 *b*) is

$$Re(\Delta Y) = \frac{\omega^2 c_g^2 R_p}{1 + \omega^2 c_g^2 R_p^2} \tag{1a}$$

$$Im(\Delta Y) = \frac{\omega c_g}{1 + \omega^2 c_g^2 R_p^2} \tag{1b}$$

with

$$\omega = 2\pi f$$

Simple inspection of these equations shows that for the unfused granule ($R_p = \infty$) $Re(\Delta Y)$ and $Im(\Delta Y)$ are both zero. When the pore forms and $\infty > R_p \gg 1/\omega C_g$, then only $Re(\Delta Y)$ will have an observable value. Finally, when the granule is completely fused ($R_p = 0$), then $Re(\Delta Y) = 0$ and $Im(\Delta Y) = \omega C_g$. The contribution of the secretory granule admittance, as R_p progresses from infinity to zero, is thus measured as a change in the total admittance of the cell reflected in the $Re(\Delta Y)$ and $Im(\Delta Y)$ projections, and can be represented vectorially as shown in Fig. 3 c.

These considerations provide an immediate explanation for the observed changes

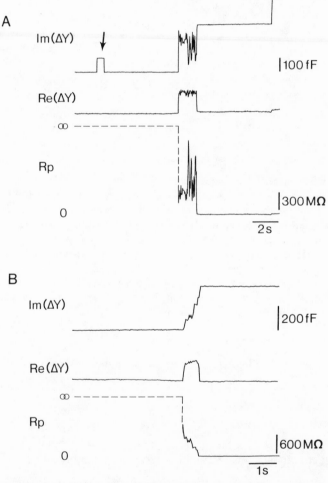

Figure 4. Reconstruction of the time course of R_p during secretory granule fusion events. R_p is calculated as $(1/\omega C_g) Re(\Delta Y)/Im(\Delta Y)$. (*A*) A calibration pulse (*arrow*) gives the scale (100 fF) for the $Im(\Delta Y)$ axis. The $Re(\Delta Y)$ axis has the same arbitrary scale. As shown, R_p fluctuates widely until its value drops to zero (F-to-S transition). (*B*) R_p calculated for another fusion event under similar conditions.

in $Re(\Delta Y)$ that are always associated with flicker. The flicker state (F) corresponds to a granule that has a connecting pore with a finite resistance R_p, whose value can fluctuate widely. The pore sometimes fails to expand and collapses, at which time R_p returns back to infinity. This process can occur several times before a complete fusion is observed (Fig. 1 *b*).

The Time Course of Fusion

Secretory granules can remain in the flicker state (F) for long periods of time before complete fusion (S state, Fig. 1 *a*). In this case a measurement of the time course of R_p can be obtained by analyzing both $Re(\Delta Y)$ and $Im(\Delta Y)$ components of the phase-detector output.

Fig. 4 shows two granule fusion events where the granules remained in the F state for a long time before completely fusing with the membrane. The upper and middle records in each part of the figure correspond to the two outputs of the phase detector measuring the imaginary $[Im(\Delta Y)]$ and real $[Re(\Delta Y)]$ components of the current. The lower trace corresponds to R_p calculated as $R_p = (1/\omega C_g) Re(\Delta Y)/Im(\Delta Y)$ (see Eq. 1a, b). As can be seen in Fig. 4 *A*, a long capacitance flicker (1.47 s) developed and finally produced a 270-fF step. During this time the conductance changed to a different value and fluctuated within a small range compared with the fluctuations in the capacitance record. R_p for this granule fusion event fluctuated from 280 to 1,276 $M\Omega$ and developed unpredictably until it decreased to zero. This erratic pattern was observed in several fusion events. Fig. 4 *B* illustrates a faster granule fusion where capacitance flicker was not as evident as in the previous example. In this case, the calculated time course for R_p followed a decreasing pattern (from 900 to 0 $M\Omega$), although erratic oscillations were also observed. In most cases however, the transition through the F state is too fast to be recorded with our present time resolution (33 Hz). An example of this is shown in the second fusion event of Fig. 4 *A*.

A Model for the Flicker State

Chandler and Heuser (1980), using quick-freezing and freeze-fracture techniques to study exocytosis in mast cells, showed that the fusion of a secretory granule began with the formation of a pore between the granule and the plasma membrane. They also showed that fusion pores could be formed even as the rest of the plasma membrane and granule membrane remained separated by as much as 0.1 μm.

Our measurements extend these observations as they are able to follow the pore resistance as a function of time. As proposed above, it is likely that the flicker state corresponds to the step where the granules are linked by this pore to the extracellular medium but have not yet undergone complete fusion. If this view is correct, it means that the pore structures observed by Chandler and Heuser are sometimes reversible and long lived. The question then is what are the mechanisms that allow the pore to change its resistivity over such wide ranges and what causes its final expansion?

Breckenridge and Almers (1987*b*), interpreting experiments similar to those presented here, proposed that the pore was formed by an aggregation of protein subunits in a ring-like structure enclosing a water-filled channel that, like the gap junction channel, bridges two membranes: the granule and plasma membranes (Fig.

5 A). Furthermore, they proposed that the addition of subunits would cause the pore to dilate while the kinetics of the aggregation/removal reaction for the subunits would cause the pore to fluctuate in diameter, causing R_p to fluctuate (Fig. 5 A). Presumably, the transition from the F to the S state would be caused by the addition of a large number of subunits, enlarging the pore dimensions.

If the pore was formed by a ring of protein subunits, it might be observable as an increased density of proteins or intramembranous particles (IMP) near the mouth of the pore. Satir et al. (1973) reported such ring-like formation of proteins near the mouth of fusing mucocysts in Tetrahymena. However, the experiments of Chandler

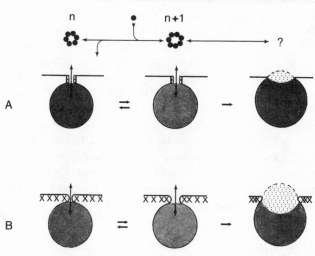

Figure 5. Two different explanations of the fluctuating values of R_p during the flicker state. (A) The pore is lined by protein subunits in a way similar to ionic channels. In the flicker state, the addition or removal of subunits causes the pore to change its size, changing the pore resistance. A final enlargement of the pore causes the transition from the fluctuating flicker state to the secretory state. (B) The pore is simply membrane lined. The surface tension driving its expansion is countered by a mesh of cytoskeletal elements. The balance between these forces will be associated with fluctuations in pore dimensions causing R_p to fluctuate. In this case, swelling of the granule core will cause the irreversible expansion of the pore, which will drive the granule to the secretory state.

and Heuser (1980) in mast cells demonstrated that the density of IMPs was not different in the mouth of the pore than in the rest of the plasma membrane.

Alternatively, after formation of the pore by an unknown structure the pore could then develop as a purely membrane-lined pore (Fig. 5 B). We picture such pores as having a structure similar to that of the sarcolemmal bridges that form between a cell and a patch-pipette when, after formation of a gigaseal, the patch pipette is withdrawn from the cell surface (i.e., see Fig. 10 d in Hamill et al., 1981). Upon formation of the water-filled pore, a large surface tension will be generated and will tend to rapidly expand the pore. Its expansion could be resisted by the mesh of cytoskeletal elements that underly the plasma membrane (marked as **X** in Fig. 5 *B*). This equilibrium of forces, slightly unbalanced by the thermal motion of the granule (or perhaps other forces), could cause the long-lived, fluctuating pores that are described in Figs. 1 *b*, 2, and 4. In contrast to the subunit aggregation model (see above) the final expansion of the pore (F-to-S transition) would be driven by the swelling of the granule core (see below).

What Is the Driving Force of Fusion?

It has been long recognized that secretory materials are stored in the granules in a highly condensed form. Upon fusion of the granule, extensive swelling of the granule content takes place in a manner not unlike that of ion exchange resins (Tam and Verdugo, 1981; Verdugo, 1984). A good example is the secretion of mucus by goblet cells. Highly condensed mucins undergo a large post-exocytotic volume expansion that is regulated by a Donnan exchange of water and ions between the mucin network and the extracellular medium and perhaps involving a polymer gel phase transition of the secretory product (Verdugo, 1986; Verdugo et al., 1987). Similarly, in mast cells a condensed network of heparin and histamine undergoes a volumetric expansion after exocytosis (Breckenridge and Almers, 1987*a*; Zimmerberg et al., 1987). These observations may provide an explanation for the transition between the F state and the S state. If the granule remains in the F state long enough, a limited exchange of water and ions will take place between the granule core and the extracellular medium, which causes a localized swelling of the granule near the pore. The localized swelling that begins after the fusion pore is formed (Chandler and Heuser, 1980) may be the force that will expand the pore (Zimmerberg et al., 1987) against the cytoskeletal elements that constrain it, causing the irreversible transition from the F to the S state. In other words, the cause of irreversible fusion would not simply be the formation of a water-filled pore between the extracellular medium and the granule, but rather the swelling of the granule core after enough water and ions have entered through the fusion pore.

In the protein subunit model (Satir et al, 1973; Breckenridge and Almers, 1987*a,b*) it is difficult to see how the force generated by the swelling of the granule core could participate in the expansion of a pore, driven only by the aggregation of protein subunits.

Membrane "Pinch-Off" Events

Several types of membrane retrieval mechanisms have been described, such as coated pit endocytosis and pinocytosis. Other phenomena, such as bleb formation and extrusion, also result in a net loss of plasma membrane. All of these events share in common a step in which a fraction of the plasma membrane "pinches off" and becomes separated. The final stages of these processes typically involve a narrowing of the connection between the membrane being separated and the plasma membrane. This narrow connection appears similar to the fusion pore discussed in this work. It is interesting to ask then how different is this narrowing pore from an expanding exocytotic pore.

In normal nondegranulating mast cells, spontaneous decreases of the surface membrane area can sometimes be observed (Fernandez et al., 1984). This phenomenon can be monitored by following the cell membrane capacitance. These experiments show that during a membrane pinch-off event, the capacitance [$Im(\Delta Y)$], instead of decreasing instantaneously, often decreases slowly. Simultaneously, a large increase in the $Re(\Delta Y)$ component can be observed (Fig. 6). By analogy with exocytosis, the data of Fig. 6 can be interpreted by an equivalent circuit similar to that of Fig. 3 *b*; however, unlike exocytosis, R_p goes from zero to infinity for the fully separated membrane. This can be appreciated in Fig. 6 (bottom trace). The range of values of R_p is similar to that observed during exocytosis (compare with Fig. 4 *B*), but with its time course reversed. These observations suggest that the initial stages of exocytosis and the final stages of membrane pinch-off have in common the formation of a high resistance, water-filled pore of similar characteristics. It is tempting to speculate that the structures that initiate and stabilize these pores are similar.

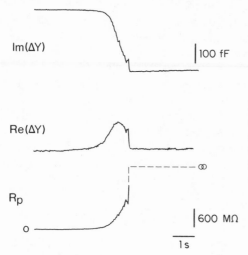

Figure 6. Formation of a narrowing pore during the final stages of membrane "pinch off." R_p is calculated as in Fig. 4. In contrast to exocytosis, R_p progresses from zero to infinity for the fully separated membrane.

Conclusion

In a remarkable sequence of photographs, Chandler and Heuser (1980, Figs. 4–7) showed the possible sequence of events that lead to the fusion of secretory granules with the plasma membrane of mast cells. The phase-detector measurements discussed here confirm and extend many of their views. In particular, they provide a detailed account of the time evolution of the fusion pore. Although the nature of the pore and the forces controlling its expansion remain controversial, the various models make predictions that can be tested with the methods presented. An intriguing result is the possible similarity between exocytotic fusion pores and the collapsing pores observed during membrane pinch-off events.

A major question still remains unexplored: How are the fusion pores formed? Perhaps the small network of fibrils that contact the granule membrane, observed in the early stages of granule fusion (Chandler and Heuser, 1980), relates to this question.

Acknowledgments

We thank Dr. Jack Kaplan, Chaya Joshi, and Nancy Fidler for comments on the manuscript.

This work was supported by National Institutes of Health grant GM-38857-01.

Address reprint request to Dr. J.M. Fernandez, Department of Physiology, University of Pennsylvania, Philadelphia, PA 19104-6085.

References

Breckenridge, L. J., and W. Almers. 1987a. Final steps in exocytosis observed in a cell with giant secretory granules. *Proceedings of the National Academy of Sciences.* 84:1945–1949.

Breckenridge, L. J., and W. Almers. 1987b. Currents through the fusion pore that forms during exocytosis of a secretory vesicle. *Nature.* 328:814–817.

Chandler, D. E., and J. Heuser. 1980. Arrest of membrane fusion events in mast cells by quick freezing. *Journal of Cell Biology.* 86:666–674.

Cole, K. S. 1968. Membrane Ions and Impulses. University of California Press, Berkeley. 569 pp.

Fernandez, J. M., E. Neher, and B. D. Gomperts. 1984. Capacitance measurements reveal stepwise fusion events in degranulating mast cells. *Nature.* 312:453–455.

Gillespie, J. I. 1979. The effect of repetitive stimulation on the passive electrical properties of the presynaptic terminal of the squid giant synapse. *Proceedings of the Royal Society of London Biological Sciences.* 206:293–306.

Hamill, O. P., A. Marty, E. Neher, B. Sakmann, and F. J. Sigworth. 1981. Improved patch-clamp techniques for high-resolution current recording from cells and cell free membrane patches. *Pfleugers Archiv European Journal of Physiology.* 391:85–100.

Jaffe, L. A., S. Hagiwara, and R. T. Kado. 1978. The time course of cortical vesicle fusion in sea urchin eggs observed as membrane capacitance changes. *Developmental Biology.* 67:243–248.

Joshi, C., and J. M. Fernandez. 1988. Capacitance measurements: an analysis of the phase detector technique used to study exocytosis and endocytosis. *Biophysical Journal.* 53:885–892.

Lindau, M., and J. M. Fernandez. 1986. A patch clamp study of histamine-secreting cells. *Journal of General Physiology* 88:349–368.

Neher, E., and A. Marty. 1982. Discrete changes of cell membrane capacitance observed under conditions of enhanced secretion in bovine adrenal chromaffin cells. *Proceedings of the National Academy of Sciences.* 79:6712–6716.

Peres, A., and G. Bernardini. 1985. The effective membrane capacity of xenopus eggs: its relation with membrane conductance and cortical granule exocytosis. *Pfluegers Archiv European Journal of Physiology.* 404:266–272.

Satir, B., C. Schooley, and P. Satir. 1973. Membrane fusion in a model system. *Journal of Cell Biology.* 56:153–176.

Tam, P. Y., and P. Verdugo. 1981. Control of mucus hydration as a Donnan equilibrium process. *Nature.* 292:340–342.

Verdugo, P. 1984. Hydration kinetics of exocytosed mucins in cultured secretory cells of the rabbit trachea: a new model. *Ciba Foundation Symposia.* 109:212–221.

Verdugo, P. 1986. Polymer phase transition: a novel mechanism of product release in secretion. *Biophysical Journal.* 49:231a. (Abstr.)

Verdugo, P., M. Aitken, L. Langley, and M. J. Villalon. 1987. Molecular mechanism of product storage and release in mucin secretion. II. The role of extracellular Ca^{++}. *Biorheology.* 24:625–634.

Zimmerberg, J., M. Curran, F. S. Cohen, and M. Brodwick. 1987. Simultaneous electrical and optical measurements show that membrane fusion precedes secretory granule swelling during exocytosis of beige mouse mast cells. *Proceedings of the National Academy of Sciences.* 84:1585–1589.

Phosphoinositide Cycle
and Protein Kinase C

Chapter 28

Inositol Lipids and Phosphates in Erythrocytes and HL60 Cells

R. H. Michell, C. E. King, C. J. Piper, L. R. Stephens, C. M. Bunce, G. R. Guy, and G. Brown

Departments of Biochemistry and Immunology, The University of Birmingham, Birmingham, United Kingdom

Cell Physiology of Blood © 1988 by The Rockefeller University Press

Introduction

In recent years we have seen acceptance of the idea that the hydrolysis of phosphatidyl-inositol 4,5-bisphosphate [PtdIns $(4, 5)P_2$] by phosphoinositidase C (PIC) is a major and widespread receptor-coupled transmembrane signaling mechanism (Michell et al., 1981; Berridge, 1987). The basic mechanisms involved in the generation and cellular actions of two second messengers, Ins $(1, 4, 5)P_3$ and 1,2-diacylglycerol, from PtdIns $(4, 5)P_2$ are summarized in Fig. 1, together with the pathways of further metabolism of these messenger molecules. Much of the extensive recent work in this field is discussed in four recently published or forthcoming multiauthor collections (Michell and Putney, 1987; Putney, 1987; Berridge and Michell, 1988; Michell et al., 1988).

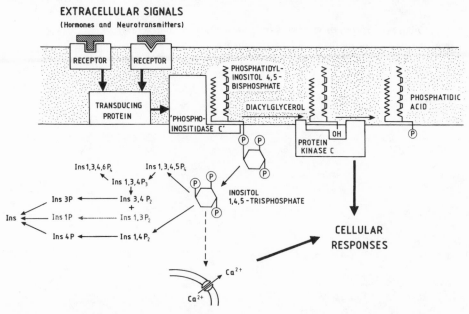

Figure 1. A summary of the basic mechanism thought to be involved in the widespread signaling mechanism that uses PIC-catalyzed hydrolysis of PtdIns $(4, 5)P_2$ as its central reaction. Ins $(1, 4, 5)P_3$, one of the two second messengers formed from PtdIns $(4, 5)P_2$, causes the release of Ca^{2+} from an intracellular membrane-enclosed pool, while the other (1, 2-diacylglycerol) activates one or more of the isoforms of protein kinase C. Also shown on this scheme is the relatively complex pattern of pathways for the metabolism of Ins $(1, 4, 5)P_3$ that has recently been described.

In this paper we will discuss briefly two important biochemical problems that have recently been highlighted by studies of inositol lipid and inositol phosphate metabolism in a variety of cells, including those of hemopoietic lineages. The first is the fact that cells contain two or more metabolic pools of inositol phospholipids that do not freely interchange over periods of a few hours. Moreover, when receptors cause the activation of PIC within intact cells there is selective hydrolysis of an inositol phospholipid pool that is, at least sometimes, only a small fraction of the total inositol lipid in the cells. Notable samples of this pooling occur (*a*) in blowfly salivary glands, where 5-hydroxytryptamine initiates hydrolysis of only a small inositol lipid pool that has a very rapid basal metabolic turnover even in the unstimulated gland (Fain and Berridge, 1979); and (*b*) in WRK1 mammary tumor cells, where a rapidly metabolized lipid pool consisting of metabolically interconvertible PtdIns and PtdIns $(4, 5)P_2$ has its turnover accelerated by vasopressin acting through V_1 receptors (Monaco, 1982; Monaco and Woods, 1983; Koreh and Monaco, 1986; Monaco, 1987). To this previously described metabolic compartmentation can now be added the surprising observation that metabolically distinct pools of phosphatidylinositol 4-phosphate (PtIns4*P*), of PtdIns $(4, 5)P_2$, and of phosphatidate (PtdOH) coexist in the plasma membrane of mammalian erythrocytes, the simplest of all eukaryote cells (see below).

The second topic we shall discuss is the recent discovery of numerous inositol polyphosphates in mammalian cells. Analysis of this area started as an exploration of the metabolism of the novel second messenger Ins $(1, 4, 5)P_3$, but has quickly expanded to encompass inositol polyphosphates synthesized by other metabolic pathways. Below we describe preliminary studies of the composition and metabolism of the complex mixture of inositol polyphosphates present in the human promyeloid cell line HL60, whose most notable features are the very substantial intracellular concentrations of some of these compounds.

Inositol Phospholipid Pools in Intact Human Erythrocytes

Erythrocytes are structurally the simplest of mammalian cells; they consist of a biconcave discoid envelope of plasma membrane that encloses a soluble cytoplasmic protein solution that is largely hemoglobin. This shape is maintained by a complex membrane-associated cytoskeletal web at the inner surface. As is described elsewhere in this book, anionic phospholipids, including the phosphoinositides, may exist not only as constituents of the lipid bilayer of the membrane but also as mediators of dynamic molecular associations between cytoskeletal proteins and between the cytoskeleton and the cytoplasmic leaflet of the lipid bilayer. Intracellular organelles are absent, except in sickle cells or drug-induced stomatocytes, which invaginate a small proportion of their plasma membrane to form endocytic vesicles. Lipid metabolism is therefore very limited, comprising a modest repertoire of reactions involving deacylation/reacylation of preexisting lipids, and the removal and replacement of the monoester phosphate groups of PtdOH, PtdIns4*P*, and PtdIns $(4, 5)P_2$.

Given the simple structure of human erythrocytes, we initiated studies to follow labeling with ^{32}P under steady-state conditions of the monoester phosphate groups of PtdOH, PtdIns4*P*, and PtdIns $(4, 5)P_2$, and of the γ phosphate of ATP, their immediate metabolic precursor. We expected that the results would show a steady progression of each of the lipid phosphate groups to isotopic equilibrium with the γ phosphate of ATP. To facilitate this kinetic analysis, we developed a novel incubation medium for red cells in which ^{32}P$_i$ enters the cells and equilibrates with ATP very quickly; this was achieved by essentially eliminating from the extracellular medium the many anions that compete with P$_i$ for entry on the nonspecific anion transporter that is the major entry route of this ion. Using this method, cells achieved steady-state labeling of the β and γ phosphates of ATP, and also of 2,3-bisphosphoglycerate, their most abundant intracellular phosphate, in approximately 3 h. Thereafter, cells remained metabolically stable for 4 h; inositol lipid and PtdOH levels were constant throughout these incubations (King et al., 1987).

The final steady state–specific radioactivities achieved by the monoester phosphate groups of PtdOH, PtdIns4P, and PtdIns $(4, 5)P_2$ in these experiments were consistently only one-quarter to one-third of the steady-state–specific radioactivity of the γ phosphate of ATP in the same cells (Fig. 2; taken from King et al., 1987). Thus the majority of the molecules of these three lipids in red cell suspensions were metabolically inert over the 7-h incubation period. One possible explanation of this observation was a loss of lipid-metabolizing ability as erythrocytes aged in the circulation, but this was eliminated by demonstrating that density-separated young and old cells metabolized these lipids in an identical manner (King et al., 1987). Another possibility is that the metabolically active and inert lipid pools are in the inner leaflet and the outer leaflet of the lipid bilayer, respectively. We tested this idea by

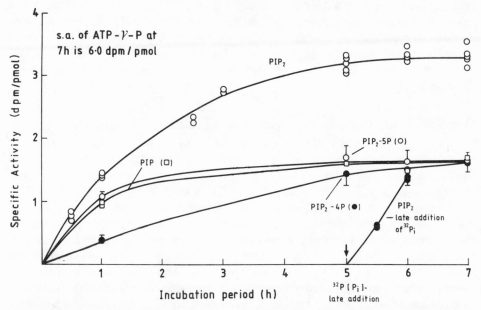

Figure 2. Incorporation of ^{32}P$_i$ into the phospholipids of human erythrocytes. Experimental details are given in King et al. (1987). Briefly, cells were equilibrated in a HEPES-rich incubation medium designed to allow rapid phosphate entry, and a tracer quantity of ^{32}P$_i$ was then added at either time 0 or 5 h later. Rapid and similar incorporation at either time indicated that lipid turnover was sustained at an undiminished rate during the entire incubation period. The concentrations and total specific radioactivities of PtdIns4P, PtdIns $(4, 5)P_2$, and ATP, and also the specific radioactivities of the monoester phosphate of the lipids and the γ phosphate of ATP, were then determined.

specifically activating the endogenous PIC of erythrocytes. This enzyme is at the cytoplasmic surface of the membrane and can be activated if intracellular [Ca^{2+}] is greatly elevated by the use of an ionophore such as A23187. Under conditions in which approximately half of the PtdIns $(4, 5)P_2$ of the cells was degraded by this PIC, there was equal attack upon each of the metabolic pools. We therefore tentatively conclude that: (*a*) the metabolically active and inert pools of PtdOH, PtdIns4P, and PtdIns $(4, 5)P_2$ exist throughout the erythrocyte population; and (*b*) both of these pools are located at the cytoplasmic surface of the membrane where they are partially accessible to the membrane PIC.

In 1986, Muller et al. had also concluded that the inositol lipids of erythrocytes are not metabolically homogeneous. They analyzed the initial rates of labeling of the γ phosphate of ATP and of the monoester phosphates of PtdIns4P and PtdIns $(4, 5)P_2$ after the addition of a trace of $^{32}P_i$ to human erythrocytes incubated at steady state in a "traditional" Cl^--based medium. In this medium, P_i enters much more slowly than in the new medium devoid of competing anions. Muller et al. modeled their data in terms of the following metabolic sequence consisting of the two kinase/phosphatase cycles:

$$\text{PtdIns} \rightleftarrows \text{PtdIns4}P \rightleftarrows \text{PtdIns } (4, 5)P_2.$$
(Cycle 1)   (Cycle 2)

They found that the closest fit between their experimental data and the kinetic model was obtained when only a small fraction of the total PtdIns $(4, 5)P_2$ and about one-half of the PtdIns4P were admitted to the calculations; when all of the lipids present in the cells were included in the calculations, the predicted kinetic behavior of lipid labeling was outside the statistical limits set by their data.

The slow entry of P_i into cells in the medium used by Muller et al. (1986) limited the temporal separation between ATP and lipid labeling. This separation was greatly increased when initial rate kinetic experiments of a similar type were undertaken using the new medium designed to facilitate rapid P_i entry (King, C. E., P. T. Hawkins, L. R. Stephens, and R. H. Michell, unpublished results). Kinetic modeling of the data obtained in these experiments was then undertaken, with the fraction of the total cellular PtdIns4P and PtdIns $(4, 5)P_2$ participating in metabolic turnover set at a range of values from 5 to 100%. Models were also considered in which the metabolically active fractions of PtdIns4P and of PtdIns $(4, 5)P_2$ were varied independently. The best fit between the experimental and modeled data was obtained by assuming that one-quarter of the PtdIns4P and one-quarter of the PtdIns $(4, 5)P_2$ were in the metabolically active pool. The active fraction of PtdIns $(4, 5)P_2$ could be varied between 20 and 30% of the total without any appreciable loss of fit. Good fit was rapidly lost if the active fraction of PtdIns4P was set at any value appreciably different from 25%. The flux rates through cycles 1 and 2 of the above pathway in healthy human red cells could then be calculated, assuming that one-quarter of the PtdIns4P and one-quarter of the PtdIns $(4, 5)P_2$ undergo rapid metabolic turnover of their monoester phosphate groups. The values obtained were 2.1 nmol/ml packed cells per h for the interconversion of PtdIns and PtdIns4P and 5.7 nmol/ml packed cells per h for the PtdIns4P/PtdIns $(4, 5)P_2$ cycle. An expenditure of <10 nmol ATP γ phosphate per milliliter packed cells per hour for "maintenance" turnover of PtdIns4P and PtdIns $(4, 5)P_2$ and probably of a similar amount for the turnover of PtdOH, represents only a very small fraction of the total ATP consumption for maintenance purposes in the healthy erythrocyte, the majority of which is used for maintenance of ion gradients.

To summarize, the first evidence for metabolic heterogeneity of inositol phospholipids in human erythrocytes came from Muller et al. (1986), but with estimates of the fractions of cellular PtdIns4P and PtdIns $(4, 5)P_2$ that were in the metabolically active lipid pools that were of only modest precision. We have now obtained two further estimates, one by a comparison of the specific activities of ATP and these two lipids when steady-state labeling had been achieved and the other by modeling the early kinetics of ATP and lipid labeling into the kinase/phosphatase cycles that interconvert PtdIns, PtdIns4P, and PtdIns $(4, 5)P_2$. These methods have given estimates that are more precise and also concordant. They indicate that approximately one-quarter of erythrocyte PtdIns4P and PtdIns (4, 5)P_2 molecules are metabolically active, with half-times for turnover of ~40 and 80 min, respectively. The remaining three-quarters show no significant metabolic turnover during a 7-h incubation period.

The physical nature of this metabolic pooling within a single leaflet of a continuous lipid envelope remains unknown. The most reasonable speculations are probably those that envisage long-lived noncovalent interactions with specific proteins, possibly those of the cytoskeleton. However, none of the interactions reported thus far would be expected to show long-term stability. In recent experiments, we searched for such stable interactions by analyzing the PtdIns (4, 5)P_2 that can be extracted from labeled erythrocyte ghosts by graded treatment with Triton X-100. However, we obtained no evidence for selective extraction of either the metabolically active or inert lipid pool under conditions that achieved only partial extraction of PtdIns (4, 5)P_2. We will next investigate whether we can reproduce the pooling seen in intact cells by using labeled ATP as a phosphate donor for PtdIns4P and PtdIns (4, 5)P_2 synthesis in permeabilized erythrocyte preparations.

Inositol Polyphosphates in HL60 Cells

One of the first receptors that was shown to cause PtdIns (4, 5)P_2 hydrolysis, with intracellular accumulation of InsP_3, was the $f$met-leu-phe receptor of neutrophils and DMSO-differentiated HL60 cells (reviewed by Cockcroft, 1986). Initially it was assumed that the Ins (1, 4, 5)P_3 formed in these and other stimulated cells had only one fate, namely stepwise hydrolysis, via InsP_2 and Ins$P$, to free inositol (Storey et al., 1984). More recently, however, it has become clear that Ins (1, 4, 5)P_3 is also metabolized by an alternative set of pathways that start with its phosphorylation to Ins (1, 3, 4, 5)P_4 (summarized in Fig 1; see Irvine et al., 1986; Shears et al., 1987 a, b, c). In addition, the observation of substantial quantities of highly phosphorylated derivatives of inositol in the GH4 pituitary cell line by Heslop et al. (1985) has led to the discovery of these compounds in a variety of mammalian cells.

The HL60 cell line can be used to study changes in inositol phospholipid and inositol phosphate levels and metabolism that may occur during both differentiation and responses to receptor activation. In preliminary experiments these cells were labeled to equilibrium with [^{3}H]inositol in order to identify and quantitate their various inositol-containing constituents, with particular emphasis on the inositol polyphosphates. HL60 cells were grown in RPMI 1640 medium containing 20% fetal calf serum and [^{3}H]inositol for 9 d, during which time they grew exponentially with a doubling time of ~19 h. Dimethyl sulfoxide (DMSO) (1.25% vol/vol) was added to some cells after 4 d in order to stimulate their differentiation towards neutrophils, a transition that includes the expression of $f$met-leu-phe receptors. Cell division halted during the first 3 d, and 83% of the previously nonphagocytic cells had acquired the ability to phagocytize complement-coated yeast cells by day 5. At this stage, some of the cells were stimulated with $f$met-leu-phe for 35 s. Cell samples were killed with perchloric acid, and the acid extracts were neutralized and used for analysis of inositol phosphates by HPLC on a Partisil SAX anion-exchange column. The precipitates were treated with acid chloroform–methanol to extract labeled inositol lipids.

Analysis of the specific radioactivity of the inositol in the spent growth medium and of phosphatidylinositol, the predominant inositol phospholipid, showed that equilibrium labeling of the lipids was attained within ~3 d of the addition of [³H]inositol to the medium of the exponentially growing cells. Regular dilution during cell growth with identical radioactive medium ensured equilibrium labeling even of cell constituents that turn over slowly. Cell volumes were assessed by rounding up the cells by incubation in RPMI medium containing 10% FCS and 10% human AB serum for 30 min at 37°C, and measuring their diameters using a light microscope fitted with a reticle. Exponentially growing cells had a volume of slightly <1 pl, which approximately halved during differentiation in DMSO for 5 d (Table I). Since the cells had been equilibrium-labeled with [³H]inositol of known specific radioactivity, we were able to calculate the absolute concentrations of the various inositol phosphates present in these cells (Table I). These values are averaged over the entire cell volume, and much higher concentrations will occur at certain intracellular sites if there is appreciable intracellular compartmentation of inositol phosphates.

TABLE I
Cell Volumes and Inositol Phosphate Concentrations in HL60 Cells

Inositol phosphate	Approximate concentrations of inositol phosphates in cells		
	Exponentially growing (0.97 pl)*	DMSO differentiated (0.53 pl)	DMSO differentiated + *f*met-leu-phe (0.53 pl)
	μM		
Ins*P*	19	22	35
Ins*P*₂	23	10	14
Ins*P*₃ (und.)	2.1	0.8	0.7
Ins(1,3,4)*P*₃	1.4	1.1	3.6
Ins(1,4,5)*P*₃	0.4	0.4	1.6
Ins*P*₄	4.1	1.6	4.6
Ins*P*₅	26	25	29
Ins*P*₆	42	12	21

HL60 cells were labeled to equilibrium with [³H]inositol both during exponential growth and during growth followed by 5 d of differentiation towards neutrophils in 1.25% (vol/vol) DMSO. Some of the DMSO-differentiated cells were then stimulated for 35 s with *f*met-leu-phe.
*Cell volume shown in parentheses.

The most striking characteristics of the inositol phosphate patterns revealed by HPLC of the acid extracts of HL60 cells were the great variety of compounds present and the substantial concentrations of some of these compounds. The sample HPLC chromatogram shown in Fig. 3 is of an extract from cells that had been induced to differentiate and then stimulated with *f*met-leu-phe, since cells treated in this way most clearly showed the presence of Ins (1, 4, 5)*P*₃ and Ins (1, 3, 4)*P*₃. Table I lists the approximate concentrations of the individual inositol phosphate species present in HL60 cells under three sets of conditions.

Ins (1, 4, 5)*P*₃ and other inositol trisphosphates, the molecules whose function and metabolism have been most thoroughly investigated in the past, were among the least abundant of the inositol phosphates detected in these cells (Table I). Three isomeric inositol trisphosphates were present. The first to elute has not been identified, and the second and third were tentatively identified as Ins (1, 3, 4)*P*₃ and Ins (1, 4, 5)*P*₃, respectively, since they coeluted with authentic samples of these two InsP_3s and the concentrations of both increased after stimulation with *f*met-leu-phe (Table I). Assuming that all of the material eluting in the final peak was Ins $(1, 4, 5)P_3$, then the concentration of this key Ca^{2+}-mobilizing intracellular second messenger in control DMSO-differentiated cells rose about fourfold during 35 s of stimulation.

Less highly phosphorylated inositol phosphates, possible metabolites of the highly phosphorylated species, were more abundant, with major peaks in the InsP_2 and Ins$P$ regions of the chromatogram. One major peak corresponded to Ins1P or Ins3P (these are enantiomers), and the second, more slowly eluting Ins$P$ peak has not been identified but may have been Ins4P. Although it is not clear from the low resolution HPLC run shown in Fig. 3, the main InsP_2 peak eluted slightly later than a standard

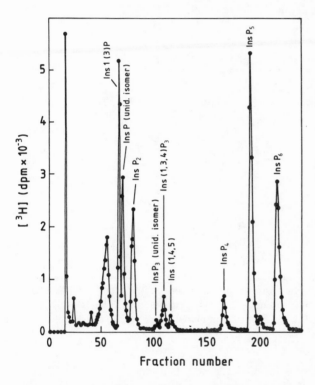

Figure 3. Separation by HPLC on a Partisil 10SAX column of the acid-soluble inositol-containing constituents of HL60 cells that had been equilibrium labeled with [^{3}H]inositol during exponential growth, differentiated in DMSO in the continued presence of [^{3}H]inositol, and stimulated with *f*met-leu-phe for 35 s. The arrows indicate the elution positions of standards of particular inositol polyphosphate isomers.

sample of Ins $(1, 4)P_2$; this separation is shown in greater detail in Fig. 4 *A*. The major InsP_2 isomer present in these cells is yet to be definitively identified, but its elution position would be consistent with Ins $(3, 4)P_2$, the major InsP_2 product of Ins $(1, 3, 4)P_3$ metabolism (Shears et al., 1987*a*, *b*).

Three InsP_4 isomers are well established either as intracellular components present in mammalian cells or as metabolites of Ins $(1, 4, 5)P_3$, or as both. Ins $(1, 3, 4, 5)P_4$ is formed directly from Ins $(1, 4, 5)P_3$; Ins $(1, 3, 4, 6)P_4$ is formed by the phosphorylation of the Ins $(1, 3, 4)P_3$ derived from Ins $(1, 3, 4, 5)P_4$; and Ins $(3, 4, 5, 6)P_4$ is of unknown origin. The major InsP_4 present in unstimulated HL60 cells cochromatographed with Ins $(3, 4, 5, 6)P_4$ rather than Ins $(1, 3, 4, 5)P_4$ (Fig. 4, *B* and *C*). The isomeric form that contributes most to the rise in total InsP_4 on brief stimulation (Table I) has not been identified, but is most likely to be Ins $(1, 3, 4, 5)P_4$.

InsP_5 and InsP_6 were particularly abundant (Fig. 3 and Table I) and took several days to come close to metabolic equilibrium with the added [^{3}H]inositol even in exponentially growing cells (data not shown). The concentrations of these compounds were such that they must constitute an appreciable, and very metabolically stable fraction of the total organic phosphate pool in HL60 cells. If they are free in the cytosol, they might be expected to exert important effects, for example on the

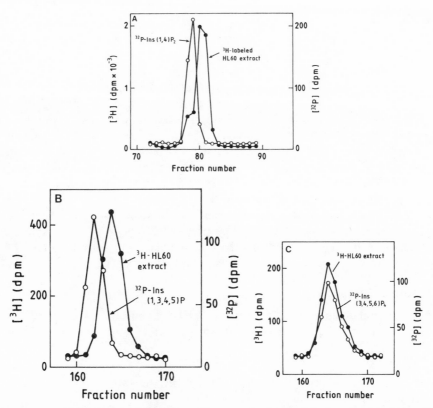

Figure 4. Partial HPLC chromatograms to illustrate the detailed elution characteristics of the major InsP_2 and InsP_4 species present in labeled HL60 extracts. (*A*) The separation of the major InsP_2 of HL60 cells from a standard sample of Ins $(1, 4)P_2$. The major InsP_4 species in extracts of unstimulated HL60 cells was clearly separable from Ins $(1, 3, 4, 5)P_4$ (*B*) but ran in a position coincident with Ins $(3, 4, 5, 6)P_4$ [$\equiv$L-Ins $(1, 4, 5, 6)P_4$] (*C*).

intracellular buffering of bivalent cations. The major InsP_5 species present co-chromatographed with Ins $(1, 3, 4, 5, 6)P_5$ (data not shown), which was recently discovered as a major inositol polyphosphate of other mammalian cells (Stephens et al., 1987). In the absence of standard samples of other InsP_5 isomers, this identification must remain tentative.

Not only were there rapid changes in the concentrations of Ins $(1, 4, 5)P_3$ and its metabolites in differentiated HL60 cells upon stimulation, but also there were slower changes in the concentrations of some inositol phosphates during differentiation. In particular, there was a large decrease in the concentration of $InsP_6$ from its initially high value (Table I). Other compounds that decreased their concentrations, but from initially lower starting values, were $InsP_4$ and $InsP_2$ (isomers unidentified), and possibly also the unidentified isomer of $InsP_3$. The mechanism underlying these changes are yet to be investigated. However, given the slow turnover of $InsP_6$, as indicated by its slow approach to equilibrium labeling during cell growth, the decline in its concentration during differentiation must imply either a marked decrease in the rate of synthesis or a substantial stimulation of degradation.

Conclusion

In conclusion, the preliminary results shown here indicate that HL60 cells grown in a culture medium of relatively high inositol content and in the presence of fetal calf serum contain a great variety of inositol phosphates, including large concentrations of metabolically stable $InsP_5$ and $InsP_6$. However, these growth conditions pose two disadvantages for any detailed analysis of the changes that may accompany differentiation or follow receptor stimulation. First, the serum may contain growth factors or other components that themselves regulate the metabolism of inositol lipids and/or inositol phosphates. Secondly, it is extremely expensive to use such a medium to achieve a high enough specific radioactivity in the [^{3}H]inositol to allow adequate labeling of the minor inositol phosphate constituents of cells. To circumvent these problems, we have now adapted HL60 cells to grow in a serum-free RMPI medium containing 1 mg inositol/liter supplemented only with insulin, transferrin, and selenium dioxide. The doubling time for the cells is slightly longer under these conditions (28 h), but they still differentiate satisfactorily in response to DMSO (French, P. J., J. A. Creba, and C. M. Bruce, unpublished data).

Acknowledgments

We are grateful to the Medical Research Council, the Leukaemia Research Fund, and the Royal Society for providing financial support for this work.

References

Berridge, M. J. 1987. Inositol trisphosphate and diacylglycerol: two interacting second messengers. *Annual Reviews of Biochemistry.* 56:159–193.

Berridge, M. J., and R. H. Michell, editors. 1988. Inositol Lipids and Transmembrane Signalling. *Philosophical Transactions of the Royal Society of London Series B.* 320:235–436.

Cockcroft, S. 1986. Phosphoinositides and neutrophil activation. *In* Phosphoinositides and Receptor Mechanisms. J. W. Putney, Jr., editor. Alan R. Liss, Inc., New York. 287–310.

Fain, J. N., and M. J. Berridge. 1979. Relationship between PtdIns synthesis and recovery of 5-hydroxytryptamine responsiveness in blowfly salivary glands. *Biochemical Journal.* 180:655–661.

Heslop, J. P., R. F. Irvine, A. H. Tashjian, and M. J. Berridge. 1985. Inositol tetrakis and pentakisphosphates in GH4 cells. *Journal of Experimental Biology.* 119:395–401.

Irvine, R. F., A. J. Letcher, J. P. Heslop and M. J. Berridge. 1986. The inositol tris/tetrakis phosphate pathway. Demonstration of Ins $(1, 4, 5)P_3$ 3-kinase activity in animal tissues. *Nature.* 320:631–634.

King, C. E., L. R. Stephens, P. T. Hawkins, G. R. Guy, and R. H. Michell. 1987. Multiple metabolic pools of phosphoinositides and phosphatidate in human erythrocytes incubated in a medium that permits rapid transmembrane exchange of phosphate. *Biochemical Journal.* 244:209–217.

Koreh, K., and M. Monaco. 1986. The relationship of hormone-sensitive and hormone-insensitive phosphatidylinositol to phosphatidylinositol 4, 5-bisphosphate in the WRK1 cell. *Journal of Biological Chemistry.* 261:88–91.

Michell, R. H., A. H. Drummond, and C. P. Downes, editors. 1988. Inositol Lipids in Cell Signalling. Academic Press, London. In press.

Michell, R. H., C. J. Kirk, L. M. Jones, C. P. Downes, and J. A. Creba. 1981. The stimulation of inositol lipid metabolism that accompanies calcium mobilization in stimulated cells: defined characteristics and unanswered questions. *Philosophical Transactions of the Royal Society of London, Series B.* 296:123–137.

Michell, R. H., and J. W. Putney, Jr., editors. 1987. Inositol Lipids in Cellular Signalling. *Cold Spring Harbor Current Communications in Cell Biology.*

Monaco, M. 1982. The phosphatidylinositol cycle in WRK1 cells: evidence for a separate, hormone-sensitive phosphatidylinositol pool. *Journal of Biological Chemistry.* 257:2137–2139.

Monaco, M. 1987. Inositol metabolism in WRK1 cells: relationship of hormone-sensitive to -insensitive pools of phosphoinositides. *Journal of Biological Chemistry.* 262:13001–13006.

Monaco, M., and D. Woods. 1983. Characterization of the hormone-sensitive phosphatidylinositol pool in WRK1 cells. *Journal of Biological Chemistry.* 258:15125–15129.

Muller, E., H. Hegewald, K. Jaroszewicz, G. A. Cumme, H. Hoppe, and H. Frunder. 1986. Turnover of phosphomonoester groups and compartmentation of polyphosphoinositides in human erythrocytes. *Biochemical Journal.* 235:775–783.

Putney, J. W., Jr., editor. 1986. Phosphoinositides and Receptor Mechanisms. Alan R. Liss, Inc., New York.

Shears, S. B., J. B. Parry, E. K. Y. Tang, R. F. Irvine, R. H. Michell, and C. J. Kirk. 1987*a*. Metabolism of Ins (1, 3, 4, 5)P_4 by rat liver, including the synthesis of a novel isomer of inositol tetrakisphosphate. *Biochemical Journal.* 246:139–147.

Shears, S. B., D. J. Storey, A. J. Morris, A. J. Cubitt, J. B. Parry, R. H. Michell, C. J. Kirk. 1987*b*. Dephosphorylation of Ins (1, 4, 5)P_3 and Ins (1, 3, 4)P_3. *Biochemical Journal.* 242:393–402.

Shears, S. B., C. J. Kirk, and R. H. Michell. 1987*c*. The pathway of Ins $(1, 3, 4)P_3$ dephosphorylation in liver. *Biochemical Journal.* 248:977–980.

Stephens, L., P. T. Hawkins, N. Carter, and C. P. Downes. 1987. Independent pathways of inositol phosphate metabolism in mammalian and avian cells. *In* Inositol Lipids in Cellular Signaling. R. H. Michell and J. W. Putney, Jr., editors. Cold Spring Harbor Laboratory, Cold Spring Harbor, NY. 11–14.

Storey, D. J., S. B. Shears, C. J. Kirk, and R. H. Michell. 1984. Stepwise dephosphorylation of Ins (1, 4, 5)P_3 in liver. *Nature.* 312: 374–376.

Chapter 29

Protein Kinase C and Its Associated Substrates in the Human Erythrocyte

H. Clive Palfrey and Ahmad Waseem

*Department of Pharmacological and Physiological Sciences,
University of Chicago, Chicago, Illinois*

Introduction

The responses of many cells to extracellular signals are mediated by a variety of second messengers, several of which interact with protein phosphorylation systems. The classic example of such a system is the hormone-sensitive adenylate cyclase (cAMP)-dependent protein kinase (cAMP-PK) cascade. More recently, the second messenger roles of Ca^{2+} and certain products of lipid hydrolysis have come into focus. Elevations in cellular free Ca^{2+} can activate a number of enzyme systems by binding to the small intracellular Ca receptor protein calmodulin (CaM; for review see Manalan and Klee, 1984). Ca/CaM-regulated enzymes include at least five distinct protein kinases and a protein phosphatase. Diacyglycerols (DAGs) that are produced by hydrolysis of phospholipids by receptor-activated phospholipase C–type enzymes (Michell et al., 1988) are the putative activators of another set of protein kinases, collectively termed protein kinase C (PK-C). This enzyme appears to be important in the regulation of numerous processes in diverse tissues (for reviews see Nishizuka, 1986; Neidel and Blackshear, 1986; Hannun and Bell, 1986). The enzyme, which is widespread in animal tissues, also has an absolute requirement for phospholipid and Ca^{2+} for activity. Because phospholipids and diacylglycerols are found primarily in membranes, it is widely assumed that physiological activation of PK-C only occurs in an appropriate membrane environment (Hannun and Bell, 1986).

As in the case of the cAMP and Ca/CaM systems, it is thought that the ultimate effects of extracellular stimuli that act via diacylglycerol production and PK-C activation are mediated by multiple tissue-specific substrates for PK-C (Nishizuka, 1986). Although much is known about PK-C itself, its target substrates are only beginning to be characterized in depth. As PK-C becomes membrane bound when activated it seems likely that many of its important substrates will turn out to be membrane-associated proteins. These may include integral membrane proteins such as receptors or transporters as well as peripheral proteins such as components of the membrane skeleton. Examples of receptors that are phosphorylated by PK-C include those for epidermal growth factor (e.g., Hunter et al., 1984) and transferrin (May et al., 1985). Many instances of alterations in neuronal excitability caused by PK-C activators have come to light recently (e.g., DeRiemer et al., 1985), and it is probable that many of these effects will eventually turn out to be due to phosphorylation of specific channel proteins. In the cytoskeleton, proteins such as myosin light chains (Nishikawa et al., 1983) and vinculin (Werth et al., 1983) have been shown to be substrates for PK-C. The human erythrocyte has a well-defined membrane skeleton; thus it became of interest to assess the possible consequences of PK-C activation on the phosphorylation of skeletal components in these cells.

Erythrocyte Protein Kinase C

Several studies have established the presence of PK-C in red cells (e.g., Palfrey and Waseem, 1985; Faquin et al., 1986; Cohen and Foley, 1986). A histone kinase activity that is stimulated by phosphatidylserine and high micromolar Ca^{2+} is found in human erythrocyte cytosol (Palfrey and Waseem, 1985). The $[Ca^{2+}]$ requirement for this activity is considerably reduced in the presence of active phorbol esters or diacylglycerol, a characteristic of PK-C from other sources. Phorbol esters bind tightly at the diacylglycerol site on the enzyme and influence the affinity of the Ca^{2+} and phosphatidylserine binding site(s) (Hannun and Bell, 1986). Several investigators have shown that multiple genes for PK-C exist in mammalian cells (e.g., Coussens et al., 1986; Knopf et al., 1986; Kikkawa et al., 1987) and that this diversity is reflected in chromatographically distinct species of the enzyme. In particular, hydroxyapatite chromatography is able to distinguish at least three forms of the kinase from brain (Huang et al., 1986; Kikkawa et al., 1987; Jaken and Kiley, 1987) that may have different Ca^{2+} requirements (Jaken and Kiley, 1987). When a partially purified human erythrocyte PK-C preparation is applied to hydroxyapatite, its elution pattern suggests that it is largely a type II enzyme (Fig. 1; nomenclature from Kikkawa et al., 1987). Other cell types may also contain predominantly a single form of the enzyme (e.g., murine fibroblasts contain only type III; McCaffrey et al., 1987).

We have used phorbol esters to probe the activity of PK-C in the intact red cell. These compounds rapidly penetrate the cell membrane and preferentially bind to and

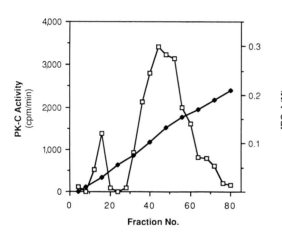

Figure 1. Hydroxyapatite chromatography of erythrocyte PK-C. The cytoplasmic fraction from 2 U of red cells was chromatographed on DEAE-cellulose and the peak PK-C-containing fractions were pooled and adsorbed to a 10-ml column of hydroxyapatite-Ultrogel (IBF Biotechnics, Savage, MD). Kinase activity (—□—) was eluted with a linear gradient from 0 to 0.3 M ammonium phosphate (—■—) and 10-μl aliquots were assayed by conventional methods. For comparison, rat brain type II kinase elutes at ~0.1 M phosphate from a hydroxyapatite HPLC column (Kikkawa et al., 1987).

activate PK-C in many cells (for review see Neidel and Blackshear, 1986). Incubation of red cells in 12-*O*-tetradecanoyl phorbol 13-acetate (TPA) or phorbol dibutyrate causes a rapid and complete translocation of PK-C from the cytoplasm to a membrane-bound form (Palfrey and Waseem, 1985; and Fig. 2). This process is complete in 5 min at 37°C but can be considerably slowed in the cold. After brief TPA treatment ghosts are prepared by conventional hypotonic lysis; PK-C remains membrane bound, even in the continued presence of EGTA, provided that ghosts are maintained at 4°C. At 37°C the enzyme rapidly dissociates from the membrane, but can be prevented from doing so by addition of further TPA. At 4°C, membrane associated PK-C can only be dissociated by treatment with nonionic detergent but not by high or low ionic strength buffers or by manipulations of $[Ca^{2+}]$. This suggests that the enzyme is bound primarily to bilayer components, presumably phosphatidylserine, which in erythrocytes is preferentially located at the inner face of the bilayer. The enzyme is active in the membrane-associated form at very low $[Ca^{2+}]$, i.e., in the presence of 1–5 mM EGTA. Both Ca^{2+} and phospholipid dependence of enzyme activity is restored when a nonionic detergent soluble fraction is prepared from ghosts that contain PK-C.

Longer term treatment of intact red cells with active phorbol esters results in a gradual disappearance of PK-C activity from the membrane (Fig. 2). This is accompanied by a reduction in the phosphorylation of endogenous PK-C substrates in isolated membranes. Enzyme activity does not reappear in the cytosol and a phospholipid- and Ca^{2+}-independent form of PK-C (that can be derived in vitro by Ca^{2+}-dependent proteolysis of PK-C, and is termed PK-M [Kishimoto et al., 1983]) is not

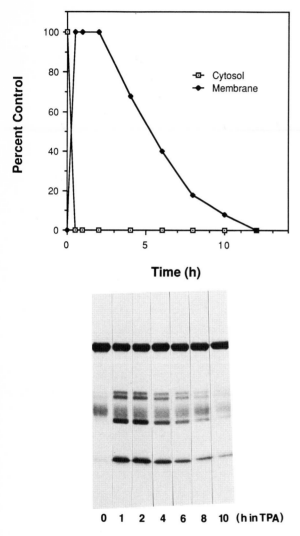

Figure 2. Translocation and downregulation of PK-C in erythrocytes. (*Top*) Red cells were treated with TPA (1 μM) at time 0 and, at various times thereafter, ghost and cytoplasmic fractions were prepared and assayed for the presence of PK-C (abscissa: expressed as percent maximal activity in cytosol or membrane fraction). (*Bottom*) Autoradiogram showing phosphorylation of membrane proteins at different times after TPA treatment of cells. Samples of ghosts were taken at the same times as those in the top panel and phosphorylated in vitro by the addition of γ-[^{32}P]ATP (Palfrey and Waseem, 1985). Five major membrane skeletal substrates for PK-C are apparent in the 1-h sample and gradually disappear on prolonged TPA treatment as PK-C activity declines.

found in the erythrocyte under these conditions. The downregulation phenomenon has been noted in many cell types and results from proteolysis of the membrane-associated species of the enzyme (Ballester and Rosen, 1985; Young et al., 1987; Woodgett and Hunter, 1987). The nature of the protease activity responsible for this breakdown has not been identified. Preincubation of red cells with a number of protease inhibitors (e.g., leupeptin, phenylmethylsulfonyl fluoride, iodoacetate) has no effect on the phorbol ester–induced loss of PK-C activity. The ability to produce cells devoid of PK-C has been useful in several instances to define the role of this enzyme in responses to extracellular hormones that may activate more than one second messenger system (e.g., Muldoon et al., 1987); it should be noted however that a form of PK-C that is "resistant" to downregulation may occur in some cell types (Kariya and Takai, 1987). It may be possible to use PK-C-deficient red cells to define the role of this enzyme in complex phenomena such as shape change.

Membrane Skeletal Substrates for PK-C in the Erythrocyte

The ability to prepare ghosts with endogenous PK-C still attached enabled us to investigate the membrane substrates for this kinase in vitro. Previously, Witters et al.

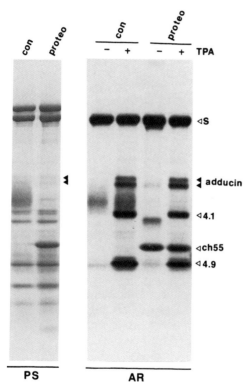

Figure 3. PK-C substrates in the erythrocyte membrane. Cells were left untreated (con) or were pretreated for 1 h with 0.25 mg/ml α-chymotrypsin at 37°C (proteo). Both sets were subsequently subdivided into a control sample or a TPA-treated sample (0.1 μM, 10 min, 37°C), after which ghosts were prepared as usual and phosphorylated as described (Palfrey and Waseem, 1985). Membrane protein stain (PS) profile at left confirms that Band 3 was cleaved to a M_r 55,000 band (Ch55) by α-chymotrypsin, revealing the two subunits of adducin at M_r 120,000 (α) and 110,000 (β). Autoradiogram (AR) at right indicates substrates for PK-C revealed by in vitro phosphorylation of membranes with γ-[^{32}P]ATP. Note that cleavage of Band 3 has no effect on the subsequent phosphorylation of the membrane by PK-C. Other major phosphoproteins are S: β-spectrin, Bands 4.1 and 4.9.

(1985) showed that the erythrocyte glucose transporter is a substrate for PK-C (the functional consequences of this phosphorylation are still apparently unknown). However, it became apparent to us that the major substrates for this kinase in red cells were membrane skeletal proteins (Palfrey and Waseem, 1985; see also Cohen and Foley, 1986; Horne et al., 1985; Faquin et al., 1986). An experiment of this type is shown in Fig. 3. Ghosts are prepared from control cells or from cells exposed to TPA for 10 min and their endogenous phosphorylation is then studied in vitro by the addition of γ-[^{32}P]ATP. As indicated above, addition of Ca^{2+} to isolated membranes containing PK-C is not essential for the full expression of enzyme activity, and even inhibits

overall membrane phosphorylation at concentrations >10 μM. Increased phosphorylation of five major proteins occurs, two of which correspond to the Band 4.1 doublet and
a third to Band 4.9 (cf., Horne et al., 1985). We confirmed this identity by isolation of
these species and recombination with purified PK-C (Palfrey and Waseem, 1985).
While proteins 4.1 and 4.9 are relatively well studied proteins whose functions are
partially understood (for review see Marchesi, 1985), the two other substrates for
PK-C (proteins of M_r 120,000 and 110,000) have been largely ignored. This is because
they are minor proteins and are not readily visible in stained gels of whole ghosts,
where they migrate just above the trailing edge of Band 3. Cleavage of the
extracellular region of Band 3 by treatment of intact red cells with chymotrypsin,
which yields a membrane-associated fragment of M_r 55,000, allows the two high Mr

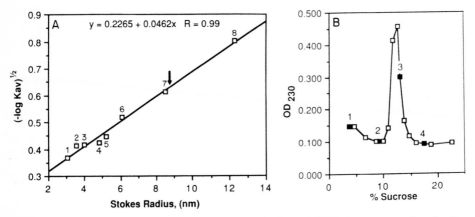

Figure 4. Hydrodynamic behavior of adducin. (*A*) Sephacryl S400 chromatography of adducin. 2 mg of adducin was applied to a 2.5 × 98 cm column equilibrated and run in 10 mM sodium
phosphate (pH 7.5), 100 mM NaCl, 1 mM dithiothreitol, 1 mM sodium azide. Calibration
proteins (with Stokes radius in nm) are: *1*, ovalbumin (3.05); *2*, BSA (3.55); *e*, glyceraldehyde
3-phosphate dehydrogenase (3.99); *4*, aldolase (4.8); *5*, catalase (5.2); *6*, ferritin (6.1); *7*,
thyroglobulin (8.5); *8*, spectrin dimer (12.3). Arrow indicates position of adducin. (*B*)
Sucrose-gradient centrifugation of adducin: 1 mg adducin was applied to a 5–20% sucrose
gradient that was centrifuged for 21 h at 270,000 *g* (SW 41 rotor). Calibration proteins ($s_{20,w}$) *1*,
cytochrome *c* (1.75); *2*, BSA (4.6); *3*, aldolase (7.4); *4*, catalase (11.3) were localized in parallel
runs.

PK-C substrates to be visualized more clearly (Fig. 3). This doublet was initially of
interest because it evidently incorporates much more phosphate per mole than either
Band 4.1 and Band 4.9. We have gone on to characterize this pair of proteins in more
detail.
      Early experiments suggested that the M_r 120,000 and 110,000 species could be
largely stripped from the membrane by dilute alkali treatment, but were not
substantially solubilized by nonionic detergents (Palfrey and Waseem, 1985), properties diagnostic of membrane skeleton proteins in the erythrocyte membrane (Yu et al.,
1973). Starting with a membrane skeleton preparation we purified these two proteins
as a complex (Waseem and Palfrey, 1988). Hydrodynamic analysis suggested that the
complex exists as a highly asymmetric heterodimer. For example, gel filtration yields
an apparent M_r of 670,000, whereas that calculated from sedimentation analysis was

M_r 253,000 (Fig. 4). While these studies were in progress, Gardner and Bennett (1986) purified a protein complex from the same source on the basis of its ability to bind CaM. We confirmed that the protein we had isolated could bind CaM by the criterion of CaM affinity chromatography (Waseem and Palfrey, 1988). Electron microscopy of the purified protein suggests that it has a flattened disc shape that is compatible with the available hydrodynamic data (Gardner and Bennett, 1986). Later, Gardner and Bennett (1987; see Bennett et al., 1988) named this protein "adducin" (from the Greek, meaning "to gather together"). They showed that adducin could form a high affinity complex with spectrin and actin, but not with either protein alone, and that once formed this complex could recruit further spectrin molecules. This model has been disputed by Mische et al. (1987), who claim that adducin alone can bundle actin filaments and that the composition of the final complex between adducin, actin, and spectrin is independent of the pathway used to assemble the complex. We are currently

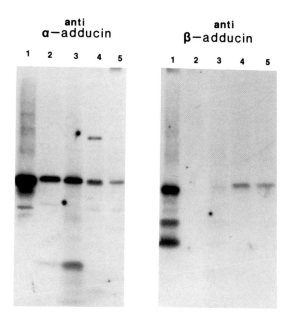

anti α—adducin

1 2 3 4 5

anti β—adducin

1 2 3 4 5

Figure 5. Proteins serologically related to α- and β-adducin are found in other cell types. Affinity-purified rabbit anti–human red cell α- and β-adducin antibodies were used to probe nitrocellulose blots of proteins from lanes *1*; human erythrocyte ghosts (5 µg); lanes *2*, platelets (50 µg); lanes *3*, lymphocytes (50 µg); lanes *4*, rat brain synaptosomal cytosol (60 µg); and lanes *5*, rat brain synaptic membrane (25 µg). Immunologically reactive proteins were localized by the ^{125}I–protein A technique. Note the poor reactivity of β-adducin antibodies with platelet and lymphocyte material and the existence of immunoreactive breakdown products of both adducin subunits.

assessing these competing claims and are also investigating the possibility that in other cell types adducin may bind to other cytoskeletal elements.

We have investigated the relationship between the α (M_r 120,000) and β (M_r 110,000) subunits of adducin in some detail. A limited degree of homology is seen in peptide maps of the iodinated subunits (Gardner and Bennett, 1986, Waseem and Palfrey, 1988), but polyclonal antibodies to the individual subunits do not cross-react with the other, which indicates divergent regions within the two proteins. Homologues of both α-adducin and β-adducin are found in other tissues, including red cells from other species (Waseem and Palfrey, 1988) and other blood cell types (Fig. 5). In both human platelets and lymphocytes, however, only very weak cross-reactivity with a β-adducin–like protein is found. In rat brain synaptic material (synaptic membranes and synaptosomal cytosol) α-adducin and β-adducin cross-reacting polypetides appear to comigrate, which may represent overlapping or identical proteins in this tissue. We have also found an α-adducin, but not a β-adducin–like protein in cultured human fibroblasts (Waseem et al., 1988). The finding that adducin may have a widespread distribution is of interest in the context of previous studies that have documented the existence of other erythrocyte membrane skeletal proteins in various tissues (for review see Bennett, 1985). Whether the structural interrelationships and functioning of these proteins in different cells resemble those in the erythrocyte membrane skeleton remains to be determined.

The existence of antibodies to adducin allowed us to reinvestigate the extraction behavior of the two proteins from ghost membranes by immunoblotting. In contrast to our previous assumption, a significant amount (15–40%) of adducin is solubilized from

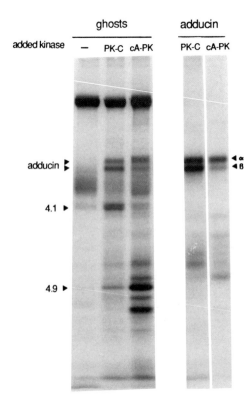

Figure 6. Phosphorylation of adducin in ghost membranes and in solution by PK-C- and cAMP-dependent protein kinase. Normal ghosts were prepared and incubated with γ-[^{32}P]ATP in the presence of 1 mM EGTA (*left lane*) or with added PK-C (5 μg/ml bovine brain enzyme with 0.2 mM Ca and 50 nM TPA present) or cAMP-PK catalytic subunit (2 μg/ml with EGTA present). Adducin was purified from membrane skeletons as described (Waseem and Palfrey, 1987), and 8 μg was incubated either with PK-C (as above but with 50 μg/ml phosphatidylserine also present) or cAMP-PK catalytic subunit (as above). Phosphoproteins were separated by SDS–7.5% PAGE and identified by autoradiography.

the membrane by nonionic detergent (Waseem and Palfrey, 1988). This suggests that adducin may interact with nonskeleton elements in the membrane. The fraction of material released by nonionic detergent appears to increase after treatment of red cells with phorbol esters, suggesting that PK-C-mediated phosphorylation of adducin modulates this interaction. Blotting of [^{125}I]adducin to whole ghost membranes in an attempt to find adducin-binding proteins, did not reveal any significant interactions, and we have no direct evidence at present what the nature of the nonskeletal association might be. It is interesting to note, however, that Wolf and Sahyoun (1986) have shown that a protein of M_r 115,000 present in both erythrocyte and synaptic membranes can bind phosphatidylserine as assayed by a blotting procedure. If this protein is one of the adducin subunits then it could indicate that adducin binds directly

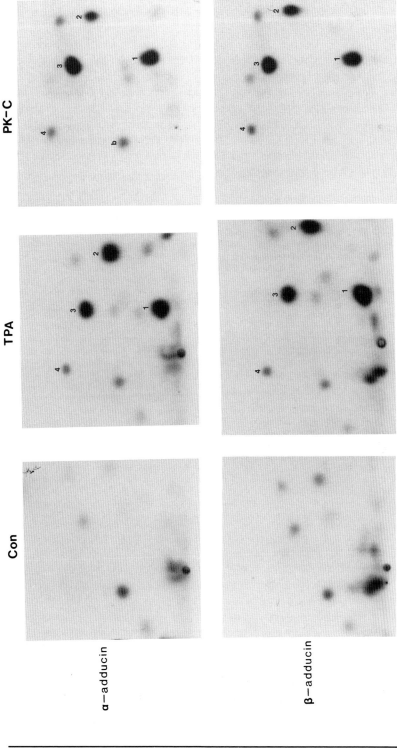

Figure 7. Phosphorylated peptides of α- and β-adducin. Adducin was phosphorylated in ^{32}P-prelabeled intact red cells either without addition or after the addition for 1 h of 1 μM TPA (four left panels.) In the right two panels purified adducin was phosphorylated in vitro with added PK-C (see Fig. 6 legend). After phosphorylation and electrophoresis the α- and β-subunit bands were excised from gels and proteolysed with trypsin. The tryptic phosphopeptides were mapped in two dimensions by sequential electrophoresis (horizontal) and chromatography (vertical). The positions of the major PK-C-stimulated phosphorylated peptides, revealed by autoradiography, are numbered *1–4*. Note their correspondence in the in vivo and in vitro labeled material. Modified from Waseem and Palfrey, 1988.

to the inner face of the bilayer. In a similar context, there are indications that Band 4.1 is also a phosphatidylserine-binding protein (Rybicki et al., 1987).

The phosphorylation of adducin can be stimulated by both phorbol esters and cAMP analogues in intact cells (Ling et al., 1986; Waseem and Palfrey, 1988). Correspondingly, addition of either PK-C or cAMP-dependent protein kinase to both red cell ghosts and purified adducin results in the phosphorylation of α- and β-adducin subunits (Fig. 6). Both are phosphorylated equally in response to PK-C, but β-adducin is preferred by cAMP-dependent protein kinase. Peptide maps of $[^{32}P]\alpha$- and β-adducin show multiple phosphorylated peptides (Fig. 7). The major stimulated phosphorylation sites on both subunits have identical mobilities after either TPA treatment of intact cells or PK-C catalyzed phosphorylation of the isolated protein (Fig. 7). These results indicate that the adducin subunits probably exhibit considerable homology around their sites of phosphorylation.

Perspective

The human erythrocyte contains several identified protein kinases and many of the targets for these enzymes are membrane skeletal proteins (Table I). Of these, PK-C

TABLE I
Protein Kinases of the Human Erythrocyte:

Kinase	Location	Membrane substrates
cAMP-PK	Membrane/cytosol (type I) Cytosol (type II)	Ankyrin, adducin, bands 4.1, 4.9, 7
Casein kinase	Membrane/cytosol (type I) Cytosol (type II)	Spectrin, ankyrin, glycophorin, bands 3, 4.1
Tyrosine kinase	Membrane	band 3
PK-C	Cytosol/membrane	Adducin, bands 4.1, 4.9, glucose transporter

appears to be the only enzyme that is activated by Ca^{2+}. Our attempts to find a CaM-dependent kinase in human erythrocytes have been negative, despite claims to the contrary (e.g., Cohen and Foley, 1986). The effects of phosphorylation on the function of erythrocyte membrane skeleton proteins are only now beginning to come to light. For example, Eder et al. (1986), have shown that phosphorylation of Band 4.1 by casein kinase leads to a reduced interaction of the protein with spectrin. In this regard the functional consequences of adducin phosphorylation, particularly with respect to its interaction with spectrin, actin, and other nonskeleton components, are clearly of interest. The multiplicity of interactions that have already been demonstrated for adducin suggest that it may have an important regulatory role in the control of membrane and cytoskeletal behavior in many cell types. As yet, no first messenger has been conclusively shown to be coupled to a second messenger and its dependent protein kinase in the human red cell. Attempts to find hormones that may activate the PI cycle in human red cells have been unsuccessful (e.g., Chandra Sekar and Hokin, 1986); thus, a physiological stimulus that could lead to DAG formation and PK-C activation has not been found. However, there is a possibility that pathological rises in intracellular Ca^{2+}, perhaps brought on by shear stress or membrane leakiness, could lead to DAG formation, as red cells possess a Ca^{2+}-sensitive phospholipase C that cleaves phosphoinositides (Downes and Michell, 1981). Despite this disadvantage the red cell remains a most amenable system to assess the effects of phosphorylation on membrane skeleton behavior.

Acknowledgments

We thank Karen Cho for technical assistance.

This work was supported by National Institutes of Health grant GM-33546.

References

Ballester, R., and O. M. Rosen. 1985. Fate of immunoprecipitable protein kinase C in GH3 cells treated with PMA. *Journal of Biological Chemistry.* 260:15194–15199.

Bennett, V. 1985. The membrane skeleton of human erythrocytes and its implications for more complex cells. *Annual Review of Biochemistry.* 54:273–304.

Bennett, V., J. Davis, K. Gardner, and J. P. Steiner. 1988. The spectrin-based membrane skeleton: extensions of the current paradigm. *In* Cell Physiology of Blood. R. B. Gunn and J. C. Parker, editors. The Rockefeller University Press, New York, NY. *Society of General Physiologists Series.* 43:101–109.

Chandra Sekar, M., and L. E. Hokin. 1986. Phosphoinositide metabolism and cGMP levels are not coupled to the muscarinic cholinergic receptor in human erythrocytes. *Life Sciences* 39:1257–1262.

Cohen, C. M. and S. Foley. 1986. Phorbol ester and Ca^{2+}-dependent phosphorylation of human red cell membrane skeletal proteins. *Journal of Biological Chemistry.* 261:7701–7709.

Coussens, L., P. J. Parker, L. Rhee, T. L. Yang-Fang, E. Chen, M. D. Waterfield, U. Francke, and A. Ullrich. 1986. Multiple distinct forms of bovine and human protein kinase C suggest diversity in cellular signalling pathways. Science. 233:859–866.

DeRiemer, S. A., J. A. Strong, K. A. Albert, P. Greengard, and L. K. Kaczmarek 1985. Enhancement of calcium current in Aplysia neurons by protein kinase C. Nature. 313:313–316.

Downes, C. P., and R. H. Michell. 1981. The polyphosphoinositide phosphodiesterase of erythrocyte membranes. *Biochemical Journal.* 198:133–140.

Eder, P. S., C. J. Soong, and M. Tao. 1986. Phosphorylation reduces the affinity of protein 4.1 for spectrin. *Biochemistry.* 25:1764–1770.

Faquin, W. C., S. B. Chahwala, L. C. Cantley, and D. Branton. 1986. Protein kinase C of erythrocytes phosphorylates Bands 4.1 and 4.9. *Biochimica Biophysica Acta.* 887:142–149.

Gardner, K., and V. Bennett. 1986. A new erythrocyte membrane-associated protein with calmodulin-binding activity. *Journal of Biological Chemistry.* 261:1339–1348.

Gardner, K., and V. Bennett. 1987. Modulation of spectrin-actin assembly by erythrocyte adducin. *Nature.* 328:359–362.

Hannun, Y., and R. M. Bell. 1986. Mechanism of activation of protein kinase C: role of diacylglycerol and calcium second messengers. *In* Cell Calcium and the Control of Membrane Transport. L. J. Mandel and D. C. Eaton, editors. The Rockefeller University Press, New York, NY. *Society of General Physiologists Series.* 40:229–240.

Horne, W. C., T. L. Leto, and V. Marchesi. 1985. Differential phosphorylation of multiple sites in protein 4.1 and protein 4.9 by phorbol ester-activated and cyclic AMP-dependent protein kinases. *Journal of Biological Chemistry.* 260:9073–9076.

Huang, K.-P., H. Nakabayashi, and F. L. Huang. 1986. Isozymic forms of rat brain Ca^{2+}-activated and phospholipid-dependent protein kinase. *Proceedings of the National Academy of Sciences.* 83:8535–8539.

Hunter, T., N. Ling, and J. A. Cooper. 1984. Protein kinase C phosphorylation of the EGF receptor at a threonine residue close to the cytoplasmic face of the plasma membrane. *Nature.* 311:480–483.

Jaken, S. and S. C. Kiley. 1987. Purification and characterization of 3 types of protein kinase C from rabbit brain cytosol. *Proceedings of the National Academy of Sciences* 84:4418–4422.

Kariya, K., and Y. Takai. 1987. Distinct functions of down-regulation-sensitive and -resistant types of protein kinase C in rabbit aorta smooth muscle cells. *FEBS Letters.* 219:119–124.

Kikkawa, U., Y. Ono, K. Ogita, T. Fuju, Y. Asaoka, K. Sekiguchi, Y. Kosaka, K. Igarishi, and Y. Nishizuka. 1987. Identification of the structures of multiple subspecies of protein kinase C expressed in rat brain. *FEBS Letters.* 217:227–231.

Kishimoto, A., N. Kajikawa, M. Shiota, and Y. Nishizuka. 1983. Proteolytic activation of calcium-activated, phospholipid-dependent protein kinase by calcium neutral protease. *Journal of Biological Chemistry.* 258:1156–1164.

Knopf, J. L., M.-H. Lee, L. A. Sultzman, R. W. Kriz, C. R. Loomis, R. M. Hewick, and R. M. Bell. 1986. Cloning and expression of multiple protein kinase C cDNAs. *Cell.* 46:491–502.

Ling, E., and V. Sapirstein. 1984. Phorbol ester stimulates the phosphorylation of rabbit erythrocyte Band 4.1. *Biochemical and Biophysical Research Communications* 120:291–298.

Ling, E., K. Gardner, and V. Bennett. 1986. Protein kinase C phosphorylates a recently identified membrane skeleton-associated calmodulin-binding protein in human erythrocytes. *Journal of Biological Chemistry.* 261:13875–13878.

Manalan, A. and C. B. Klee. 1984. Calmodulin. *Advances in Cyclic Nucleotide and Protein Phosphorylation Research.* 18:227–278.

Marchesi, V. 1985. Stabilizing infrastructure of cell membranes. *Annual Review of Cell Biology.* 531–562.

May, W. S., N. Sahyoun, S. Jacobs, M. Wolf, and P. Cuatrecasas. 1985. Mechanism of phorbol diester-induced regulation of surface transferrin receptor involves the action of activated protein kinase C and an intact cytoskeleton. *Journal of Biological Chemistry.* 260:9419–9426.

McCaffrey, P. G., M. Rosner, U. Kikkawa, K. Sekiguchi, K. Ogita, K. Ase, and Y. Nishizuka. 1987. Characterization of protein kinase C from normal and transformed murine fibroblasts. *Biochemical and Biophysical Research Communications.* 146:140–146.

Mische, S. M., M. Mooseker, and J. S. Morrow. 1987. Erythrocyte adducin: a calmodulin-regulated actin-bundling protein that stimulates spectrin-actin binding. *Journal of Cell Biology.* 105:2837–2846.

Michell, R. H., C. E. King, C. E. Piper, L. R. Stephens, C. M. Bunce, G. R. Guy, and G. Brown. 1988. Inositol lipids and phosphates in erythrocytes and HL60 cells. *In* Cell Physiology of Blood. R. B. Gunn and J. C. Parker, editors. The Rockefeller University Press, New York, NY. *Society of General Physiologists Series.* 43:000–000.

Muldoon, L. L., G. A. Jamieson, Jr., A. C. Kao, H. C. Palfrey, and M. L. Villereal. 1987. Mitogen stimulation of Na/H exchange: differential involvement of protein kinase C. *American Journal of Physiology.* 253:C219–229.

Neidel, J. E., and P. J. Blackshear. 1986. Protein kinase C. *In* Phosphoinositides and Receptor Mechanisms. J. Putney, editor. Alan R. Liss Inc., New York, NY. 47–88.

Nishikawa, M., H. Hidaka, and R. S. Adelstein. 1983. Phosphorylation of smooth muscle heavy meromyosin by calcium-activated, phospholipid-dependent protein kinase. *Journal of Biological Chemistry.* 258:14069–14073.

Nishizuka, Y. 1986. Studies and perspectives on protein kinase C. *Science.* 233:305–313.

Palfrey, H. C., and A. Waseem. 1985. Protein kinase C in the human erythrocyte. *Journal of Biological Chemistry.* 260:16021–16029.

Rybicki, A., R. Heath, B. Lubin, and R. S. Schwartz. 1988. Human erythrocyte protein 4.1 is a phosphatidylserine binding protein. *Journal of Clinical Investigation.* 81:255–260.

Waseem, A., and H. C. Palfrey. 1988. Erythrocyte adducin: comparison of the α- and β-subunits. *European Journal of Biochemistry.* In press.

Waseem, A., K. Cho, and H. C. Palfrey. 1988. Human fibroblast α-adducin is phosphorylated by protein kinase C in response to mitogens. *FASEB Journal.* 2:A349. (Abstr.)

Werth, D. K., J. E. Niedel and I. Pastan. 1983. Vinculin: a cytoskeletal substrate of protein kinase C. *Journal of Biological Chemistry.* 258:11423–11426.

Witters, L. A., C. A. Vater, and G. E. Lienhard. 1985. Phosphorylation of the glucose transporter *in vitro* and *in vivo* by protein kinase C. *Nature.* 315:777–778.

Wolf, M., and N. Sahyoun. 1986. Protein kinase C and phosphatidylserine bind to Mr 110,000/115,000 polypeptides enriched in cytoskeletal and postsynaptic density preparations. *Journal of Biological Chemistry.* 261:13327–13332.

Woodgett, J. R., and T. Hunter. 1987. Immunological evidence for two physical forms of protein kinase C. *Molecular and Cellular Biology.* 7:85–96.

Young, S., P. J. Parker, A. Ullrich, and S. Stabel. 1987. Down-regulation of protein kinase C is due to an increased rate of degradation. *Biochemical Journal.* 244:775–779.

Yu, J., D. A. Fischman, and T. L. Steck. 1973. Selective solubilization of proteins and phospholipids from red blood cell membranes by nonionic detergents. *Journal of Supramolecular Structure.* 1:233–240.

Chapter 30

Biochemical and Molecular Events Controlled by Lymphokine Growth Factors

William L. Farrar, Annick Harel-Bellan, and Douglas K. Ferris

Laboratory of Molecular Immunoregulation, Biological Response Modifier Program, Division of Cancer Treatment, and Program Resources, Inc., National Cancer Institute, Frederick Cancer Research Facility, Frederick, Maryland

Introduction

The proliferation and differentiation of hematopoietic cells of both lymphoid and myeloid lineages are governed by various polypeptide growth factors known as lymphokines. Binding of a given lymphokine by its specific high affinity cell surface receptor initiates a signal transduction system that traverses the cell membrane, eliciting a variety of cytosolic and nuclear changes related to growth or differentiation.

For a cell to grow and divide into identical daughter cells, a number of events must take place. Obviously there must be an increase in cellular components, DNA must be duplicated, organelles must be constructed, new genes specifically related to growth are activated, and cellular protein must be roughly doubled. These events take place in a spacial and temporal pattern that is precisely organized.

First, a complementary identification between the secondary structures of the external ligand, informational substance, and its molecular "antennae" (receptors) must occur in a way that allows specific recognition. Several obligatory requirements are necessary before a membrane protein can be seriously classified as a receptor. The membrane binding protein must exhibit saturable and displaceable binding of the ligand within physiological concentrations. The interactions between ligand and receptor must be transformed into another chemical form, second messengers, which now gives "new" biochemical instructions to the cell. Finally, a ligand receptor interaction must modify the biological response of the organism that expresses the particular receptor.

The hematopoietic growth factors multi–colony-stimulating factor (multi-CSF or IL-3), granulocyte/macrophage colony-stimulating factor (GM-CSF), granulocyte colony-stimulating factor (G-CSF), and interleukin 2 (IL-2) specifically control the production and proliferation of distinct leukocyte series. Each growth factor is structurally unique and acts on respective cell surface receptors. The biochemical mechanisms by which lymphokine receptors affect growth or differentiation has only recently been addressed. Here, we will summarize data from experiments that have studied lymphokine-initiated signal transduction and effects on gene expression. These studies focus on lymphokine activation of kinase systems, identification of substrates, and the consensus expression of nuclear protooncogenes common to all the major blood cell types. Furthermore, we provide evidence that part of the biochemical response to lymphokines involves an extraordinarily ancient set of genes, collectively referred to as heat shock genes. This suggests that the cellular response to growth factors incorporates elements evolved from primitive stress-response stimuli.

Results and Discussion

Alterations in the phosphorylation state of both membrane and cytosolic proteins are among the earliest changes that can be measured after growth factor binding in a variety of mammalian cell types (Hunter and Cooper, 1985). Modification in amounts of covalently bound phosphate may alter the conformation, intracellular location, or enzymatic activity of a protein. In fact, there are few if any major cellular activities that are not regulated at some level by reversible phosphorylation of structural proteins or enzymes. Several growth factor receptors possess intrinsic protein kinase activity and many hormones and growth factors rapidly activate membrane and/or cytosolic protein kinase activities (Kolata, 1983; Hunter, 1984). For these reasons, reversible

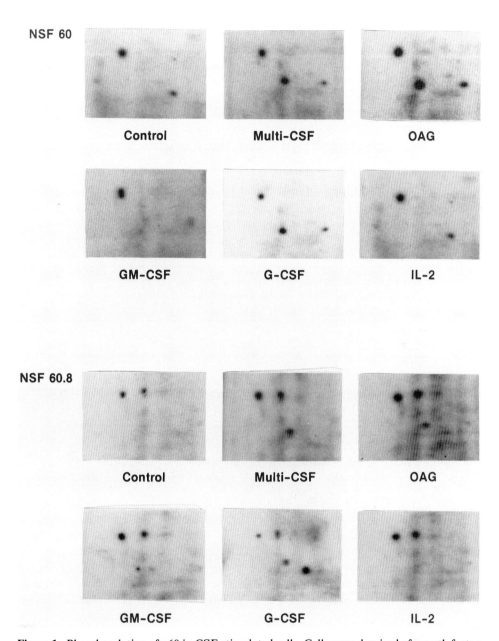

Figure 1. Phosphorylation of p68 in CSF-stimulated cells. Cells were deprived of growth factor and preequilibrated with [^{32}P]P$_i$. P$_i$-loaded cells were stimulated for 10 min with 1,000 U of IL-2/ml, 10^5 of multi-CSF, GM-CSF, or G-CSF/ml or 1 μg of OAG/ml, and then phosphoproteins were separated by two-dimensional electrophoresis (nonequilibrated pH gel electrophoresis and SDS-PAGE). The resolved phosphoproteins were revealed by autoradiography.

phosphorylation is thought to be one of the principal mechanisms by which external signals are transduced to the nucleus.

p68 Phosphorylation

While cells of different mammalian species, or even cells from different tissues in a single species, may synthesize a few unique proteins, by far the majority of proteins found in any particular cell are present in most cells of the organism and homologous proteins are found throughout the animal phyla. Thus, in examining the protein phosphorylation changes that are observed after growth factor stimulation of cells, attention is usually focused on those that are common in several cell types in the belief that these may be of more fundamental importance.

Previous work in our laboratory has shown that a 68-kD protein (p68) is rapidly phosphorylated in two myeloid and two lymphoid cell lines after stimulation with appropriate growth factors, interleukin 2 (IL-2), multi-CSF or IL-3, and G-CSF (Evans and Farrar, 1987a; Evans et al., 1987). Phosphorylation of p68 was also induced in all four cell lines by synthetic diacylglycerol, a direct activator of protein kinase C (PK-C). In all cases, p68 was only phosphorylated on threonine residues and

TABLE I

Comparison of the Effects of Physiological Growth Factors and Pharmacological Kinase Activators on Nuclear Protooncogene Expression in Lymphoid and Myeloid Cells

Oncogene	GF	PK-C	PK-A	GF+ PK-C	GF+ PK-A	PK-C+ PK-A
c-*fos*	+	+	+	+	+ +	+ + + +
c-*myc*	+ + + +	+ +	−	+ + + +	+/−	−
c-*myb*	+ + +	ND	+ +	+ + +	+ +	ND
ODC	+ +	+	−	+ +	−	ND

GF, growth factor; PK-C, activation of protein kinase C by PMA; PK-A, activation of protein kinase A by 8-bromo-cAMP. The number of +'s indicates the level of mature message detected on a Northern blot, performed using standard procedures.

tryptic peptide analysis indicated that the same residues were phosphorylated in response to growth factors and diacylglycerol. Fig. 1 shows a two-dimensional analysis of phosphoproteins stimulated by CSFs in two separate CSF-dependent cell lines, NSF60 (multi-CSF-dependent) and NSF60.8 (GM-CSF- or G-CSF-dependent). Treatment of NSF60 with either multi-CSF or the diacylglycerol analogue 1-oleoyl-2-acetylglycerol (OAG) stimulated phosphorylation of a 68-kD substrate (center). G-CSF also stimulated p68 phosphorylation and GM-CSF or IL-2 did not. IL-2 or GM-CSF do not provoke a biological response with this particular cell line. The NSF60.8 cell line proliferates in response to multi-CSF, GM-CSF, or G-CSF. With this cell line only multi-CSF and G-CSF induced phosphorylation of p68. IL-2 only induced phosphorylation in lymphoid cell lines that possess high affinity receptors (Evans and Farrar, 1987a). It therefore appears that phosphorylation of p68 is a common consequence of growth stimulation by IL-2, multi-CSF, and G-CSF. Interestingly, one of the cell lines (NSF60.8) can grow in response to either G-CSF or GM-CSF, yet p68 phosphorylation was only induced by G-CSF. This indicates the presence, within a single cell type, of two alternate signal transduction pathways leading to growth.

S6 Phosphorylation

The proliferation of cells requires the synthesis of a large amount of protein. Most of the newly synthesized protein in actively growing cells represents increased synthesis of proteins that are also expressed in cells arrested in G_1 by growth factor deprivation while a few proteins, such as histone H1 and some protooncogenes, are only synthesized in growing cells at specific stages of the cell cycle. The increase in protein synthesis following serum or growth factor stimulation of quiescent cells is primarily due to an increased rate of translation initiation. Work in our laboratory and others has shown that shortly after growth factor stimulation, the S6 protein of the 40S ribosomal

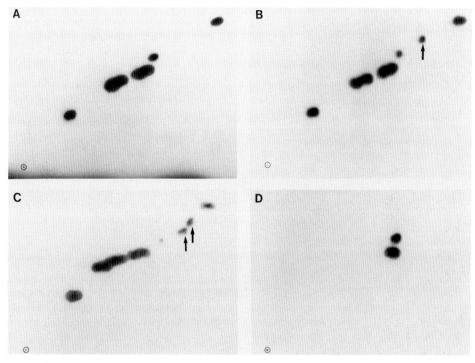

Figure 2. Peptide maps of S6 proteins phosphorylated in situ and in vitro. S6 protein was purified from intact cells or in vitro kinase reactions and subjected to complete tryptic digestion. The resultant peptides were separated by two-dimensional thin-layer electrophoresis and analyzed by autoradiography. (*A*) S6 peptides from IL-2-stimulated cells; (*B*) S6 peptides from OAG-stimulated cells; (*C*) S6 peptides after cell-free S6 kinase phosphorylation; (*D*) S6 peptides after cell-free protein kinase C phosphorylation.

subunit is phosphorylated at multiple sites (Thomas et al., 1982; Evans and Farrar, 1987b). Thomas and co-workers have shown that the phosphorylation of S6 is correlated with, and precedes, the formation of polysomes and initiation of protein synthesis. We found that both IL-2 and OAG, a synthetic direct activator of PK-C, induced phosphorylation of S6 with similar kinetics, and resulted in identical tryptic phosphopeptide patterns. Phosphorylation of purified S6 protein by purified PK-C in vitro, however, resulted in a quite different phosphopeptide map. In fact, we and others found that a distinct S6 protein kinase activity was present in extracts of growth factor–stimulated cells. While examining lymphocyte cytosol for PK-C activity, another kinase that was phospholipid- and Ca^{++}-independent was discovered. This kinase could equally phosphorylate the S6 protein and was clearly biochemically distinct from PK-C.

Fig. 2 shows the results of a comparative study between PK-C and the S6 kinase obtained from the IL-2-dependent CT6 cell line. Fig. 2 *A* shows the phosphopeptide map of S6 obtained from CT6 cells stimulated in situ with IL-2. Opposite (panel *B*) is a very similar digest pattern seen when intact cells are treated with OAG. However, when S6 protein was phosphorylated in vitro with either purified PK-C (panel *D*) or S6 kinase (panel *C*), only S6 kinase phosphorylated the ribosomal protein in a manner analogous to the physiological stimulant, IL-2. Our results indicated that while IL-2 or OAG stimulation caused rapid activation of PK-C, the protein kinase directly responsible for S6 phosphorylation was separate from, and possibly regulated by, PK-C.

Heat Shock Proteins and Lymphocyte Growth

Since heat shock proteins (HSPs) were first reported in fruit flies there has been a slowly growing interest in these ubiquitous, highly conserved proteins. It is now widely

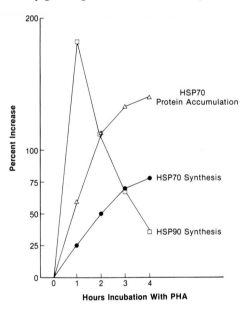

Figure 3. Mitogen stimulation of heat shock protein synthesis in human T lymphocytes. Resting human T lymphocytes were stimulated with PHA for the indicated times and then pulse labeled with [^{35}S]methionine. The labeled proteins were separated on a 10% SDS gel, transferred to nitrocellulose, immunostained, and autoradiographed. Data are expressed as percent increase in HSP70 and HSP90 synthesis rates and protein accumulation at 0–4 h after PHA stimulation.

believed that in addition to providing protection from thermal and other stresses, these proteins are also involved in cell growth and differentiation. This belief arises from the observations that: (*a*) expression of several heat shock proteins is tightly coupled to development in fruit flies (Craig et al., 1983), in amphibian oocytes (Bienz, 1984*a,b*), and in mammals (Barnier et al., 1987); (*b*) in mammals, HSP70 (which is 50% homologous to the *Escherichia coli* dnaK gene product [Bardwell and Craig, 1984]) translocates to the nucleus during S phase (Milarski and Morimoto, 1986) and has been found in association with p53 (Pinhasi-Kimhi et al., 1986), a nuclear proto-oncogene; (*c*) HSP90 has been found in association with steroid receptors and with pp60 src (Schuh et al., 1985).

We have examined the synthesis of HSP70 and HSP90 during the growth of human T lymphocytes. Proliferation of T lymphocytes requires at least two sequential extracellular signals, first an activation signal by antigen (which can be mimicked in situ by nonspecific polyclonal activators such as phytohemagglutinin [PHA]), and secondly, stimulation by IL-2, which allows the activated cells to proceed through the cell cycle. Fig. 3 shows a graphic representation of the patterns of HSP90 protein synthesis and HSP70 protein synthesis and accumulation. We found that HSP90 protein synthesis was rapidly and transiently increased in response to PHA, while HSP70 protein synthesis did not rise until later and appeared to be part of a generalized increase in protein biosynthesis. We also found that HSP70 steady-state mRNA was not increased by PHA.

After IL-2 stimulation of IL-2-dependent cells that had been deprived of the growth factor, HSP70 steady-state mRNA and protein synthesis were specifically increased, while HSP90 protein synthesis was delayed. These observations indicated that synthesis of HSP70 and HSP90 is correlated with growth in human T lymphocytes and that both transcriptional and posttranscriptional regulation of HSP70 occurs. The rapid increase in HSP90 protein synthesis after PHA may indicate an important role for this protein in the activation and differentiation of lymphocytes, while specific stimulation of HSP70 mRNA and protein synthesis by IL-2 favors a role in IL-2-directed growth (Ferris et al., 1988).

Gene Activation by Lymphokines

Since growth lymphokines stimulate cell cycle progression, it is obvious that the signal transduced from the membrane must alter or induce the expression of genes that biochemically promote cell cycle or differentiative function. Some of the transcriptional activation by IL-2 or CSFs have been recently identified. Most notable are the members of the nuclear protooncogene family, c-*fos*, c-*myc*, and c-*myb*. All the lymphokine growth factors tested thus far (IL-2, multi-CSF, G-CSF, GM-CSF) stimulate the sequential transcription of these genes in their respective cell lines (Cleveland et al., 1987; Harel-Bellan and Farrar, 1987). Moreover, the sequential expression of c-*fos* followed by c-*myc* is clearly a consensus nuclear event shared by fibroblastic cell lines responding to serum-derived growth factors (Greenberg and Ziff, 1984). Other gene transcripts, such as HSP70 or ornithine decarboxylase (Farrar et al., 1988), are also "consensus genes" activated in diverse tissues responding to proliferative signals.

Effects of Pharmacological Agonists of Protein Kinases on Activation Gene Expression

The two unique protein kinases involved in signal transduction by several hormones, the Ca^{++}/lipid-dependent PK-C and the cAMP-dependent protein kinase A, are also connected with the consensus genes induced by lymphokines. The tumor promoter phorbol myristate acetate (PMA), a pharmacological activator of PK-C, induces c-*fos* and c-*myc* expression and ODC mRNA in growth-arrested lymphoid and myeloid cells (Farrar et al., 1986; Harel-Bellan and Farrar, 1987). Direct activation of protein kinase A by analogues of cAMP such as 8-Br-cAMP, which strongly inhibit the proliferation of lymphoid and myeloid cells, have differential effects on the consensus gene family. It results in superinduction of c-*fos* mRNA, whereas most of the other genes are suppressed by addition of protein kinase A agonists. For example, IL-2-induced c-*myc* and ODC gene expression in T cell lines is inhibited by 8-Br-cAMP (Farrar et al., 1987; Harel-Bellan and Farrar, 1987).

It is clear from a number of studies that activators of PK-C generally mimick growth factor modulation of gene transcription. This does not exclude that growth factors activate discrete pathways unrelated to PK-C or unaccessible to pharmacological activation. In all probability, other pathways remain to be discovered. The ability of cAMP analogues to suppress growth and inhibit the expression of c-*myc* mRNA accumulation may be one mechanism by which this pathway affects the proliferative response to growth factors. Inhibition of c-*myc* expression by interferon or complementary antisense oligonucleotides (Harel-Bellan et al., 1988) also reduces the proliferative response to hematopoietic growth factors. Given the universal association of c-*myc* mRNA expression with the proliferative action of growth factors, the negative control of c-*myc* by cAMP-dependent intracellular pathways or others may provide clues to the function of the c-*myc* protein.

The complex mechanisms by which extracellular signals control the function of the cell still remain largely unknown. We have identified a small number of the biochemical and molecular changes observed when either myeloid or lymphoid growth factors stimulate their target cells. All the growth factors apparently regulate the phosphorylation of cytosolic proteins. One of these, 40S ribosomal S6 protein, is in fact phosphorylated by a kinase that is distinct from the one initially stimulated by diacylglycerol. This suggests a concept long sought, namely, that intrakinase cascades are activated in a manner whereby the activation of one system affects others "downstream." We can envision a wave of kinase activities occurring to control the large number of proteins whose biochemical actions are regulated by apparent phosphorylation. As described here and elsewhere (Evans and Farrar, 1987*b*), the phosphorylation status of the S6 protein was proportional to the protein chain elongation of the polysomes.

Among the consensus nuclear events described for all the lymphokine growth factors and serum-related proteins (PDGF, etc.) is the regulation of the expression of the nuclear protooncogene family. Their conservation between tissue types and transcriptional responsiveness to a broad spectrum of growth stimulants heavily indicates a critical role in the control of growth or differentiation. Phorbol esters, which promote tumor development and growth in normal cells, also stimulate c-*myc* expression. cAMP analogues, which in many instances reverse transformation and inhibit normal cell growth, inhibit stable steady-state c-*myc* mRNA accumulation. More recently, we have shown that inhibition of c-*myc* biosynthesis with antisense oligonucleotides suppresses the proliferative response of lymphocytes to IL-2 (Harel-Bellan et al., 1988). Whether the inhibition of c-*myc* biosynthesis effects subsequential gene expression or DNA polymerase activity is currently under investigation. The development of oligomer antisense technology, however, will provide a useful tool to delete certain inducible proteins from the intracellular response and allow the determination of their role in the complex coordination of intracellular events associated with growth factor initiation of cell cycle progression.

Summary

The polypeptide hormones that govern the proliferation and differentiation of the mature immune system and hematopoiesis are collectively referred to as lymphokines. We have examined a number of biochemical and molecular events stimulated by several unique lymphokines that exhibit proliferative activity on lymphoid and myeloid cell lines. IL-2 and several members of the colony-stimulating factors (multi-CSF, G-CSF, and GM-CSF) stimulate a similar pattern of cellular phosphorylation, including the prominent phosphorylation of a 68-kD substrate present in numerous distinct lineage cell lines. The 68-kD substrate is phosphorylated by protein kinase C on threonine residues and is primarily cytosolic. Another kinase system activated by either physiological ligand or synthetic diacylglycerol phosphorylated the 40S ribosomal S6 protein in a dose-dependent manner. The increased phosphorylation of S6 protein was associated with enhanced chain elongation in vitro. The kinase responsible for the in situ phosphorylation, however, was not protein kinase C but another physicochemically distinct Mg^{++}-dependent enzyme (termed S6 kinase). These studies suggested that although protein kinase C was activated by diacylglycerol, another kinase, S6 kinase, was the effector enzyme involved in the phosphorylation of the 40S protein. IL-2 and all other lymphokines tested stimulated the transcription of the nuclear protooncogenes c-*fos*, c-*myc*, and c-*myb*, as well as a member of the heat shock family of proteins, HSP 70. Phorbol esters also stimulated similar gene expression; however, cAMP analogue inhibited phorbol ester– or ligand-induced c-*myc* expression. cAMP agonists are antiprolifeative to all the growth factors tested.

References

Bardwell, J. C., and E. A. Craig. 1984. Major heat shock gene of Drosophila and the *Excherichia coli* heat inducible dnaK gene are homologous. *Proceedings of the National Academy of Sciences.* 81:848–852.

Barnier, V. B., O. Bensaude, M. Morange, and C. Babinet. 1987. Mouse 89 kD heat shock protein. *Experimental Cell Research.* 170:186–194.

Bienz, M. 1984*a*. Developmental control of heat shock response in Xenopus. *Proceedings of the National Academy of Sciences.* 81:3138–3142.

Bienz, M. 1984*b*. Xenopus hsp 70 genes are constitutively expressed in injected oocytes. *EMBO Journal.* 11:2477–2483.

Cleveland, J. L., U. R. Rapp, and W. L. Farrar. 1987. Role of c-myc and other genes in IL2 regulated CT6 T lymphocytes and their malignant variants. *Journal of Immunology.* 138:3495–3504.

Craig, E. A., T. D. Ingolia, and L. J. Manseau. 1983. Expression of *Drosophila* heat shock cognate genes during heat shock and development. *Development Biology.* 99:418–426.

Evans, S. W., and W. L. Farrar. 1987*a*. Identity of common phosphoprotein substrates stimulated by interleukin 2 and diacylglycerol suggests a role of protein kinase C for IL2 signal transduction. *Journal of Cellular Biochemistry.* 34:47–59.

Evans, S. W., and W. L. Farrar. 1987*b*. Interleukin 2 and diacylglycerol stimulate phosphorylation of 40 S ribosomal S6 protein. *Journal of Biological Chemistry.* 262:4624–4630.

Evans, S. W., D. Rennick, and W. L. Farrar. 1987. Identification of a signal-transduction pathway shared by haematopoietic growth factors with diverse biological specificity. *Biochemistry Journal.* 244:683–691.

Farrar, W. L., S. W. Evans, F. W. Ruscetti, E. Bonvini, H. A. Young, and M. B. Sparks. 1986. Biochemical mechanisms of interleukin 2 regulation of lymphokine growth. *In* Role of

Leukocytes in Host Defense. J. J. Oppenheim and D. Jacobs, editors. Alan R. Liss, Inc., New York, NY. 75–82.

Farrar, W. L., S. W. Evans, U. R. Rapp, and J. L. Cleveland. 1988. Effects of anti-proliferative cyclic adenosine 3'5' monophosphate on interleukin 2 stimulated gene expression. *Journal of Immunology.* 139:2075–2080.

Farrar, W. L., M. Vinocour, J. L. Cleveland, and A. Harel-Bellan. 1988. Regulation of ornithine decarboxylase activity by IL-2 and cyclic AMP. *Journal of Immunology.* 141:967–971.

Ferris, D. K., A. Harel-Bellan, R. Morimoto, W. Welch, and W. L. Farrar. 1988. Mitogen and lymphokine stimulation of heat shock proteins in T lymphocytes. *Proceedings of the National Academy of Sciences.* 85:3850–3854.

Greenberg, M. E., and E. B. Ziff. 1984. Stimulation of 3T3 cells induces transcription of the c-fos proto-oncogene. *Nature.* 311:433–438.

Harel-Bellan, A., and W. L. Farrar. 1987. Modulation of proto-oncogene expression by colony stimulating factors. *Biochemical and Biophysical Research Communications.* 148:1001–1008.

Harel-Bellan, A., D. Ferris, M. Vinocous, J. Holt, and W. L. Farrar. 1988. Specific inhibition of *c-myc* protein biosynthesis using an antisense synthetic deoxy-nucleotide in human T lymphocytes. *Journal of Immunology.* 140:2431–2435.

Hunter, T. 1984. The proteins of oncogenes. *Scientific American.* 251:70–79.

Hunter, T. and J. A. Cooper, 1985. Protein tyrosine kinases. *Annual Review of Biochemistry.* 54:897–930.

Kolata, G. 1983. Is tyrosine the key to growth control. *Science.* 219:377–378.

Milarski, K. L., and R. I. Morimoto. 1986. Expression of human HSP70 during the synthetic phase of the cell cycle. *Proceedings of the National Academy of Sciences.* 83:9517–9521.

Pinhasi-Kimhi, O., D. Michalovitz, A. Ben-Zeev, and M. Oren. 1986. Specific interaction between the p53 cellular tumor antigen and the major heat shock proteins. *Nature.* 320:182–184.

Schuh, S., W. Yonemoto, J. Brugge, V. J. Bauer, R. M. Riehl, W. P. Sullivan, and D. O. Toft, 1985. A 90,000-dalton binding protein common to both steroid receptors and the Rous sarcoma virus transforming protein, pp60v-src. *Journal of Biological Chemistry.* 260:14292–14296.

Thomas, G. J., J. Martin-Perez, M. Siegmann, and A. M. Otto. 1982. Effect of serum, EGF, $PDGF_{2a}$ and insulin on S6 phosphorylation and the initiation of protein and DNA synthesis. *Cell.* 30:235–240.

Chapter 31

Initiation and Termination of Ca^{2+} Signals: Studies in Human Platelets

Timothy J. Rink and Stewart O. Sage

Smith Kline & French Research Ltd., Welwyn, Herts., and Physiological Laboratory, Cambridge, United Kingdom

Cell Physiology of Blood © 1988 by The Rockefeller University Press

Introduction

Despite the discovery of increasing numbers of intracellular messengers over the past decade, Ca^{2+} ions still hold a central place in the activation cascade in many cell types, and probably play a key regulating role in all eukaryotic cells. Indeed some of the other known, putative, cytosolic messenger substances act partly or mainly by influencing Ca^{2+} entry into or exit from the cytosol, e.g., cyclic AMP (cAMP), inositol tris, and tetrakis phosphates.

Among the most important questions in cell activation, therefore, are: how are cytosolic Ca^{2+} signals set up, and how are they ended? In many cells in which rapid intracellular $[Ca^{2+}]$ ($[Ca^{2+}]_i$) transients are the critical trigger, the most important mechanism is rapid Ca^{2+} entry through voltage-gated Ca^{2+} channels. Even in these cells, as with smooth and cardiac muscle, there is also an important component of Ca^{2+} discharge from internal stores. In many nonexcitable cells—i.e., those in which action potentials are not a feature of activation—there can also be rapid Ca^{2+} entry and discharge from internal stores. In the first two sections of this chapter we outline the current state of understanding of these processes and some recent data from our laboratory, considering these processes in human platelets. In the third section we discuss the ways in which elevated $[Ca^{2+}]_i$ is restored to resting levels, with a particular focus on Ca^{2+} extrusion across the plasma membrane, including some of our own recent work and that of others.

Receptor-Mediated Discharge of Ca^{2+}

Evidence for Internal Release

The idea that stimulation, chemical or electrical, at the cell surface could activate cells by causing the discharge of internal stores of Ca^{2+} was originally based on two considerations. First, A. V. Hill (1948) demonstrated that diffusion of Ca^{2+} from the surface to the interior could not be fast enough to account for the speed of activation of skeletal muscle. Second, various cells, including platelets, could respond to stimuli assumed to act at the plasma membrane, with changes supposedly triggered by Ca^{2+}, even in the absence of external Ca^{2+}. More convincing evidence came from finding that the Ca ionophore A23187 could evoke shape change and secretion in the complete absence of external Ca^{2+} (e.g., Feinstein and Fraser, 1975). Since the primary effect of the ionophore is to translate Ca^{2+} across cell membranes this observation suggests the presence of an internal dischargeable store sequestered in an organelle. It was also found that, like most cells, platelets contain intracellular membrane systems analogous to sarcoplasmic reticulum that are capable of accumulating Ca^{2+} in an ATP-dependent manner, even at Ca^{2+} concentrations in the range expected in cytosol (e.g., O'Rourke, 1985). The chief Ca^{2+}-sequestering organelle in platelets is thought to be the "dense tubular system" described by electron microscopists.

The ability of cell stimulation to increase efflux of preloaded ^{45}Ca (e.g., Rink and Sage, 1987) is also consistent with an internal discharge elevating $[Ca^{2+}]_i$, followed by increased Ca^{2+} extrusion; however, one should also consider the possibility that an observed increased efflux may result from stimulation of a Ca^{2+} pump with consequent lowering $[Ca^{2+}]_i$.

With the invention of fluorescent Ca^{2+} indicators and a way to load them into small intact cells (Tsien et al., 1982) it became possible to analyze stimulus-evoked elevation of $[Ca^{2+}]_i$ in many cell types, including platelets. We soon found that stimulation by a range of surface ligands leads to an abrupt rise in $[Ca^{2+}]_i$ when extracellular $[Ca^{2+}]$ ($[Ca^{2+}]_0$) was reduced to <10 nM (Rink et al., 1982), and even when the cells were depolarized in high K^+ (Sage and Rink, 1986*a*), thus ensuring an outward electrochemical gradient on Ca^{2+} and precluding the possibility that Ca^{2+} entry contributed to this signal. Incidently, this observation and the finding that stimulus-evoked elevation of $[Ca^{2+}]_i$ in the absence of external calcium is not diminished when external Na^+ is replaced by a large cation such as choline or *N*-methyl-D-glucamine (Sage and Rink, 1986*a*), provides evidence that changes of the membrane potential are not important in triggering the discharge of internal Ca^{2+}.

We can calculate the minimum amount of Ca^{2+} dischargeable by receptor-mediated processes from the increase in the Ca^{2+} saturation of intracellular quin2, as $\sim$200 μmol/liter of cell water (Rink et al., 1982). The maximum discharge of $[Ca^{2+}]_i$ evoked by thrombin, which is one of the most effective platelets stimuli, is close to that which can be discharged by maximally effective concentrations of Ca^{2+} ionophores.

This finding implies that the bulk of the Ca^{2+} available for discharge in Ca^{2+}-sequestering organelles is available to receptor-mediated signaling. In platelets containing millimolar amounts of quin2 the extent of the $[Ca^{2+}]_i$ rise seen in Ca^{2+}-free medium is naturally curbed by the excess of Ca^{2+} buffering contained in the cytosol. With the more highly fluorescent indicator fura-2, signals can be obtained with dye loadings down to 20 μM, and one now finds that discharge of intracellular Ca^{2+} by thrombin or the Ca^{2+} ionophore ionomycin can raise $[Ca^{2+}]_i$ towards or beyond 1 μM (Pollock and Rink, 1986).

Linking the Receptor to Ca²⁺ Discharge

The mechanism by which an action potential causes Ca^{2+} release from the sarcoplasmic reticulum in striated muscle has long been and still remains a highly controversial and intruiging matter. No less mysterious was the linking between surface receptors and internal Ca^{2+} discharge in nonexcitable cells, but, as other accounts in this volume will testify, it is now generally believed that the missing message is inositol 1,4,5-triphosphate (InsP₃). In platelets, several pieces of the evidence required to sustain this view are available: (*a*) InsP₃ is formed within a few seconds of stimulation by many agonists, which include thrombin, platelet-activating factor, vasopressin, and thromboxin mimetics, via stimulated hydrolysis of phosphatidylinositol bisphosphate (e.g., Agranoff et al., 1983; Siess et al., 1984). (*b*) Microsomal subfractions from disrupted platelets that can accumulate ⁴⁵Ca in an ATP-dependent manner release part of the ⁴⁵Ca on exposure to InsP₃ in the micromolar range, i.e., in the range likely in stimulated cells (e.g., O'Rourke et al., 1985). (*c*) A similar effect of InsP₃ is seen with the ⁴⁵Ca taken up by platelets whose plasma membrane is permeabilized by saponin (Authi et al., 1987).

This evidence strongly implies that InsP₃ can liberate internal Ca^{2+} in platelets and is made available during platelet activation in many instances. Of course the evidence does not rule out the existence of additional or alternative mechanisms. For example, another important platelet agonist, adenosine disphosphate (ADP), is markedly less effective than thrombin in inducing the breakdown of phosphatidylinositol bisphosphate and the production of InsP₃ (Fisher et al., 1985), but is nevertheless effective at discharging internal Ca^{2+} (Sage and Rink, 1986*b*). This result might, however, be seen if ADP produced only a modest and very brief elevation of InsP₃ (see e.g., Daniel et al., 1986). Another important piece of evidence, which is presently lacking, is a kinetic analysis showing that the elevation of InsP$_3$ is fast enough to account for the observed rise in [Ca^{2+}]$_i$ in intact platelets. It would also add strength to the hypothesis to find specific blockers of the production and the effects of InsP$_3$ and demonstrate that they indeed block the stimulated discharge of Ca^{2+} in intact platelets. One way of generally inhibiting platelet function is to elevate cAMP; this has an array of actions, including a reduction in the production of InsP$_3$ and of the discharge of internal Ca^{2+} (e.g., Sage and Rink, 1985), although naturally this does not by itself demonstrate a causal link. The development of monoclonal antibodies that can interfere with the ability of InsP$_3$ to release Ca^{2+} from sequestering organelles offers an elegant way of further testing the hypothesis, although this will require the incorporation of large molecules into a resealed cell.

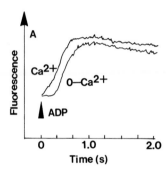

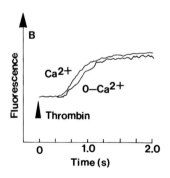

Figure 1. Kinetics of changes in [Ca^{2+}]$_i$ evoked by thrombin and ADP. Fura-2-loaded platelets were mixed with (*A*) ADP at a final concentration of 40 μM and (*B*) thrombin at a final concentration of 4 U/ml in a Hi-Tech Scientific SFA-II Rapid Kinetic Accessory mounted in a Perkin-Elmer MPF-44A spectrophotometer. Excitation was at 340 nm, emission at 500 nm. Traces are the average of 10–12 scans recorded on an Acorn BBC microcomputer. The medium contained 1 mM CaCl$_2$ (Ca^{2+}) or 1 mM EGTA (0-Ca^{2+}) as indicated.

Stopped-Flow Fluorescence Studies

We have begun to examine the question of the kinetics of the Ca^{2+} discharge by using stopped-flow fluorescence measurements of fura-2–loaded human platelets (Sage and Rink, 1986*a, b*; 1987). Typical responses to thrombin and ADP are shown in Fig. 1. In Ca^{2+}-free medium containing 1 mM EGTA, even with supramaximally effective concentrations of agonists there is an irreducible delay of ~200 ms before there is a discernible rise in signal. Thereafter the signal rises almost immediately at maximum rate and peaks within the first second. We suppose that the delay represents time for the many processes that have to occur between the rapid ($\sim$10–20 ms) mixing of ligand and platelets and the actual release of Ca^{2+} from the dense tubular system and its binding to the fura-2. The effective response time of fura-2 under these conditions is $\sim$10 ms, so instrumental and indicator lag are not responsible for the bulk of the observed delay. Presumably during this lag phase the following sequence of events takes place: ligand binding to the receptor; conformational changes in the receptor and the associated G proteins, the activation of phospholipase C; the hydrolysis of phosphatidyl inositol biphosphate (PIP_2) and production of $InsP_3$; the diffusion of this soluble message to the dense tubular system; the binding of $InsP_3$ to its receptor; conformational changes on that receptor, its coupling mechanisms to calcium channels; and finally the discharge of Ca^{2+}. Clearly, these observations require that the putative messenger $InsP_3$ reaches adequate concentrations within the first 200 ms after stimulation.

The second half of the analysis thus requires an accurate subsecond kinetic analysis of the hydrolysis of PIP_2. So far, to our knowledge, this has not been achieved for any cell type; however, a recent report shows an elevation of platelet $InsP_3$ 1 s after thrombin stimulation (Daniel et al., 1987).

The mechanism of Ca^{2+} release from the tubular system is not known. In the one study of this point on rough endoplasmic reticulum from rat liver, Muallem et al. (1985) found evidence that $InsP_3$-induced ^{45}Ca release was electrogenic.

Termination of the Calcium Signal

Termination of a Ca^{2+} signal will normally require two events: ending the processes generating the Ca^{2+} flux into the cytosol, and subsequent removal of Ca^{2+} to restore basal $[Ca^{2+}]_i$. Physiologically, removal of a stimulus will often result in shutting off the supply of Ca^{2+}, and the Ca^{2+} pumps in the plasma membrane and sequestering organelles will remove the excess until the normal pump/leak balance at resting $[Ca^{2+}]_i$ is restored. Two additional processes can help terminate the Ca^{2+} signal, even in the continued presence of a stimulus. Various kinds of inactivation or desensitization may occur at the level of the receptor or at subsequent transduction steps; and the various cellular messengers may stimulate Ca^{2+} removal.

In platelets, many agonists, including thromboxane mimetics and platelet-activating factor, show rapid desensitization so that both Ca^{2+} entry and internal release last only a few seconds. Part of this effect may be a direct consequence of receptor occupancy, but may in part be due to effects of protein kinase C.

Stimulation of protein kinase C using phorbol esters inhibits both agonist-evoked (PtdIns $(4, 5)P_2$ hydrolysis (e.g., Rittenhouse and Sasson, 1985) and rises in $[Ca^{2+}]_i$ in platelets (MacIntyre et al., 1985; Zavioco et al., 1985). Therefore, it has been suggested that the accumulation of diacylglycerol can help terminate the generation of Ca^{2+} fluxes by negative feedback on phospholipase C. A fall in the production of Ins $(1, 4, 5)P_3$ after activation of protein kinase C would be expected to result in the cessation of Ca^{2+} release from the internal stores. The influx of Ca^{2+} across the plasma membrane would similarly cease if inositol phosphates also mediate this response. The ability of diacylglycerol to inhibit Ca^{2+} flux generation may not be limited to the suppression of intermediate formation; ADP-evoked elevation of $[Ca^{2+}]_i$, whose influx component is probably independent of inositol phosphate formation, is also reduced (Drummond and MacIntyre, 1987). The Ca^{2+} channels in both the plasma and internal store membranes may thus be modulated by protein kinase C–dependent phosphorylation.

Once Ca^{2+} fluxes into the cytosol have been terminated, the restoration of $[Ca^{2+}]_i$ to basal levels can result from sequestering of Ca^{2+} into the internal stores and/or extrusion of Ca^{2+} across the plasma membrane. The sequestering of Ca^{2+} by the dense tubular system has been recognized for some time, since vesicles believed to be derived from this membrane system accumulate Ca^{2+} in vitro in an ATP-dependent manner (Robblee et al., 1973; Kaser-Glanzmann et al., 1977; Menashi et al., 1982).

Ca^{2+} Extrusion

The nature and regulation of systems capable of extruding Ca^{2+} across the plasma membrane of platelets is more controversial. Some authors have found no evidence for a transport Ca^{2+}-ATPase in purified plasma membrane preparations (e.g., Menashi et. al., 1984), but two recent papers report finding transport Ca^{2+}-ATPase in platelet surface membranes (Enyedi et al., 1986; Enouf, 1987).

We have examined the ways in which elevated $[Ca^{2+}]_i$ is restored to resting levels in human platelets loaded with fura-2 and ^{45}Ca. To avoid complications of changing Ca^{2+} influx, experiments were done with low external $[Ca^{2+}]$, <20 nM (Pollock et al., 1987) or 200 nM (Rink and Sage, 1987). After stimulation by thrombin in these conditions, $[Ca^{2+}]_i$ falls from its early peak towards basal value with half-time measured in seconds. Part of this fall is due to Ca^{2+} extrusion, evidenced by ^{45}Ca loss, and partly reflects resequestration by internal stores, which could be shown to be partially refilled from the size of the subsequent discharge produced by optimal concentration of Ca^{2+} ionophore (Pollock et al., 1987).

When $[Ca^{2+}]_i$ is initially elevated by Ca^{2+} ionophore, the subsequent fall must be due very largely to Ca^{2+} extrusion since the Ca^{2+}-sequestoring organelles are continually open "short-circuited" by the continued presence of the ionophore that translocates re-accumulated Ca^{2+} back into the cytosol. (Extruded Ca^{2+} is of course captured by the external EGTA and is not available for re-entry.)

The observed decline in $[Ca^{2+}]_i$ was surprisingly slow, having an initial drop to ~500 nM in an few tens of seconds and then having a slow phase lasting many minutes, so that $[Ca^{2+}]_i$ was still >400 nM 6 min after application of the ionophore (see Fig. 2 C). The measured loss of ^{45}Ca from the cells was similar to that seen with thrombin, but of course with ionophore present this loss reflects extrusion, as little could be resequestered into internal stores (as was shown by the lack of any elevation of $[Ca^{2+}]_i$ after subsequent application of thrombin (Pollock et al., 1987)). Indeed, application of thrombin after ionophore stimulated recovery towards basal $[Ca^{2+}]_i$, as seen in Fig. 2 B, and this reflected an enhanced extrusion, evidenced by a stimulated loss of ^{45}Ca as shown in Fig. 2 C.

The main established mechanisms for extrusion of calcium across the plasma membrane are Ca^{2+}-ATPases and Na^+/Ca^{2+} exchange carriers. We have shown that the replacement of external Na^+ with N-methyl-D-glucamine has no effect on the time course of the return of $[Ca^{2+}]_i$ to basal levels (Rink and Sage, 1987) and does not significantly affect ^{45}Ca efflux from resting platelets, or platelets stimulated with thrombin, ionomycin, or both agonists (Fig. 2 C). It therefore appears that Na^+/Ca^+ exchange does not play an important role in mediating Ca^{2+} extrusion in resting or stimulated platelets.

As mentioned above there have been reports that the platelet plasma membrane does not possess a Ca^{2+}-ATPase (Cutler et al., 1978) and that vesicles derived from platelet outer membranes do not accumulate Ca^{2+} (Menashi et al., 1982; Menashi et al., 1984).

More recently, however, there have been reports of two different Ca^{2+}-ATPase activities in platelet membrane fractions; one of these enzymes showed structural and kinetic similarities to the Ca^{2+}-ATPase of the endoplasmic reticulum, which suggests

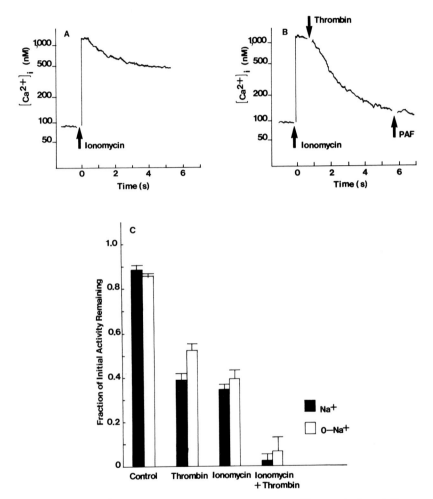

Figure 2. Stimulated Ca^{2+} efflux from fura-2-loaded platelets. Changes in fura-2 fluorescence recorded from a stirred platelet suspension containing 1 mM EGTA. (*A*) Ionomycin (1 μM) evokes a rise in $[Ca^{2+}]_i$ with an elevated plateau. (*B*) The addition of thrombin (4 U/ml) after 1 μM ionomycin restores $[Ca^{2+}]_i$ toward basal. The subsequent addition of 20 ng/ml PAF evokes no rise in $[Ca^{2+}]_i$, indicating that the internal Ca^{2+} stores have not refilled in the presence of the ionophore. (*C*) Effects of Na^+ replacement on $^{45}Ca^{2+}$ efflux from cells co-loaded with fura-2 and $^{45}Ca^{2+}$. Bars show the fraction of initial $^{45}Ca^{2+}$ activity remaining inside cells after 10 min under control conditions, and after the addition of 1 U/ml thrombin, 1 μM ionomycin, or 1 μM ionomycin with 4 U/ml thrombin. Filled bars show responses in control medium and open bars parallel responses in medium in which all Na^+ was replaced by *N*-methyl-D-glucamine. All values are means ($\pm$SEM) of triplicate determinations on four donors.

that it was derived from internal membranes, and the other showed similarities to the Ca^{2+}-ATPase of the erythrocyte plasma membrane, which suggests that it was of surface origin (Enyedi et al., 1986; Enouf et al., 1987). Our results support the existence of a Ca^{2+}-ATPase in the platelet plasma membrane.

Ca^{2+} pumping may be stimulated by Ca^{2+}-calmodulin, cAMP-dependent protein kinases and protein kinase C. Since $[Ca^{2+}]_i$ does not return to basal levels after stimulation by ionomycin, but reaches a plateau at ~500 nM, Ca^{2+}-calmodulin does not seem to be the stimulus to Ca^{2+} extrusion in platelets, at least not at $[Ca^{2+}]_i$ below ~500 nM. cAMP-dependent protein kinases do not appear to promote Ca^{2+} extrusion either, since we found that elevation of cAMP by using forskolin did not stimulate a fall in $[Ca^{2+}]_i$ (Rink and Sage, 1987). However, we found that phorbol myristate acetate mimicked the ability of thrombin to promote a fall in $[Ca^{2+}]_i$ in the presence of ionomycin (Pollock et al., 1987) and to stimulate $^{45}Ca^{2+}$ efflux (Rink and Sage, 1987).

It therefore appears that protein kinase C stimulates a Ca^{2+}-ATPase in the platelet plasma membrane, as has been reported in other cell types (e.g., Lagast et al., 1984). Hence, diacylglycerol appears to be a bidirectional regulator of platelet activity, acting synergistically with Ca^{2+} to evoke aggregation and secretion (Nishizuka, 1984; Rink et al., 1983) and helping to terminate the Ca^{2+} signal in platelets in at least two ways: by stimulating Ca^{2+} efflux, as well as by reducing agonist-evoked Ca^{2+} flux into the cytosol.

Receptor-mediated Ca^{2+} Entry

There are several lines of evidence pointing to the existence of receptor-mediated Ca^{2+} entry in human platelets. (*a*) ^{45}Ca uptake is rapidly increased on application of agonists such as thrombin or platelet-activating factor (Massini and Luscher, 1976; Lee et al., 1981). (*b*) The fluorescent signal from platelets loaded with millimolar levels of quin2 is substantially larger when they are stimulated in the presence of external Ca than in its absence (Rink et al., 1982). The simplest interpretation is that a stimulated Ca^{2+} entry adds to the Ca^{2+} entering the cytosol from internal stores. An alternative explanation could be that external Ca^{2+} is needed for efficient coupling of the receptor to internal release. This is unlikely for two reasons: the increase in $[Ca^{2+}]_i$ in platelets loaded with millimolar quin2 evoked by saturating amounts of Ca ionophore in Ca^{2+}-free medium, which presumably discharges virtually all mobilizable Ca^{2+}, is considerably smaller than that evoked by thrombin in the presence of external Ca^{2+}; and also, application of divalent cations such as Ni^{2+} or Cd^{2+}, which blocks Ca^{2+} entry through voltage-gated channels, can reduce the evoked increase in $[Ca^{2+}]_i$ in the presence of external Ca^{2+} to the level seen in Ca^{2+}-free medium (e.g., Hallam and Rink, 1985a). (*c*) We can exploit the fact that Mn^{2+} binds more avidly than does Ca^{2+} to quin2 or fura-2 and quenches the fluorescence, to show that thrombin, platelet-activating factor, or ADP can promote Mn^{2+} entry into human platelets (Hallam and Rink 1985a).

It seems most likely that Mn^{2+} enters via the route normally used by Ca^{2+}. Mn^{2+} entry is blocked by Ni^{2+}, as is the putative Ca^{2+} entry. So far there is no direct electrophysiological evidence for receptor-mediated Ca^{2+} currents in platelets, although patch clamp techniques can now be applied to very small cells or cell fragments. (Of course the absence of a measurable ligand-stimulated Ca current will not exclude a physiologically relevant receptor-mediated Ca entry in intact platelets, and finding some form of Ca^{2+} current in patch clamp experiments does not itself define physiological relevance for that Ca^{2+} entry.)

The Role of Membrane Potential

We do not think that changes in membrane potential play an important part in initiating Ca^{2+} entry in platelets. Imposed depolarization with high K^+ or with gramicidin does not elevate $[Ca^{2+}]_i$ (e.g., Sage and Rink, 1986a; Sage and Rink, unpublished observations). In normal Na^+ medium, thrombin, platelet-activating factor, or ADP do produce a small depolarization as indicated by potential-sensing fluorescent probes (Pipili, 1985), but only by some 5–10 mV, i.e., less than that usually needed to promote voltage-gated Ca^{2+} entry. More important, replacement of Na by large organic cations such as choline converts the small depolarization into a small hyperpolarization with minimal effects on the extent of $[Ca^{2+}]_i$ elevation observed in quin-2–loaded platelets in the presence of external Ca^{2+} (Sage and Rink, 1986a). Organic Ca^{2+} antagonists such as nifedipine and D600 have little effect on $[Ca^{2+}]_i$ signals over the concentration range in which they block voltage-gated channels in cardiac and smooth muscle (e.g., Hallam and Rink, 1985b). Human platelets lack nitrendipine binding sites (Motulski et al., 1983). However, K^+ depolarization substantially reduces the component of agonist-evoked $[Ca^{2+}]_i$ rise attributable to Ca^{2+} entry (Sage and Rink, 1986a).

We cannot yet fully interpret this finding, but it might reflect one of a number of mechanisms: K^+/Ca^{2+} competition at an external site; a reduced driving force for Ca^{2+} entry; an inactivation of Ca^{2+} entry that is dependent on depolarization (even though activation does not require depolarization); an effect of K^+ or membrane depolarization on the coupling of the receptor to Ca^{2+} entry. K^+ depolarization does not have a general effect in reducing signal transduction since the internal discharge of Ca^{2+} is barely affected. The finding that gramicidin, which depolarizes platelets, also reduces the $[Ca^{2+}]_i$ signal attributable to Ca^{2+} entry (Sage, S. O., and T. J. Rink, unpublished observations) may indicate that it is membrane depolarization rather than elevated external K^+ per se that is responsible for the effect. However, gramicidin will rapidly exchange internal potassium for sodium and so there will be a major change in the ionic gradients across the membrane. We need a sodium-specific ionophore that can cause a substantial depolarization of a membrane without significant alterations of the ionic environment, to isolate the effect membrane potential.

Mechanisms for Receptor-mediated Ca²⁺ Entry

We believe that the evidence outlined above strongly suggests that there is a significant component of stimulated Ca^{2+} entry in activated platelets which results from receptor occupation, not from changes in voltage. There are many possible mechanisms by which this entry could be triggered and several modes of Ca^{2+} entry to consider. Perhaps the most likely entry mechanism is the opening of some aqueous pore in the membrane, a channel, that is at least somewhat selective for Ca^{2+} and might well permit the transfer of other cations as well.

Another possibility is an exchanger mechanism whereby Ca^{2+} may be exchanged for the exit of another cation or combined with the entry of an anion. An exchanger might be capable of carrying electric current if the charges do not balance, or could be electroneutral. At present we do not have the evidence to separate these possibilities.

Among the mechanisms that can be proposed for triggering the Ca entry are the following: (*a*) A direct effect of ligand binding on a Ca^{2+}-passing channel, analogous to the influence of ACh on the nicotinic receptor; (*b*) receptors may be coupled to ionic channels via GTP-binding proteins in or attached to the inner face of the membrane; (*c*) A diffusable second messenger produced by receptor occupation may serve to promote Ca^{2+} entry. Irvine and Moor (1986) have suggested that inositol tetrakis phosphate might serve this role; (*d*) a subset of the mechanism in *c* is for Ca^{2+} released from the internal store to trigger Ca^{2+} entry, as has been suggested for neutrophils (Von-Tscharner et al., 1986); (*e*) yet another is one in which Ca^{2+} entry occurs direct from the external medium through some form of coupling device, possibly analogous to a gap junction, into the Ca-sequestering organelles of the dischargeable internal store. In this class of models some consequence of the depletion of the internal store by InsP$_3$ results in direct filling of the store from the external medium which, in the continued open state of the InsP$_3$-dependent channels of the internal membrane, can lead to Ca entry into the cytosol. (Putney, 1986; Merritt and Rink, 1987). Of course, these mechanisms need not be mutually exclusive and more than one may be operating simultaneously or sequentially during stimulation by any given ligand. There is relatively little direct evidence for or against these mechanisms in human platelets, but there are some clues. Application of large concentrations of fluoride, in the presence Al^{3+} ions, does evoke an increase in $[Ca^{2+}]_i$, part of which is attributable to Ca^{2+} entry.

Since AlF_4^- is thought to directly activate GTP-binding proteins that couple receptors to transduction processes this may constitute evidence that a G protein is involved in some form of Ca^{2+} entry (e.g., Brass et al., 1986). It has recently been shown that inositol tetrakis phosphate is rapidly formed in thrombin-stimulated human platelets (Daniel et al., 1987), so a role for this putative mediator of Ca^{2+} entry is not excluded.

Subsecond Kinetics

We have sought to examine the processes of Ca^{2+} entry, and to see whether certain models could be eliminated or rendered unlikely, by examining the rapid kinetics of $[Ca^{2+}]_i$ elevation in the presence of external Ca^{2+} compared with those in the absence of Ca^{2+}. Typical results for thrombin and ADP, two agonists that we expected to act through different mechanisms (Rink and Hallam, 1984), are shown in Fig. 1. With thrombin, the signal in the presence of external Ca^{2+} occurs slightly, but reproducibly, earlier than that seen in the absence of external Ca^{2+}, and is slightly larger. The simplest interpretation of this finding is that a component of the elevation of $[Ca^{2+}]_i$ is attributable to an early Ca^{2+} entry that precedes the internal discharge. It may be argued that the presence of EGTA to chelate residual Ca^{2+} may interfere with the coupling mechanisms and thereby produce the observed lag in the elevation of $[Ca^{2+}]_i$ seen in Ca^{2+}-free medium. Arguing against this idea, although not completely excluding it, is our recent finding that external Ni^{2+} produces very similar changes in the time course to those seen with EGTA (Sage and Rink, 1987).

If our interpretation is correct, this experimental observation appears to exclude mechanisms in which the entry of Ca^{2+} is consequent upon internal discharge; i.e., mechanisms that depend on Ca^{2+} activation or Ca^{2+} entry, or depend on depletion of Ca^{2+} from the internal store.

A much more striking difference in the signal with and without external Ca^{2+} is seen with ADP. The elevation of $[Ca^{2+}]_i$ produced in the presence of external Ca^{2+} occurs with no measurable lag (i.e., no more than the dead-time of $\sim$20 ms). As with thrombin, the longer lag seen in Ca^{2+}-free medium containing EGTA is also produced by application of Ni^{2+} (Sage and Rink, 1987). The rapidity of the $[Ca^{2+}]_i$ increase seen with ADP in the presence of external Ca^{2+} initially suggested to us that ADP might be working by a direct effect of ligand binding on a Ca^{2+} channel, whereas other ligands, such as thrombin, might act through a diffusable second messenger. However, a further examination of responses to ADP showed that a lag phase did appear with suboptimal concentrations of this agonist, which indicates some slower step in the activation cascade. However, for any given extent of elevation of $[Ca^{2+}]_i$ in the presence of external Ca^{2+} the lag is much less with ADP than with thrombin, suggesting a different coupling mechanism. Possibly the ADP receptor is coupled to the Ca^{2+} entry mechanism via a GTP-binding coupling protein, while thrombin acts more slowly via production of a diffusable second messenger.

Conclusions

In human platelets there is good evidence for receptor-mediated Ca^{2+} entry. With most stimuli this happens with an irreducible lag of $\sim$200 ms, earlier than the discharge of internal stores. With one agonist, ADP, the entry of Ca^{2+} appears to start with no measurable delay ($<$20 ms). This observation suggests that there are two distinct coupling mechanisms. Little is known of the actual entry mechanism, save that it appears not to be voltage gated and may be electrogenic (i.e., driven by an electric field as well as by a concentration gradient). We have no good experimental evidence to support or refute the idea that $InsP_4$ may mediate Ca^{2+} entry.

As in many cells, there is now good evidence to support the hypothesis that the main trigger for discharge of internal Ca^{2+} stores is mediated by rapid formation of $InsP_3$. However, it has not yet been shown that this messenger is formed fast enough to account for speed of the $[Ca^{2+}]_i$ response—rising maximally within 300 ms in our stopped-flow experiments.

Termination of a Ca^{2+} signal has many components, including: shutting off the extra supply of Ca^{2+} to the cytosol-entry and internal release; resequesterisation into internal stores; and extrusion of Ca^{2+} to the external medium. Recent evidence from our own laboratory and others suggests that despite previous reports to the contrary, platelets do have a plasma membrane transport Ca-ATPase that is regulated by protein kinase C. We find no evidence for an important contribution of Na:Ca exchange.

Acknowledgments

We thank Ms. M. Fitzgerald and Mrs. B. Leigh for help in preparing this manuscript.

S. O. Sage was a Medical Research Council Scholar.

References

Agranoff, B. W., P. Murthy, and E. B. Seguin. 1983. Thrombin-induced phosphodiesteric cleavage of phosphatidylinositol bisphosphate in human platelets. *Journal of Biological Chemistry.* 258:2076–2078.

Authi, K., E. J. Hornby, B. J. Evenden, and N. Crawford. 1987. Inositol 1,4,5-triphosphate (IP₃) induced rapid formation of thromboxane B₂ in saponin-permeabilised human platelets: mechanism of IP₃ action. *FEBS Letters* 213:95–101.

Brass, L. F., M. Laposata, H. S. Banga, and S. E. Rittenhouse. 1986. Regulation of the phosphoinositide hydrolysis pathway in thrombin-stimulated platelets by a pertussis toxin-sensitive guanine nucleotide-binding protein. *Journal of Biological Chemistry.* 261:16838–16847.

Cutler, L., G. Rodan, and M. Feinstein. 1978. Cytosolic localisation of adenylate cyclase and of calcium ion, magnesium ion-activated ATPases in the dense tubular system of human platelets. *Biochimica Biophysica Acta.* 542:357–371.

Daniel, J. L., C. A. Dangelmaier, M. Selak, and J. B. Smith. 1986. ADP stimulates IP₃ formation in human platelets. *FEBS Letters.* 206:299–303.

Daniel, J., C. Dangelmaier, and J. Smith. 1987. Formation and metabolism of inositol 1,4,5-triphosphate in human platelets. *Biochemistry Journal.* 246:109–114.

Drummond, A. H., and D. E. MacIntyre. 1987. Inositol lipid metabolism and calcium flux. *In* Platelets in Biology and Pathology III D. E. MacIntyre and J. L. Gordon, editors. Elsevier Science Publishers B.V., Amsterdam. 373–431.

Enouf, J., R. Bredoux, N. Bourdeau, and S. Levy-Toledano. 1987. Two different Ca²⁺ transport systems are associated with plasma and intracellular human platelet membranes. *Journal of Biological Chemistry.* 262:9293–9297.

Enyedi, A., B. Sarkadi, Z. Foldes-Papp, S. Monostory, and G. Gardos. 1986. Demonstration of two destinct calcium pumps in human platelet membrane vesicles. *Journal of Biological Chemistry.* 261:9558–9563.

Feinstein, M. B., and C. Fraser. 1975. Human platelet secretion and aggregation induced by Ca²⁺ ionophores. Inhibition by PGE1 and dbcAMP. *Journal of General Physiology.* 66:561–581.

Fisher, G. J., S. Bakshian, and J. J. Baldassare. 1985. Activation of human platelets by ADP causes a rapid rise in cytosolic free Ca²⁺ without hydrolysis of phosphatidylinositol-4,5-biphosphate. *Biochemical Biophysical Research Communications.* 129:958–964.

Hallam, T. J., and T. J. Rink. 1985*a*. Agonists stimulate divalent cation channels in the plasma membrane of human platelets. *FEBS Letters.* 186:175–179.

Hallam, T. J., and T. J. Rink. 1985*b*. Responses to adenosine diphosphate in human platelets loaded with the fluorescent calcium indicator quin2. *Journal of Physiology.* 368:131–146.

Hill, A. V. 1948. On the time required for diffusion and its relation to processes in muscle. *Proceedings of the Royal Society (London) Series B.* 135:446–453.

Irvine, R. F., and R. M. Moor. 1986. Micro-injection of inositol 1,3,4,5-tetrakisphosphate activates sea urchin eggs by a mechanism dependent on external Ca²⁺. *Biochemistry Journal.* 240:917–920.

Kaser-Glanzmann, R., M. Jakabova, J. N. George, and E. F. Luscher. 1977. Stimulation of Ca²⁺ uptake in platelet membrane vesicles by cAMP and protein kinases. *Biochimica Biophysica Acta.* 466:429–440.

Lagast, H., T. Pozzan, F. A. Waldvogel, and P. D. Lew. 1984. Phorbol myristate acetate stimulates ATP-dependent calcium transport by the plasma membrane of neutrophils. *Journal of Clinical Investigation.* 73:878–883.

Lee, T., B. Malone, M. L. Blank, and F. Snyder. 1981. PAF stimulates Ca²⁺ influx into rabbit platelets. *Biochemical and Biophysical Research Communications.* 102:1262–1268.

MacIntyre, D. E., A. McNicol, and A. H. Drummond. 1985. Tumor-promoting phorbol esters inhibit agonist-induced phosphatidate formation and Ca²⁺ flux in human platelets. *FEBS Letters.* 180:160–164.

Massini, P., and E. F. Luscher. 1976. On the significance of the influx of Ca²⁺ into stimulated human blood platelets. *Biochimica Biophysica Acta.* 436:652–663.

Menashi, S., C. Davis, and N. Crawford. 1982. Calcium uptake associated with an intracellular membrane fraction prepared from human blood platelets by high-voltage, free-flow electrophoresis. *FEBS Letters.* 140:298–302.

Menashi, S., K. S. Authi, F. Carey, and N. Crawford. 1984. Characterisation of the calcium-sequestering process associated with human platelet intracellular membranes isolated by free-flow electrophoresis. *Biochemistry Journal.* 222:413–417.

Merritt, J. E., and T. J. Rink. 1987. Rapid increases in cytosolic free calcium in response to muscarinic stimulation of rat parotid ascinar cells. *Journal of Biological Chemistry.* 262:4958–4960.

Motulsky, H. J., M. D. Snavely, P. J. Hughes, and P. A. Insel. 1983. Interaction of verapamil and other calcium channel blockers with α- and β-adrenergic receptors. *Circulation Research.* 52:226–231.

Muallem, S., M. Schoeffield, S. Pandol, and G. Sachs. 1985. Inositol triphosphate modification of ion transport in rough endoplasmic reticulum. *Proceedings of the National Academy of Sciences.* 82:4433–4437.

Nishizuka, Y. 1984. The role of protein kinase-C in cell surface signal transduction and tumour promotion. *Nature.* 308:693–698.

O'Rourke, F. A., S. P. Halenda, G. B. Zavioco, and M. B. Feinstein. 1985. Inositol 1,4,5-triphosphate releases Ca²⁺ from a Ca²⁺-transporting membrane vesicle fraction derived from human platelets. *Journal of Biological Chemistry.* 260:956–962.

Pipili, E. 1985. Platelet membrane potential: simultaneous measurements of diS-C3-(5) fluorescence and optical density. *Thrombosis and Haemostasis.* 54:645–649.

Pollock, W. K. and T. J. Rink. 1986. Thrombin and Ionomycin can raise platelet cytosolic Ca²⁺ stores: studies using fura-2. *Biochemical and Biophysical Research Communications.* 139:308–314.

Pollock, W. K., S. O. Sage, and T. J. Rink. 1987. Stimulation of Ca²⁺ efflux from fura-2 loaded platelets activated by thrombin or phorbol myristate acetate. *FEBS Letters.* 210:132–136.

Putney, J. W., Jr. 1986. A model for receptor-regulated calcium entry. *Cell Calcium* 7:1–12.

Rink, T. J., S. W. Smith, and R. Y. Tsien. 1982. Cytoplasmic free Ca²⁺ in human platel: Ca²⁺ thresholds and Ca²⁺-dependent activation for shape change and secretion. *FEBS Letters.* 148:21–26.

Rink, T. J., and A. Sanchez, and T. J. Hallam. 1983. Diacylglycerol and phorbol ester stimulate secretion without raising cytoplasmic free calcium in human platelets. *Nature.* 305:317–319.

Rink, T. J. and T. J. Hallam. 1984. What turns platelets on? *Trends in Biochemical Sciences.* 9:215–219.

Rink, T. J. and S. O. Sage. 1987. Stimulated calcium efflux from fura-2-loaded human platelets. *Journal of Physiology.* 393:513–524.

Rittenhouse, S. E. and J. P. Sasson. 1985. Mass changes in myoinositol triphosphate in human platelets stimulated by thrombin. Inhibitory effects of phorbol ester. *Journal of Biological Chemistry*. 260:8657–8660.

Robblee, L., D. Shepro, and F. Belamarich. 1973. Calcium uptake and associated ATPase activity of isolated platelet membranes. *Journal of General Physiology*. 61:462–481.

Sage, S. O., and T. J. Rink. 1985. Inhibitor by forskolin of cytosolic calcium rise, shape change and aggregation in quin2-loaded human platelets. *FEBS Letters*. 188:135–140.

Sage, S. O., and T. J. Rink. 1986a. Effects of ionic substitution on $[Ca^{2+}]_i$ rises evoked by thrombin and PAF in human platelets. *European Journal of Pharmacology*. 128:99–107.

Sage, S. O., and T. J. Rink. 1986b. Kinetic differences between thrombin-induced and ADP-induced calcium influx and release from internal stores in fura-2-loaded human platelets. *Biochemical and Biophysical Research Communications*. 136:1124–1129.

Sage, S. O. and T. J. Rink. 1987. The kinetics of changes in intracellular calcium concentration in fura-2-loaded human platelets. *Journal of Biological Chemistry*. 267:16364–16369.

Siess, W., P. Weber, and E. Lapetina. 1984. Activation of phospholipase C is dissociated from arachidonate metabolism during platelet shape change induced by thrombin or platelet-activating factor. Epinephrine does not induce phospholipase C activation or platelet shape change. *Journal of Biological Chemistry*. 259:8286–8292.

Tsien, R. Y., T. Pozzan, and T. J. Rink. 1982. Calcium homeostasis in intact lymphocytes: cytoplasmic free calcium monitored with a new, intracellularly trapped fluorescent indicator. *Journal of Cell Biology*. 94:325–334.

Von-Tscharner, V., B. Prod-Hom, M. Baggrolini, and H. Reuter. 1986. Ion channels in human neutrophils activated by a rise in free cytosolic calcium concentration. *Nature*. 324:369–372.

Zavioco, G. B., S. P. Halenda, R. I. Sha'afi, and M. B. Feinstein. 1985. Phorbolmyristate acetate inhibits thrombin-stimulated Ca^{2+} mobilization and phosphatidylinositol 4,5-biphosphate hydrolysis in human platelets. *Proceedings of the National Academy of Sciences*. 82:3859–3862.

List of Contributors

John W. Adamson, Department of Medicine, Division of Hematology, University of Washington School of Medicine, Seattle, Washington

Peter Agre, Department of Medicine, Johns Hopkins Hospital, Baltimore, Maryland

Qais Al-Awqati, Department of Medicine, Columbia University College of Physicians and Surgeons, New York, New York

Guillermo Alvarez de Toledo, Department of Physiology, University of Pennsylvania, Philadelphia, Pennsylvania

Matthew P. Anderson, Department of Physiology and Biophysics, University of Texas Medical Branch, Galveston, Texas

Mona M. Andersson, Whitehead Institute for Biomedical Research, Cambridge, Massachusetts

Andrew Asimos, University of Pittsburgh School of Medicine, Pittsburgh, Pennsylvania

R. Eugene Bailey, College of Medicine, State University of New York Health Science Center, Syracuse, New York

Vann Bennett, The Howard Hughes Medical Institute and Department of Biochemistry, Duke University Medical Center, Durham, North Carolina

Ellen M. Bifano, Department of Pediatrics, State University of New York Health Science Center, Syracuse, New York

David Bodine, National Heart, Lung, and Blood Institute, National Institutes of Health, Bethesda, Maryland

Janet K. Boyles, The Gladstone Foundation Laboratories, University of California, San Francisco, San Francisco, California

Jesper Brahm, Department of General Physiology and Biophysics, The Panum Institute, University of Copenhagen, Copenhagen, Denmark

Robert Briddell, Hematology/Oncology Section, Department of Medicine, Indiana University School of Medicine, Indianapolis, Indiana

Timothy Browder, National Heart, Lung, and Blood Institute, National Institutes of Health, Bethesda, Maryland

G. Brown, Department of Immunology, University of Birmingham, Birmingham, United Kingdom

Edward Bruno, Hematology/Oncology Section, Department of Medicine, Indiana University School of Medicine, Indianapolis, Indiana

C. M. Bunce, Department of Immunology, University of Birmingham, Birmingham, United Kingdom

Michael D. Cahalan, Department of Physiology and Biophysics, University of California, Irvine, Irvine, California

John C. Cambier, DNAX Research Institute of Molecular and Cellular Biology Inc., Palo Alto, California

Siu-Wah Chung, National Heart, Lung, and Blood Institute, National Institutes of Health, Bethesda, Maryland

Thomas M. Chused, Laboratory of Immunology, National Institute of Allergy and Infectious Diseases, Bethesda, Maryland

Lynn M. Crespo, Department of Pharmacology, Osteopathic Medical Center, North Miami Beach, Florida

Theresa A. Davies, Department of Biochemistry, Boston University School of Medicine, Boston, Massachusetts

Jonathan Davis, The Howard Hughes Medical Institute and Department of Biochemistry, Duke University Medical Center, Durham, North Carolina

T. M. Dexter, Department of Experimental Haematology, Paterson Institute for Cancer Research, Christie Hospital, Withington, Manchester, United Kingdom

Cynthia Dunbar, National Heart, Lung, and Blood Institute, National Institutes of Health, Bethesda, Maryland

Judith M. Dunn, Department of Biochemistry, Boston University School of Medicine, Boston, Massachusetts

Douglas V. Faller, Dana Farber Cancer Institute, Boston, Massachusetts

William L. Farrar, Laboratory of Molecular Immunoregulation, National Cancer Institute, Frederick Cancer Research Facility, Frederick, Maryland

Julio M. Fernandez, Department of Physiology, University of Pennsylvania, Philadelphia, Pennsylvania

Douglas K. Ferris, Program Resources, Inc., National Cancer Institute, Frederick Cancer Research Facility, Frederick, Maryland

Fred D. Finkelman, Department of Medicine, Uniform Services University of the Health Sciences, Bethesda, Maryland

Joan E. B. Fox, The Gladstone Foundation Laboratories, University of California, San Francisco, San Francisco, California

Jeffrey C. Freedman, Department of Physiology, State University of New York Health Science Center, Syracuse, New York

Otto Fröhlich, Department of Physiology, Emory University School of Medicine, Atlanta, Georgia

Wendy Furuya, Division of Cell Biology, Hospital for Sick Children, Toronto, Ontario, Canada

J. T. Gallagher, Department of Medical Oncology, Paterson Institute for Cancer Research, Christie Hospital, Withington, Manchester, United Kingdom

Elaine K. Gallin, Department of Physiology, Armed Forces Radiobiology Research Institute, Bethesda, Maryland

Kevin Gardner, Department of Cell Biology and Anatomy, The Johns Hopkins University School of Medicine, Baltimore, Maryland

I. M. Glynn, University of Cambridge, Cambridge, United Kingdom

Sheryl M. Greenberg, Division of Hematology, Brigham and Women's Hospital, Boston, Massachusetts

David Greenblatt, Laboratory of Microbial Immunity, National Institute of Allergy and Infectious Diseases, Bethesda, Maryland

Sergio Grinstein, Division of Cell Biology, Hospital for Sick Children, Toronto, Ontario, Canada

Robert B. Gunn, Department of Physiology, Emory University School of Medicine, Atlanta, Georgia

G. R. Guy, Department of Biochemistry, University of Birmingham, Birmingham, United Kingdom

Annick Harel-Bellan, Laboratory of Molecular Immunoregulation, National Cancer Institute, Frederick Cancer Research Facility, Frederick, Maryland

John H. Hartwig, Hematology-Oncology Unit, Massachusetts General Hospital, Boston, Massachusetts

William O. Haston, Research Department, Pharmaceutical Division, CIBA-Geigy Corporation, Summit, New Jersey

Doris A. Herzlinger, Department of Medicine, Columbia University College of Physicians and Surgeons, New York, New York

C. Heyworth, Department of Experimental Haematology, Paterson Institute for Cancer Research, Christie Hospital, Withington, Manchester, United Kingdom

Ronald Hoffman, Hematology/Oncology Section, Department of Medicine, Indiana University School of Medicine, Indianapolis, Indiana

William C. Horne, Department of Pathology, Yale University Medical Center, New Haven, Connecticut

Paul A. Janmey, Hematology-Oncology Unit, Massachusetts General Hospital, Boston, Massachusetts

Michael L. Jennings, Department of Physiology and Biophysics, University of Texas Medical Branch, Galveston, Texas

Stefan Karlsson, National Heart, Lung, and Blood Institute, National Institutes of Health, Bethesda, Maryland

C. E. King, Department of Biochemistry, University of Birmingham, Birmingham, United Kingdom

Patricia A. King, Department of Physiology and Biophysics, University of Vermont, Burlington, Vermont

Ron R. Kopito, Department of Biological Science, Stanford University, Stanford, California

Richard S. Lewis, Department of Physiology and Biophysics, University of California, Irvine, Irvine, California

Harvey F. Lodish, Whitehead Institute for Biomedical Research, Cambridge, Massachusetts

Li Lu, Hematology/Oncology Section, Department of Medicine, Indiana University School of Medicine, Indianapolis, Indiana

Sally J. McCormick, Department of Physiology and Biophysics, University of Texas Medical Branch, Galveston, Texas

Leslie C. McKinney, Department of Physiology, Armed Forces Radiobiology Research Institute, Bethesda, Maryland

Robert H. Michell, Department of Biochemistry, University of Birmingham, Birmingham, United Kingdom

David G. Nathan, Children's Hospital Research Center, Boston, Massachusetts

Charlotte M. Niemeyer, Dana Farber Cancer Institute, Boston, Massachusetts

Arthur Nienhuis, National Heart, Lung, and Blood Institute, National Institutes of Health, Bethesda, Maryland

Terri S. Novak, Department of Physiology, State University of New York Health Science Center, Syracuse, New York

Makio Ogawa, Veterans Administration Medical Center, Charleston, South Carolina

H. Clive Palfrey, Department of Pharmacological and Physiological Sciences, University of Chicago, Chicago, Illinois

John C. Parker, Department of Medicine, University of North Carolina, Chapel Hill, North Carolina

C. J. Piper, Department of Biochemistry, University of Birmingham, Birmingham, United Kingdom

I. L. O. Ponting, Department of Experimental Haematology, Paterson Institute for Cancer Research, Christie Hospital, Withington, Manchester, United Kingdom

Promod R. Pratap, Department of Physiology, State University of New York Health Science Center, Syracuse, New York

John T. Ransom, Department of Molecular Immunology, Syntex, Inc., Palo Alto, California

John R. Rediske, Research Department, Pharmaceutical Division, CIBA-Geigy Corporation, Summit, New Jersey

Timothy J. Rink, Smith, Kline & French Research Laboratory, Welwyn, Herts, United Kingdom

R. A. Roberts, Department of Experimental Haematology, Paterson Institute for Cancer Research, Christie Hospital, Withington, Manchester, United Kingdom

Bruce Roth, Hematology/Oncology Section, Department of Medicine, Indiana University School of Medicine, Indianapolis, Indiana

Ori D. Rotstein, Department of Surgery, University of Toronto, Toronto Western Hospital, Toronto, Ontario, Canada

Ali M. Saboori, Department of Medicine, Johns Hopkins Hospital, Baltimore, Maryland

Stewart O. Sage, Smith, Kline & French Research Laboratory, Welwyn, Herts, United Kingdom

Bruce Seligmann, Research Department, Pharmaceutical Division, CIBA-Geigy Corporation, Summit, New Jersey

Colin A. Sieff, Dana Farber Cancer Institute, Boston, Massachusetts

Louis Simchowitz, John Cochran Veterans Administration Medical Center, St. Louis, Missouri

Elizabeth R. Simons, Department of Biochemistry, Boston University School of Medicine, Boston, Massachusetts

Barbara L. Smith, Department of Medicine, Johns Hopkins Hospital, Baltimore, Maryland

E. Spooncer, Department of Experimental Haematology, Paterson Institute for Cancer Research, Christie Hospital, Withington, Manchester, United Kingdom

Joseph P. Steiner, The Howard Hughes Medical Institute and Department of Biochemistry, Duke University Medical Center, Durham, North Carolina

L. R. Stephens, Department of Biochemistry, University of Birmingham, Birmingham, United Kingdom

John Straneva, Hematology/Oncology Section, Department of Medicine, Indiana University School of Medicine, Indianapolis, Indiana

Alan S. Waggoner, Department of Biological Sciences, Carnegie Mellon University, Pittsburgh, Pennsylvania

Ronald Walenga, Department of Pediatrics, State University of New York Health Science Center, Syracuse, New York

Ahmad Waseem, Department of Pharmacological and Physiological Sciences, University of Chicago, Chicago, Illinois

James S. Wasvary, Research Department, Pharmaceutical Division, CIBA-Geigy Corporation, Summit, New Jersey

H. Alexander Wilson, Laboratory of Microbial Immunity, National Institute of Allergy and Infectious Diseases, Bethesda, Maryland

Peter Wong, National Heart, Lung, and Blood Institute, National Institutes of Health, Bethesda, Maryland

Ken S. Zaner, Hematology-Oncology Unit, Massachusetts General Hospital, Boston, Massachusetts

Susan Zuk, State University of New York Health Science Center, Syracuse, New York

Subject Index

Actin, cortical, 125
  filaments, 111, 125
Actin-binding proteins, 111, 125
Adducin, 101, 357
Anion,
  conductance, 181
  exchange, 151, 193
  permeant, 209
  selectivity, 181
  transport, 141, 163, 181
Ankyrin, 101, 151
Antigen receptors, 241, 253
Arginine residues, 163

Band 3, 91, 101, 151, 163, 357
B lymphocytes, 241, 253
Bone marrow, 19, 47, 79

Calcium,
  cascade, 217
  influx, 241
  signals, 381
Calcium probe, intracellular, 233
Capnophorin, 141, 163
  (see Band 3)
Carboxyl groups, 163
Cations, 263
  permeant, 209
Charybdotoxin, 281
Chemotactic peptide, 233
Chloride,
  channels, 281
  conductance, 163, 217
  net flux, 181
Colony-stimulating factors, 19, 25, 47, 67, 371
Cytoskeleton, 111

DIDS (see Stilbenes)
Differentiation of stem cells, 25, 39

DNDS (see Stilbenes)
Electrical potential, 217, 253
Elliptocytosis, 91
Endocytosis, 333
Erythrocytes, 141, 151, 163, 181, 217, 345, 357
Erythrocyte membrane, 91
Erythropoietin, 57
Exocytosis, 333
Extracellular matrix, 25

Flicker phenomenon, 333
Fluorescent dyes, 209, 233, 253

Gene transfer, 79
Glycoprotein Ib, 111
Granule fusion, 333
Granulocyte/macrophage colony-stimulating
  factors, 25, 47, 67, 371
Growth factors, 25, 39, 47, 67, 79, 371
  regulation of, 57

HL60 cells, 345
Heat shock proteins, 371
Hematopoiesis, 19, 25, 39, 59, 79, 371

Inositol, lipids, 345, 381
  phosphates, 345
Interleukin
  IL-1, 25, 39
  IL-2, 371
  IL-3, 25, 39, 47, 67, 79

Leukocytes, 125
Lymphokines, 371

Macrophages, 125, 315
Megakaryocytopoiesis, 67
Membrane,
  capacitance, 333
  pinch-off events, 333
  potentials, 209, 217, 233, 253, 265, 381
Membrane skeleton,
  platelet, 111
  proteins, 357
  spectrin based, 357
  substrates for protein kinase C, 357
Molecular zipper mechanism, 163

Na,K-ATPase,
  conformational changes, 1
  occluded ion forms, 1
Net transport, 181
Neutrophils, 193, 233, 303

Papain, 163
Phagosome, 303
Phorbol esters, 357, 371
Phosphoinositidase C, 345
pH regulation, 193, 303
Platelets, 67, 111, 265, 381
Potassium,
  channels, 281
  conductance, 31, 315
  efflux, 241
Protein kinase C, 101, 357, 371, 381

Proton-anion cotransport, 163
Pyropoikilocytosis, 91

Regulatory volume decrease, 281
Renal cells, 151
Retroviral vectors, 79
Rh antigen, 91

Self-renewal and differentiation, 25
Sickle cell anemia, 79
Slippage, 141, 181
Sodium/calcium exchange, 381
Sodium/hydrogen
    exchanger, 265, 303
Sodium-potassium pump, 1

Spectrin, 101
Spherocytosis, 91
Stem cells, 25, 39, 79
Stilbenes, 163, 181
Stimulus response, 265
Stromal cells, 25, 47
Sulfate net flux, 181

T cell receptor, 253
T lymphocytes, 253, 281
Thalassemia, 79
Thrombocytopoietic stimulating factor, 67
Tumor necrosis factor, 47
Tunneling mechanisms, 141, 181